Soil Ecological Risk Assessment Based on LULC

Soil Ecological Risk Assessment Based on LULC

Guest Editor

Lúcia Barão

MDPI

Basel • Beijing • Wuhan • Barcelona • Belgrade • Novi Sad • Cluj • Manchester

Guest Editor
Lúcia Barão
Instituto Superior Técnico
Universidade de Lisboa
Lisboa
Portugal

Editorial Office
MDPI AG
Grosspeteranlage 5
4052 Basel, Switzerland

This is a reprint of the Special Issue, published open access by the journal *Land* (ISSN 2073-445X), freely accessible at: https://www.mdpi.com/journal/land/special_issues/06103822MH.

For citation purposes, cite each article independently as indicated on the article page online and as indicated below:

Lastname, A.A.; Lastname, B.B. Article Title. *Journal Name* **Year**, *Volume Number*, Page Range.

ISBN 978-3-7258-7010-3 (Hbk)
ISBN 978-3-7258-7011-0 (PDF)
https://doi.org/10.3390/books978-3-7258-7011-0

Contents

About the Editor

Lúcia Barão

Lúcia Barão is an Environmental Engineer with a PhD in Biology. Her work focuses on topics covering sustainable agriculture, soil quality assessment, soil quality parameter measurement and soil modelling. Her main topics cover: (1) earth dynamics and soil biogeochemical processes; (2) link between soil nutrient cycles and soil and water quality; and (3) human impact on the soil processes and their local/global implications.

land

Article

Sustainable Soil Management: The Dynamic Impact of Combined Use of Crop Rotation and Fertilizers from Agri-Food and Sulfur Hydrocarbon Refining Processes Wastes

Angela Maffia, Federica Marra, Mariateresa Oliva, Santo Battaglia, Carmelo Mallamaci and Adele Muscolo *

Agriculture Department, Mediterranea University, 89124 Reggio Calabria, Italy; angela.maffia@unirc.it (A.M.); federica.marra@unirc.it (F.M.); mariateresa.oliva@unirc.it (M.O.); santo.battaglia@unirc.it (S.B.); carmelo.mallamaci@unirc.it (C.M.)

* Correspondence: amuscolo@unirc.it; Tel.: +39-0965-169-4364

Abstract: Sustainable agriculture increasingly relies on strategies that improve soil fertility while reducing the environmental footprint of chemical inputs. The primary objective of this research was to disentangle the individual and combined effects of crop rotation and fertilization on soil quality. This study aimed to determine whether the effectiveness of fertilization was modified by rotational practices—exploring whether these interactions were additive, antagonistic, or synergistic. This study assessed the impact of two-year open-field crop rotations—broccoli–tomato and broccoli–pepper—combined with organic and mineral fertilization on soil chemical and biological properties. Treatments included sulfur bentonite enriched with orange waste (SBO), horse manure (HM), mineral fertilizer (NPK), and an unfertilized control (CTR). Soil samples were collected after each crop cycle and analyzed for enzymatic activities (fluorescein diacetate hydrolase, dehydrogenase, catalase), microbial biomass carbon (MBC), organic matter, total nitrogen, and macro- and micronutrient content. The results showed that organic amendments, particularly SBO and HM, significantly increased microbial activity, MBC, and nutrient availability compared to NPK and CTR. Organic treatments also led to a reduction in soil pH (−12%) and a more balanced ionic profile, enhancing soil biological fertility across both rotations. By contrast, the NPK treatments favored higher nitrate and chloride concentrations (3.5 and 4.6 mg * g^{-1} dw, respectively) but did not improve biological indicators. Improvements were more pronounced in the second crop cycle, suggesting the cumulative benefits of organic amendments over time. These findings highlight the potential of combining organic fertilization with crop rotation to enhance soil health and support long-term sustainability in horticultural systems.

Keywords: soil fertility; organic amendments; crop rotation; microbial biomass; enzyme activity; sustainable agriculture

1. Introduction

Global food production must increase by 70% to meet the demands of a growing population, projected to reach 9.7 billion by 2050 [1]. This has led to the intensification of agriculture to boost productivity, driven by the widespread use of mineral fertilizers and plant protection products. However, this intensification has also contributed to a rise in soil degradation, threatening soil health. Agriculture plays a crucial role in natural resource management, but its expansion and intensification can negatively impact environmental protection and accelerate the degradation of agricultural lands [2,3]. In response, an ongoing debate has emerged regarding the need for improved fertilizer management and more sustainable soil nutrient use. This highlights the importance of integrating methods and practices that ensure adequate plant nutrition while maintaining long-term soil productivity. Sustainable soil management remains a fundamental challenge in modern agriculture, particularly in the face of growing environmental concerns and the need for efficient resource utilization. To maintain or to improve the fertility and productivity of the agricultural soil, many types of waste can be used, such as organic municipal waste, sewage sludge, waste from agricultural crops, manure from animals, and some types of industrial wastes, as a source of organic matter [4]. In previous works Muscolo et al. [5], Maffia et al. [6], and Panuccio et al. [7] used fertilizers produced from different agro-industrial wastes to improve crop yield and quality, and soil quality. The results evidenced at different extents a positive impact of all the fertilizers produced from wastes on the soil and crops. The integration of agro-industrial waste-derived fertilizers with crop rotation strategies has recently emerged as a promising approach to enhance soil fertility while reducing reliance on synthetic inputs. However, the dynamic interactions between soil amendments and cropping systems require thorough investigation to verify if their combination can be more suitable and productive than their single use, to optimize agronomic practices and ensure long-term sustainability. Crop rotation is a -time-honored agricultural practice known to improve soil health, manage pests and diseases, and enhance crop yields by diversifying the crops grown in succession on the same land [8]. By alternating crops with different nutrient requirements and rooting patterns, crop rotation helps to maintain soil nutrient balance and structure, thereby reducing the need for chemical fertilizers [9]. The utilization of agro-industrial wastes as fertilizers presents an eco-friendly alternative to conventional fertilization methods. Composting and vermicomposting agricultural and industrial wastes can improve both aboveground and belowground ecosystem services, contributing to sustainable agriculture [10]. Sulfur is a macronutrient essential for optimal plant growth and for enhancing resistance to soil-borne fungal and bacterial diseases [11]. In recent years, sulfur has gained renewed importance as a plant nutrient due to its declining availability in the soil. This reduction is attributed to the following: (1) the increased use of mono-ammonium and di-ammonium phosphate fertilizers, which contain less sulfur compared to traditional superphosphate; (2) the decreased atmospheric deposition of sulfur as a result of reduced industrial emissions; and (3) the adoption of higher-yielding crop varieties that extract greater amounts of sulfur from the soil during harvest [12].

Furthermore, sulfur is permitted in organic agriculture worldwide, making the recovery of sulfur from agricultural and industrial wastes an agronomically beneficial and environmentally sustainable strategy [13]. In the Calabria region of Southern Italy, research has shown that the application of sulfur bentonite combined with orange processing wastes—rich in valuable biomolecules such as nutrients and phenolic compounds—can improve soil fertility and promote eco-friendly agricultural practices [14–17]. Based on previous experiments, this study investigated the combined effects of crop rotation and the innovative fertilizer produced from agro-industrial wastes (sulfur bentonite plus orange waste) on soil quality and fertility. Conducted over two consecutive years, the research employed a rotational system alternating a winter crop (broccoli) with two different summer crops (tomato and pepper). Specifically, Calabrian broccoli, Big Rio tomato, and Topepo pepper were selected, as these are crops cultivated in Calabria with significant economic impact. The experiment was carried out in two distinct plots (1/2 hectare each) within the same area, each plot was further divided into multiple subplots to ensure statistical robustness. One plot featured a broccoli–pepper rotation, while the other employed a broccoli–tomato rotation. The primary objective of this research was to disentangle the individual and combined effects of crop rotation and fertilization on soil quality, with particular attention to the influence of different summer crops grown in rotation with broccoli. The study aimed to determine whether the effectiveness of fertilization was modified by rotational practices—exploring whether these interactions were additive, antagonistic, or synergistic—and whether different rotational crops exerted distinct impacts on soil fertility parameters. Additionally, the research sought to evaluate how the integration of agro-industrial waste-derived fertilizers with diversified cropping systems could contribute to the long-term improvement of soil health and function. By analyzing a range of soil physical and chemical properties under varied crop and fertilization combinations, this study offers critical insights into soil–plant dynamics, informing carbon farming strategies and supporting the transition toward sustainable, resilient, and productive agricultural systems.

2. Materials and Methods

2.1. Fertilizer Production

Sulfur bentonite plus orange waste (called pastazzo) have been produced by SBS Steel Belt Systems srl in tablets of 3/4 mm, as described by Muscolo et al. [18,19]. Sulfur (S) was mixed with bentonite (B) and orange residues (O) coming from the food industry. Elemental S was the principal component of the fertilizer [20]. The fertilizer was tested for pathogens (*total coliforms*, *faecal coliforms*, *salmonella* spp., and *Escherichia coli*) and heavy metals to avoid a pollutant impact on the soil [21]. The results evidenced an absence of pathogens and heavy metals [20]. The composition of the sulfur bentonite plus orange waste and horse manure fertilizers is reported below, in Table 1.

Table 1. Chemical characteristic of sulfur bentonite plus orange waste and horse manure fertilizers. Data are the mean of three replications ± standard deviation. Different letters in the same row indicate, significant differences among the treatments (* Tukey's test, $p \leq 0.05$).

Chemical Properties	HM	SBO
pH	7.0 ^{a*} ± 0.1	6.8 ^{a} ± 0.2
EC (mS/cm)	12 ^{a} ± 1.1	1.3 ^{b} ± 0.9
Moisture (%)	86.7 ^{a} ± 3.2	nd ± 2.9
OC (%)	22.6 ^{a} ± 1.9	6.3 ^{b} ± 2.5
TN (%)	3.0 ^{a} ± 0.6	0.9 ^{b} ± 0.3
C/N	7.3 ^{a} ± 1.9	9.3 ^{a} ± 1.7
Na^+ (mg g^{-1} dw)	1.8 ^{a} ± 0.5	0.12 ^{b} ± 0.2
NH_4^+ (mg g^{-1} dw)	0.24 ^{a} ± 0.03	0.09 ^{b} ± 0.04
K^+ (mg g^{-1} dw)	3.2 ^{a} ± 2.3	1.32 ^{b} ± 2.6
Mg^{2+} (mg g^{-1} dw)	1.9 ^{a} ± 0.4	1.41 ^{b} ± 0.7
Ca^{2+} (mg g^{-1} dw)	2.3 ^{a} ± 0.7	1.1 ^{b} ± 1.0
Cl^- (mg g^{-1} dw)	3.8 ^{a} ± 0.5	0.11 ^{b} ± 0.6
PO_4^{3-} (g g^{-1} dw)	2.8 ^{a} ± 0.4	0.3 ^{b} ± 0.3

2.2. Open Field Experiment

Extensive open-field tests were conducted over a total area of 1 hectare to evaluate the effects of fertilizer application and seasonal vegetable crop rotation on soil properties. The experiment took place at Falcone Farm in San Lorenzo (Reggio Calabria). The soil is classified as Sandy–Clay–Loam, with an alkaline pH and a low organic matter content ranging from 0.7% to 1.6%. Before the fertilizer application, the soil in the subplots underwent comprehensive analysis, with the results reported in the Tables as T0. The 1-hectare field was divided into two plots of 0.5 hectares each, which were further subdivided into subplots. Four fertilizer treatments were applied on a nitrogenous base:

- SBO (476 kg S ha^{-1})—Sulfur-based amendment

- CTR (Control)—Unfertilized soil
- NPK (20/10/10) (476 kg S ha^{-1})—Chemical fertilizer
- HM (430 kg ha^{-1})—Horse manure as an organic amendment

The experiment was arranged in a randomized complete block design and three replicates of each treatment were conducted across three subplots, with six soil samples collected per subplot for each treatment. The experiment was conducted over two consecutive years, and the results represent the average of three independent trials. Throughout the study, irrigation was maintained at 70% of field capacity to support soil vitality. Soil water content was monitored using a direct-read soil pH/moisture meter (R181). The SBO application rate was selected based on the literature data, which suggests pure sulfur application rates ranging from 2200 kg S ha^{-1} to 3300 kg S ha^{-1}, depending on the soil texture [18,22]. Broccoli was placed in the soil in October and collected in February. After broccoli harvesting, one plot rotated with the Topepo red pepper, and the other one with the tomato Big Rio F1. The tomato and pepper were harvested in July. Soil was collected after the crop harvesting, dried and sieved at 2 mm except for microbial biomass C, and analyzed for their chemical and biological properties, as reported in the soil chemical analysis paragraph.

2.3. Soil Chemical Analysis

The following soil properties were detected. Soil texture with the hydrometer method [23]; electric conductibility (EC) in 1:5 soil/water suspension, after stirring at 15 rpm for 1 h was measured with a Hanna instrument conductivity meter; the pH was determined in soil/solution ratio 1:2.5 with a glass electrode. Organic carbon was detected using the Walkley–Black method [24]. Total nitrogen (TN) was assessed using the Kjeldahl method [25]. C/N was quantified as a carbon/nitrogen ratio. Water soluble phenols were extracted and analyzed as described by Kaminsky and Müller [26], and monomeric and polyphenols were determined by the Box method [27], using tannic acid as the standard. The concentration of water-soluble phenolic compounds was expressed as tannic acid equivalents (μg TAE g^{-1}D.W.). The cation exchange capacity was analyzed using the barium chloride method [28]. Cations and anions were detected using ion chromatography (DIONEX ICS-1100) [29].

2.4. Soil Biological Analysis

The microbial biomass carbon (MBC) was assessed using the chloroform fumigation–extraction procedure [30] on fresh soil. Fumigated and unfumigated soil sample extracts were used to detect the soluble organic C [24]. Fluorescein diacetate hydrolase (FDA) activity was determined according to the method of Adam and Duncan [31]. Dehydrogenase (DHA) activity was assessed using the von Mersi and Schinner method [32]. Catalase activity (CAT) was detected by assessing the absorbance during the transformation of H_2O_2 to oxygen and water [18]. The decrease in the absorbance was measured at 240 nm, using the extinction coefficient of 39.4 M^{-1} cm^{-1}. Protease activity was detected as reported in Muscolo et al. [18]. Urease activity was determined as described in Sidari et al. [33]. Ammonium concentrations were determined at 690 nm by using a calibration curve. The results are reported as μg N-NH_4 g^{-1} d^{-1} 3 h^{-1}.

2.5. Statistical Analysis

An analysis of variance was used for all the data sets. One-way ANOVA with Tukey's honestly significant difference tests for analyzing the effects of fertilizer and crop rotation on each of the parameters measured were used. The ANOVA and *t*-test were performed using SPSS software, version 29.0. The effects were significant at $p \leq 0.01$. To analyze the relationships between the fertilizer, crop rotation, and soil parameters in the two different plots, a principal component analysis (PCA) was used.

3. Results

After the first cycle of the broccoli–tomato crop rotation, notable changes were observed in the soil chemical properties. The water content increased across all treatments compared to both the baseline control (initial soil conditions) and the unfertilized soil (Table 2). The soil pH decreased significantly only in the presence of SBO. Among the treatments, the SBO uniquely increased the concentration of water-soluble phenols (WSP), whereas the NPK was the only treatment that led to a significant reduction in WSP. Organic carbon (OC), total nitrogen (TN), cation exchange capacity (CEC), and microbial biomass carbon (MBC) increased across all treatments, including the unfertilized control (Table 2). Enzymatic activities, including fluorescein diacetate hydrolysis (FDA) and dehydrogenase (DHA), increased in the HM and SBO treatments relative to both controls. However, catalase (CAT) activity decreased with the SBO application (Figure 1). No significant differences were found in the lithium content. By contrast, potassium and magnesium levels rose compared to the control, while the calcium content declined in the soil treated with HM and SBO. Nitrate concentrations peaked with the NPK treatment, whereas sulfate concentrations were highest in the soil treated with SBO (Figure 2).

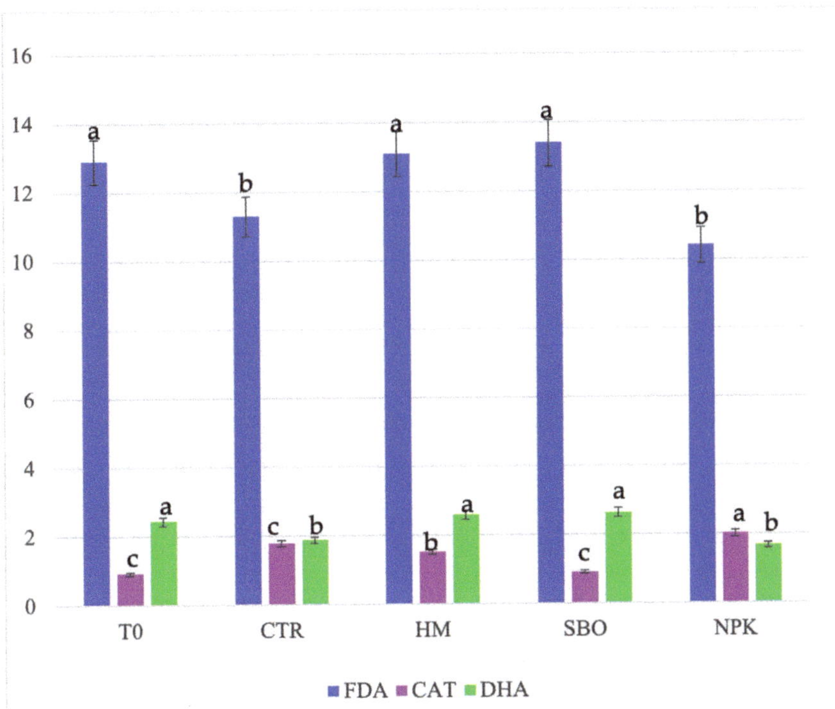

Figure 1. Fluorescein diacetate hydrolase (FDA, µg FDA/g d.s.), dehydrogenase (DHA, µg TTF/h/g d.s.), and catalase, (CAT, O_2/3 min/g d.s.) detected in soil after the second cycle of broccoli–tomato crop rotation. The data refers to the open field experiments carried on from October 2022 to July 2023. Soil samples were collected after the tomato harvest. The treatments include T0 (initial soil), CTR (unfertilized control), HM (horse manure), SBO (sulfur bentonite plus orange waste), and NPK (mineral fertilizer). Different letters indicate statistically significant differences among the treatments according to Tukey's test ($p < 0.05$).

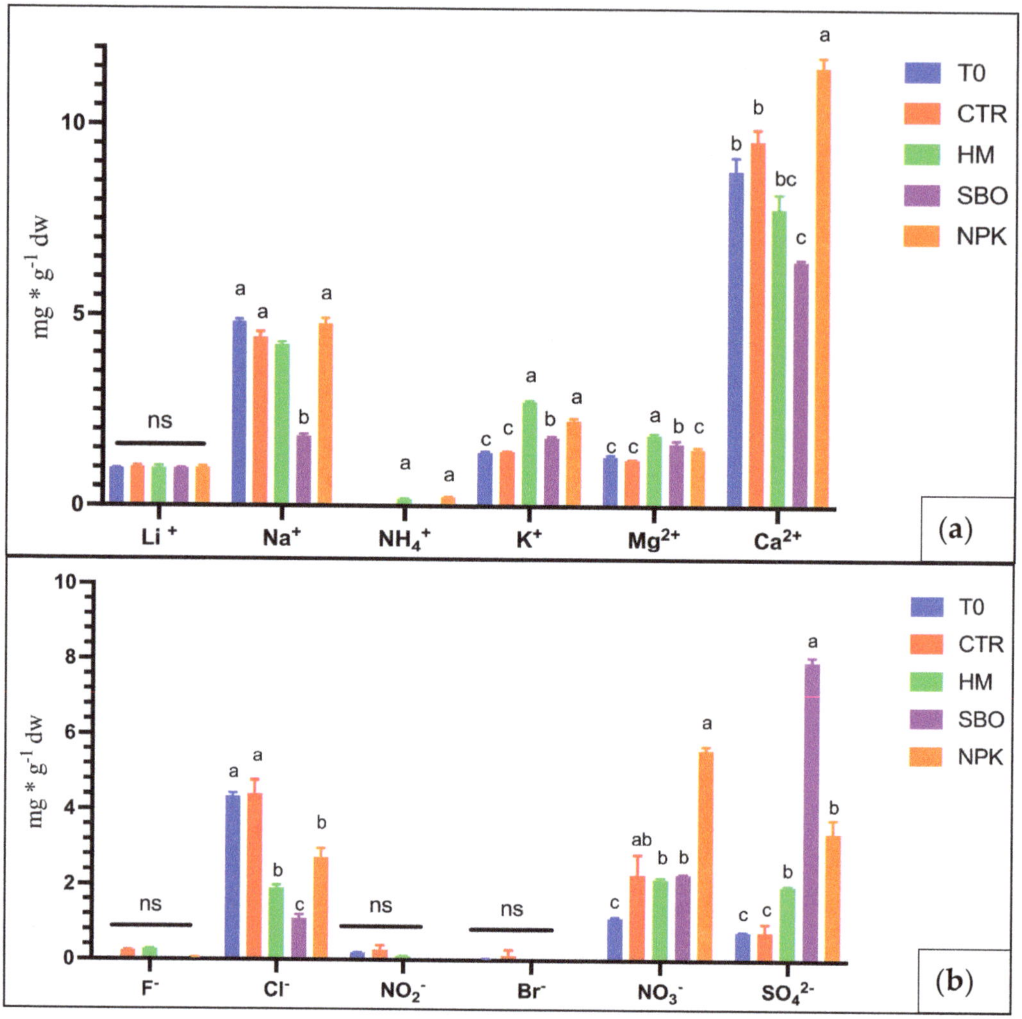

Figure 2. Cations (**a**) and anions (**b**) (mg * g^{-1} dw) in soil under the broccoli–tomato crop rotation. The first crop rotation cycle, from October 2022 to July 2023, with soil samples collected after the tomato harvest. The treatments include T0 (initial soil), CTR (unfertilized control), HM (horse manure), SBO (sulfur bentonite plus orange waste), and NPK (mineral fertilizer). Different letters indicate statistically significant differences among the treatments according to Tukey's test ($p < 0.05$). n.s. = not significant ($p > 0.05$).

In the second cycle of the broccoli–tomato rotation, electrical conductivity (EC) increased compared to the initial conditions (T0), especially in the soil treated with SBO and NPK. The pH again decreased with the SBO application (Table 3), while WSP increased in all treatments, most markedly in the unfertilized soil. FDA and DHA activities, along with OC, TN, and MBC, increased in the HM and SBO treatments relative to controls. Conversely, the CAT activity declined (Figure 3).

Table 2. Chemical and biological properties of soil after the first crop cycle of the broccoli–tomato rotation (October 2022–July 2023), with samples collected after tomato harvest. The treatments include T0 (initial soil), CTR (unfertilized control), HM (horse manure), SBO (sulfur bentonite plus orange waste), and NPK (mineral fertilizer). The measured variables include water content (WC, %), pH, electrical conductivity (EC, μS cm^{-1}), water-soluble phenols (WSP, μg TAE g^{-1} dry soil), organic carbon (OC, %), organic matter (OM, %), total nitrogen (TN, %), carbon-to-nitrogen ratio (C/N), cation exchange capacity (CEC, $cmol^{+}$ kg^{-1}), and microbial biomass carbon (MBC, μg C g^{-1}). Values are expressed as a mean ± standard deviation. Different letters within the same column indicate statistically significant differences among the treatments according to Tukey's test ($p < 0.05$).

	WC (%)	pH	EC	WSP	OC	OM	TN	C/N	CEC	MBC
T0	7.8 ± 0.2^{b}	8.16 ± 0.1^{a}	100 ± 6^{e}	23.3 ± 3^{b}	0.73 ± 0.03^{d}	1.25 ± 0.1^{d}	0.09 ± 0.01^{b}	8.1 ± 0.2^{b}	13.4 ± 0.1^{b}	433 ± 8^{e}
CTR	8 ± 0.2^{b}	8.24 ± 0.2^{a}	171 ± 9^{d}	23.2 ± 2^{b}	1.40 ± 0.02^{c}	2.40 ± 0.3^{c}	0.10 ± 0.02^{b}	14.0 ± 0.2^{a}	15.2 ± 0.1^{b}	476 ± 10^{d}
HM	13 ± 0.2^{a}	8.19 ± 0.2^{a}	232 ± 10^{a}	22.4 ± 3^{b}	1.70 ± 0.04^{b}	2.92 ± 0.1^{b}	0.15 ± 0.01^{a}	11.3 ± 0.1^{a}	17.1 ± 0.2^{a}	752 ± 12^{b}
SBO	16 ± 0.2^{a}	7.99 ± 0.1^{a}	207 ± 8^{b}	28.8 ± 2^{a}	1.98 ± 0.1^{a}	3.40 ± 0.2^{a}	0.14 ± 0.02^{a}	$14.1 \pm 0.\ 2^{a}$	17.0 ± 0.2^{a}	788 ± 11^{a}
NPK	16 ± 0.2^{a}	8.38 ± 0.2^{a}	195 ± 9^{b}	19.5 ± 3^{b}	1.67 ± 0.07^{b}	2.87 ± 0.2^{b}	0.18 ± 0.02^{a}	9.2 ± 0.01^{b}	16.5 ± 0.02^{a}	611 ± 13^{c}

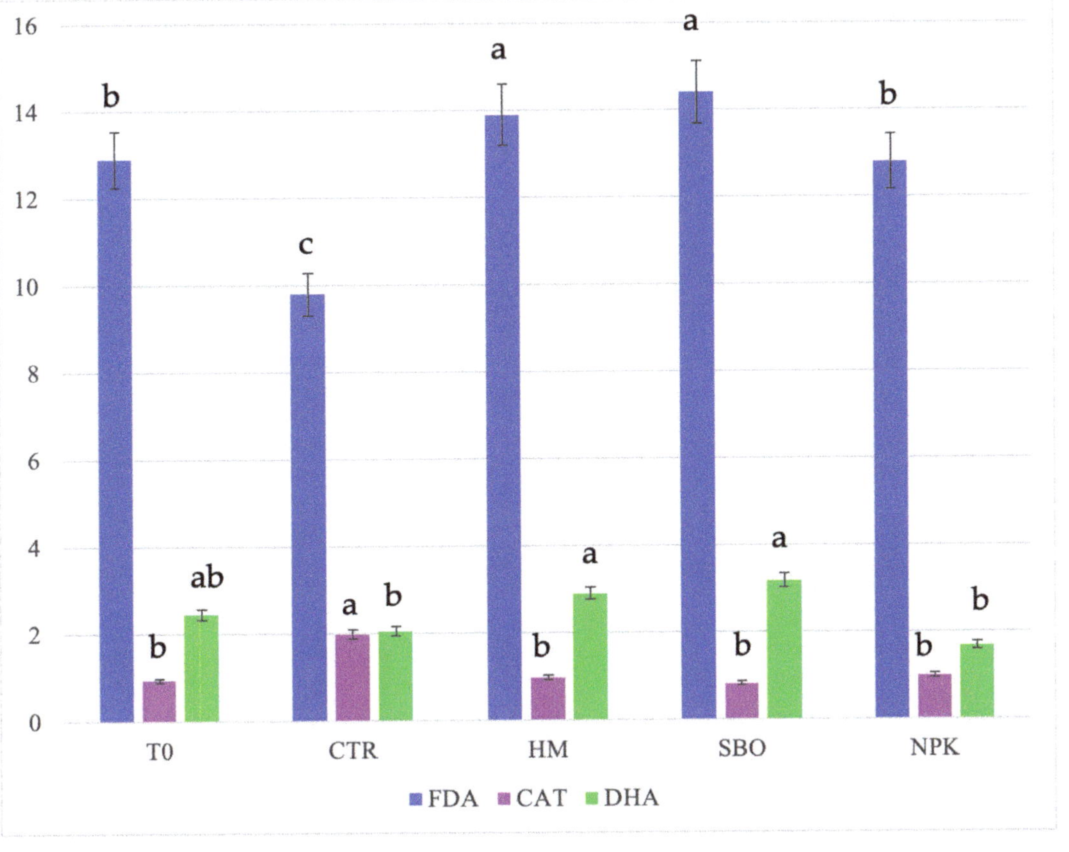

Figure 3. Fluorescein diacetate hydrolase (FDA, μg FDA/g d.s.), dehydrogenase (DHA, μg TTF/h/g d.s.), and catalase, (CAT, O_2/3 min/g d.s.) detected in soil after the second cycle of the broccoli–tomato crop rotation. The data refers to the open field experiments carried on from October 2022 to July 2023. Soil samples were collected after the tomato harvest. Different letters indicate statistically significant differences among the treatments according to Tukey's test ($p < 0.05$).

Table 3. Chemical and biological properties of soil after the second crop cycle of the broccoli–tomato rotation (October 2022–July 2023), with samples collected after tomato harvest. The treatments include T0 (initial soil), CTR (unfertilized control), HM (horse manure), SBO (sulfur bentonite plus orange waste), and NPK (mineral fertilizer). The measured variables include water content (WC, %), pH, electrical conductivity (EC, μS cm^{-1}), water-soluble phenols (WSP, μg TAE g^{-1} dry soil), organic carbon (OC, %), organic matter (OM, %), total nitrogen (TN, %), carbon-to-nitrogen ratio (C/N), cation exchange capacity (CEC, $cmol^+ kg^{-1}$), and microbial biomass carbon (MBC, μg C g^{-1}). Values are expressed as a mean ± standard deviation. Different letters within the same column indicate statistically significant differences among the treatments according to Tukey's test ($p < 0.05$).

	WC (%)	pH	EC	WSP	OC	TN	C/N	OM	CEC	MBC
T0	7.85 ± 0.2^{c}	8.16 ± 0.2^{a}	100 ± 7^{d}	23.3 ± 2^{c}	0.73 ± 0.05^{d}	0.09 ± 0.001^{b}	8.1 ± 1^{d}	1.25 ± 0.05^{d}	13.4 ± 0.5^{c}	433 ± 15^{d}
CTR	11.0 ± 0.1^{b}	8.21 ± 0.1^{a}	217 ± 9^{c}	45.1 ± 6^{a}	1.62 ± 0.1^{c}	0.08 ± 0.002^{b}	20.2 ± 3^{a}	2.40 ± 0.1^{c}	15.7 ± 0.7^{a}	498 ± 14^{d}
HM	14.0 ± 0.7^{a}	7.89 ± 0.2^{ab}	220 ± 8^{c}	30.2 ± 4^{b}	2.14 ± 0.2^{ab}	0.19 ± 0.001^{a}	11.2 ± 1^{c}	3.68 ± 0.2^{a}	16.3 ± 0.5^{a}	952 ± 15^{b}
SBO	14.0 ± 0.6^{a}	7.50 ± 0.2^{b}	310 ± 11^{b}	27.9 ± 3^{b}	2.30 ± 0.2^{a}	0.17 ± 0.001^{a}	13.5 ± 2^{b}	3.95 ± 0.3^{a}	15.0 ± 1^{a}	988 ± 10^{a}
NPK	11.0 ± 0.8^{b}	8.29 ± 0.3^{a}	370 ± 9^{a}	25.0 ± 2^{c}	1.95 ± 0.1^{b}	0.17 ± 0.002^{a}	11.4 ± 1^{c}	3.35 ± 0.2^{b}	14.4 ± 0.4^{b}	0761 ± 11^{c}

When comparing the first and second crop cycles, soil properties, anions, and cations changed within each treatment. To integrate these multidimensional changes, a principal component analysis (PCA) was conducted using all chemical, biochemical, and ionic parameters measured across both crop cycles under the broccoli–tomato rotation. The biplot (Figure 4) shows the spatial distribution of treatments and variables along the first two principal components, which together explained 59.03% of the total variance (F1: 39.49%; F2: 19.54%). Samples from the second cycle (T02, CTR2, SBO2, HM2, NPK2) generally shifted toward the positive side of F1, closely associated with improved biological and fertility indicators, including MBC, DHA, FDA, EC, TN, and CEC. By contrast, untreated or initial-cycle samples, like T02 and CTR1, were positioned on the negative side of F1, in proximity to variables such as Cl^-, Br^-, NO_2^-, and Na^+, suggesting lower microbial activity and potential ionic accumulation. This multivariate visualization supports the observed trend of progressive soil improvement under organic treatments across successive crop cycles, distinguishing them from the mineral or unfertilized plots both chemically and biologically.

In the first broccoli–pepper crop rotation cycle, the SBO treatment led to a decrease in the soil pH and an increase in EC, WSP, FDA, OC, OM, and MBC, especially when compared to the NPK treatment and the two controls. The water content increased across all treatments compared to T0 and the control (Table 4).

Table 4. Chemical analysis of soil under the broccoli–pepper crop rotation. The first cycle refers to the first crop rotation cycle, from October 2022 to July 2023, with soil samples collected after the pepper harvest. The treatments include T0 (initial soil), CTR (unfertilized control), HM (horse manure), SBO (sulfur bentonite plus orange waste), and NPK (mineral fertilizer). The measured variables include water content (WC, %), pH, electrical conductivity (EC, μS cm^{-1}), water-soluble phenols (WSP, μg TAE g^{-1} dry soil), organic carbon (OC, %), organic matter (OM, %), total nitrogen (TN, %), carbon-to-nitrogen ratio (C/N), cation exchange capacity (CEC, $cmol^+ kg^{-1}$), and microbial biomass carbon (MBC, μg C g^{-1}). Values are expressed as a mean ± standard deviation. Different letters within the same column indicate statistically significant differences among the treatments according to Tukey's test ($p < 0.05$).

	WC (%)	pH	EC	WSP	OC	TN	C/N	OM	CEC	MBC
T0	7.85 ± 0.2^{c}	8.16 ± 0.3^{a}	100 ± 2.1^{c}	23.34 ± 0.6^{b}	0.73 ± 0.5^{c}	0.09 ± 0.5^{b}	8.1 ± 0.6^{c}	1.25 ± 0.1^{c}	13.4 ± 0.4^{b}	433 ± 4.6^{c}
CTR	9.3 ± 2.1^{b}	8.16 ± 0.5^{a}	188 ± 4.1^{b}	30.96 ± 0.7^{a}	1.48 ± 0.4^{b}	0.08 ± 0.1^{b}	18.5 ± 0.3^{a}	2.54 ± 0.7^{b}	14.1 ± 0.5^{b}	499 ± 3.4^{c}
HM	16.5 ± 1.2^{a}	8.14 ± 0.7^{a}	233 ± 2.3^{ba}	26.85 ± 0.5^{b}	1.93 ± 0.1^{b}	0.18 ± 0.1^{a}	10.7 ± 0.5^{b}	3.31 ± 0.2^{a}	18.2 ± 0.5^{a}	852 ± 4.4^{b}
SBO	16.5 ± 3.1^{a}	7.94 ± 0.9^{a}	275 ± 2.1^{a}	24.34 ± 2.1^{b}	2.16 ± 0.1^{a}	0.17 ± 0.2^{a}	12.7 ± 0.4^{ab}	3.71 ± 0.1^{a}	16.9 ± 0.7^{a}	988 ± 5.6^{a}
NPK	16.00 ± 3.4^{a}	8.38 ± 0.1^{a}	295 ± 3.4^{a}	19.55 ± 1.4^{c}	1.67 ± 0.3^{b}	0.18 ± 0.1^{a}	9.2 ± 0.6^{c}	2.87 ± 0.2^{b}	15.6 ± 0.8^{b}	735 ± 5.5^{b}

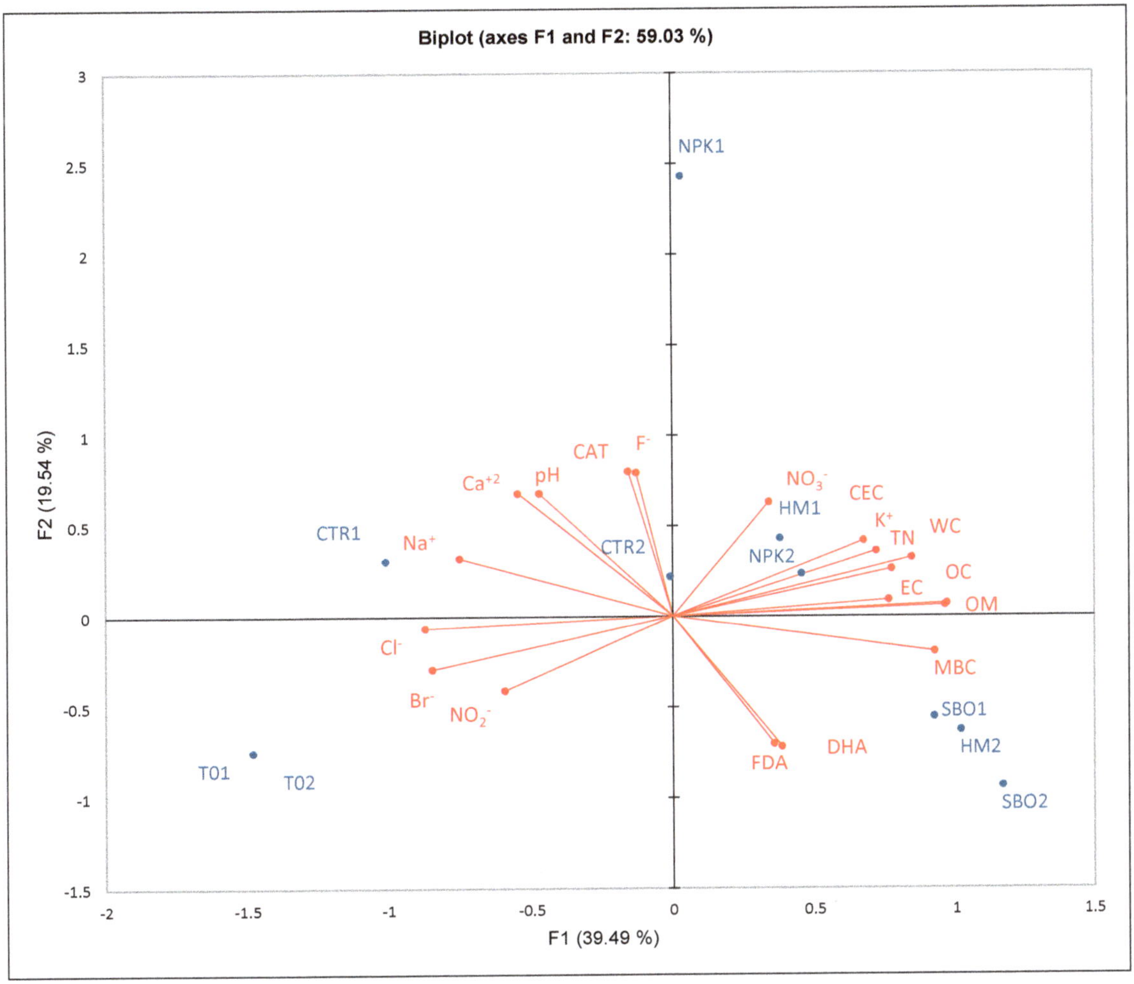

Figure 4. Principal component analysis (PCA) of chemical, biochemical, and ionic soil parameters under the broccoli–tomato crop rotation. The biplot represents soil samples collected after the first cycle (T01, CTR1, NPK1, HM1, SBO1) and the second cycle (T02, CTR2, NPK2, HM2, SBO2).

FDA activity, representing the overall microbial hydrolytic potential, was highest in the SBO-treated soil, followed closely by HM, while the lowest values were observed in the CTR and NPK treatments. DHA activity followed a similar pattern, peaking under the SBO treatment, indicative of enhanced microbial respiration and metabolic activity. By contrast, CAT activity, which reflects oxidative stress regulation, remained relatively low across all treatments, with only slight variations among them. The T0 and NPK treatments showed slightly higher CAT values compared to the organic amendments (Figure 5). The cation analysis revealed higher concentrations of K^+ and Mg^{2+} in the HM and SBO treatments, indicating improved nutrient availability from organic sources. Ca^{2+} was most abundant in the NPK and CTR treatments, while Na^+ levels were elevated in the T0 and mineral-fertilized soil (Figure 6a). Regarding anions, NO_3^- and Cl^- peaked under NPK, confirming the influence of synthetic fertilization. SO_4^{2-} and PO_4^{3-} were notably higher in the SBO-treated plots, reflecting sulfur and phosphorus contributions from the amendment (Figure 6b). Overall, organic treatments enriched the soil with key macroelements while minimizing salt accumulation. This trend continued in the second cycle: The pH decreased in the HM and SBO treatments, while EC, OC, OM, TN, CEC, and MBC increased compared to both T0 and control (Table 5).

Table 5. Chemical analysis of soil under the broccoli–pepper crop rotation. The second cycle refers to the second crop rotation cycle, from October 2023 to July 2024, with soil samples collected after the pepper harvest. The treatments include T0 (initial soil), CTR (unfertilized control), HM (horse manure), SBO (sulfur bentonite plus orange waste), and NPK (mineral fertilizer). The measured variables include water content (WC, %), pH, electrical conductivity (EC, μS cm^{-1}), water-soluble phenols (WSP, μg TAE g^{-1} dry soil), organic carbon (OC, %), organic matter (OM, %), total nitrogen (TN, %), carbon-to-nitrogen ratio (C/N), cation exchange capacity (CEC, $cmol^+$ kg^{-1}), and microbial biomass carbon (MBC, μg C g^{-1}). Values are expressed as a mean ± standard deviation. Different letters within the same column indicate statistically significant differences among the treatments according to Tukey's test ($p < 0.05$).

	WC (%)	pH	EC	WSP	OC	TN	C/N	OM	CEC	MBC
T0	7.85 ± 0.1^{c}	8.16 ± 0.1^{a}	100 ± 1.5^{b}	23.34 ± 1.5^{b}	0.73 ± 0.1^{c}	0.09 ± 0.2^{a}	8.1 ± 0.7^{c}	1.25 ± 0.1^{c}	13.4 ± 1.2^{b}	433 ± 2.5^{c}
CTR	$10 0 \pm 0.2^{b}$	8.25 ± 0.2^{a}	192 ± 2.5^{ab}	37.5 ± 0.5^{a}	1.54 ± 0.1^{b}	0.10 ± 0.3^{a}	15.4 ± 0.8^{a}	2.65 ± 0.3^{b}	13.9 ± 1.3^{b}	501 ± 2.7^{c}
HM	13.5 ± 0.2^{b}	7.97 ± 0.4^{a}	223 ± 2.7^{a}	25.50 ± 1.7^{b}	1.93 ± 0.2^{b}	0.16 ± 0.3^{a}	12.1 ± 1.1^{b}	3.31 ± 0.3^{a}	15.2 ± 1.6^{a}	879 ± 3.5^{a}
SBO	13.0 ± 0.3^{b}	7.81 ± 0.4^{a}	249 ± 3.5^{a}	32.50 ± 1.2^{a}	2.10 ± 0.2^{a}	0.13 ± 0.3^{a}	16.2 ± 1.2^{a}	3.61 ± 0.4^{a}	15.81 ± 1.2^{a}	955 ± 4.7^{a}
NPK	16.00 ± 0.3^{a}	8.38 ± 0.7^{a}	264 ± 2.3^{a}	19.55 ± 1.4^{c}	1.67 ± 0.3^{b}	0.18 ± 0.2^{a}	9.2 ± 1.2^{bc}	2.87 ± 0.1^{b}	15.6 ± 1.2^{a}	755 ± 5.6^{b}

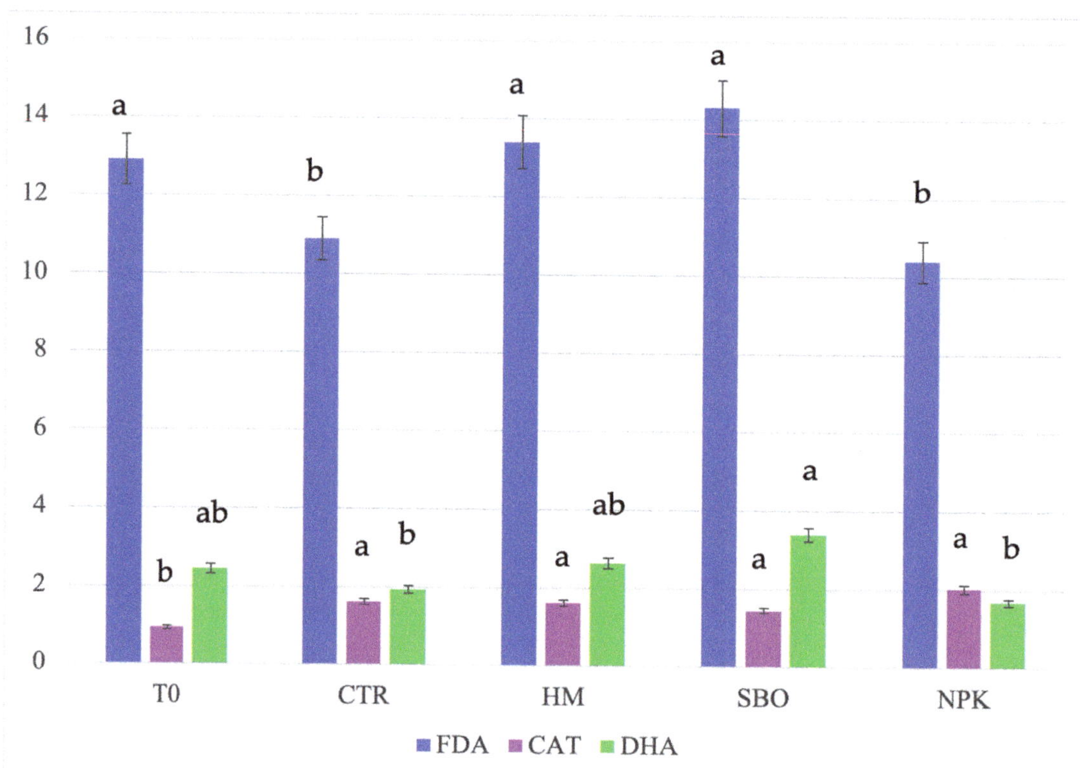

Figure 5. Fluorescein diacetate hydrolase (FDA, μg FDA/g d.s.), dehydrogenase (DHA, μg TTF/h/g d.s.), and catalase, (CAT, O_2/3 min/g d.s.) detected in soil after the first cycle of the broccoli–pepper crop rotation. The data refers to the open field experiments carried on from October 2022 to July 2023. Soil samples were collected after the tomato harvest. Different letters indicate statistically significant differences among the treatments according to Tukey's test ($p < 0.05$).

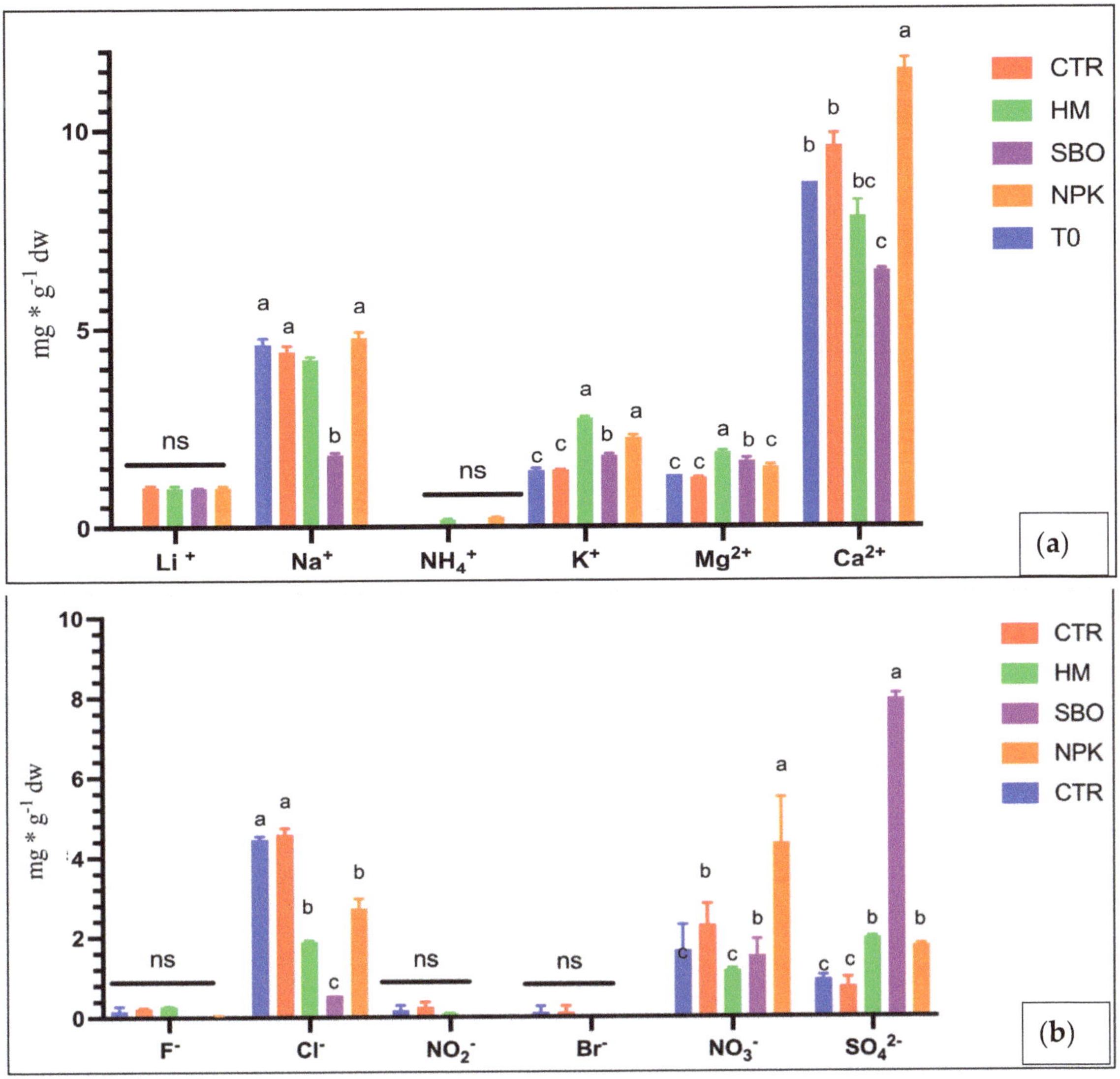

Figure 6. Cations (**a**) and anions (**b**) (mg * g^{-1} dw) in soil under the broccoli–pepper crop rotation. The first crop rotation cycle, from October 2022 to July 2023, with soil samples collected after the pepper harvest. The treatments include T0 (initial soil), CTR (unfertilized control), HM (horse manure), SBO (sulfur bentonite plus orange waste), and NPK (mineral fertilizer). Different letters indicate statistically significant differences among the treatments according to Tukey's test ($p < 0.05$). n.s. = not significant ($p > 0.05$).

Enzymatic activities measured after the second cycle of the broccoli–pepper rotation (Figure 7) showed an overall increase compared to the first cycle. FDA activity remained highest in the HM and SBO treatments, indicating sustained microbial hydrolytic capacity, followed by the T0 and NPK treatments. DHA levels were also elevated under the SBO and HM treatments, confirming enhanced microbial respiration. CAT activity remained consistently low across all treatments, with minor differences; slightly higher values were recorded in the T0 and CTR treatments suggesting limited oxidative stress variation among the treatments. Overall, the enzymatic profiles in the second cycle mirrored those of the first, with a general trend of increased microbial functional activity in organically amended soil. In the second cycle of the broccoli–pepper rotation, Ca^{2+} remained highest in the NPK-treated soil, while K^+ and Mg^{2+} increased under the HM and SBO treatments, indicating enhanced nutrient input from organic amendments, while Na^+ persisted at higher levels in T0, and NH_4^+ stayed low across all treatments. Among anions, Cl^- and NO_3^- peaked in the NPK treatment, reflecting continued mineral fertilization. SO_4^{2-} and PO_4^{3-} were elevated in the SBO and HM treatments, supporting the influence of sulfur- and phosphorus-rich amendments. Minor ions, such as NO_2^- and Br^-, remained low, with slight increases in the mineral treatments. Overall, organic inputs improved nutrient availability, while the NPK treatment favored ion accumulation (Figure 8).

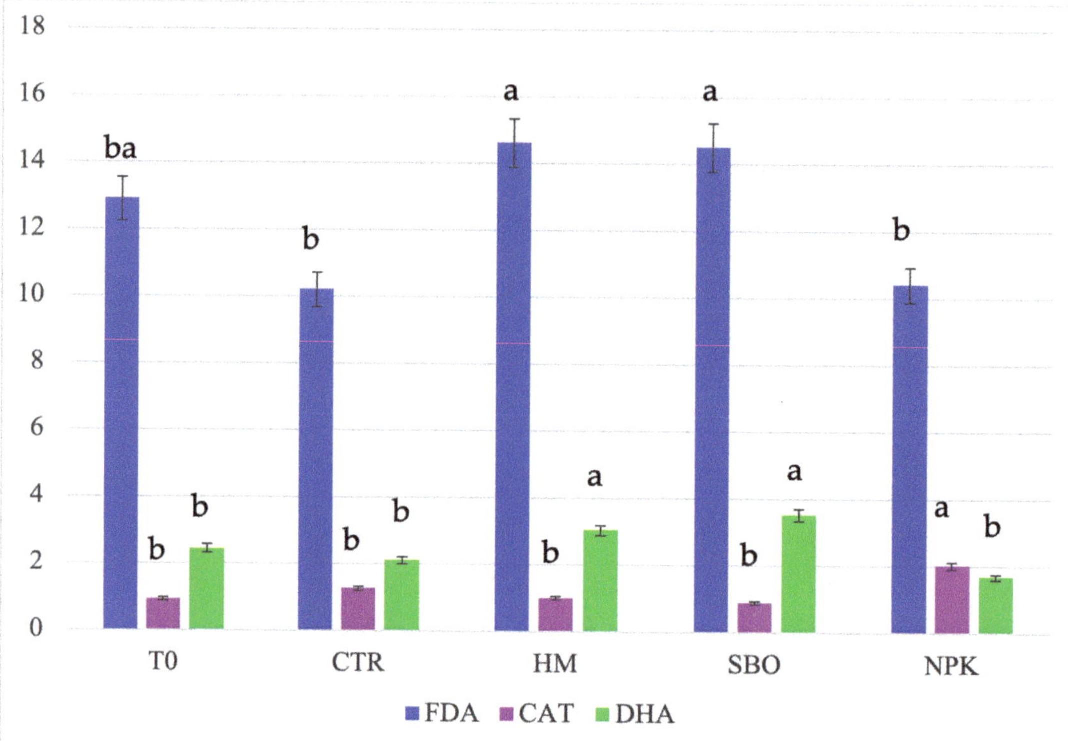

Figure 7. Fluorescein diacetate hydrolase (FDA. μg FDA/g d.s.), dehydrogenase (DHA. μg TTF/h/g d.s.), and catalase. (CAT. O_2/3 min/g d.s.) detected in soil after the second cycle of the broccoli–pepper crop rotation. The data refers to the open field experiments carried on from October 2022 to July 2023. Soil samples were collected after the tomato harvest. The treatments include T0 (initial soil), CTR (unfertilized control), HM (horse manure), SBO (sulfur bentonite plus orange waste), and NPK (mineral fertilizer). Different letters indicate statistically significant differences among the treatments according to Tukey's test ($p < 0.05$).

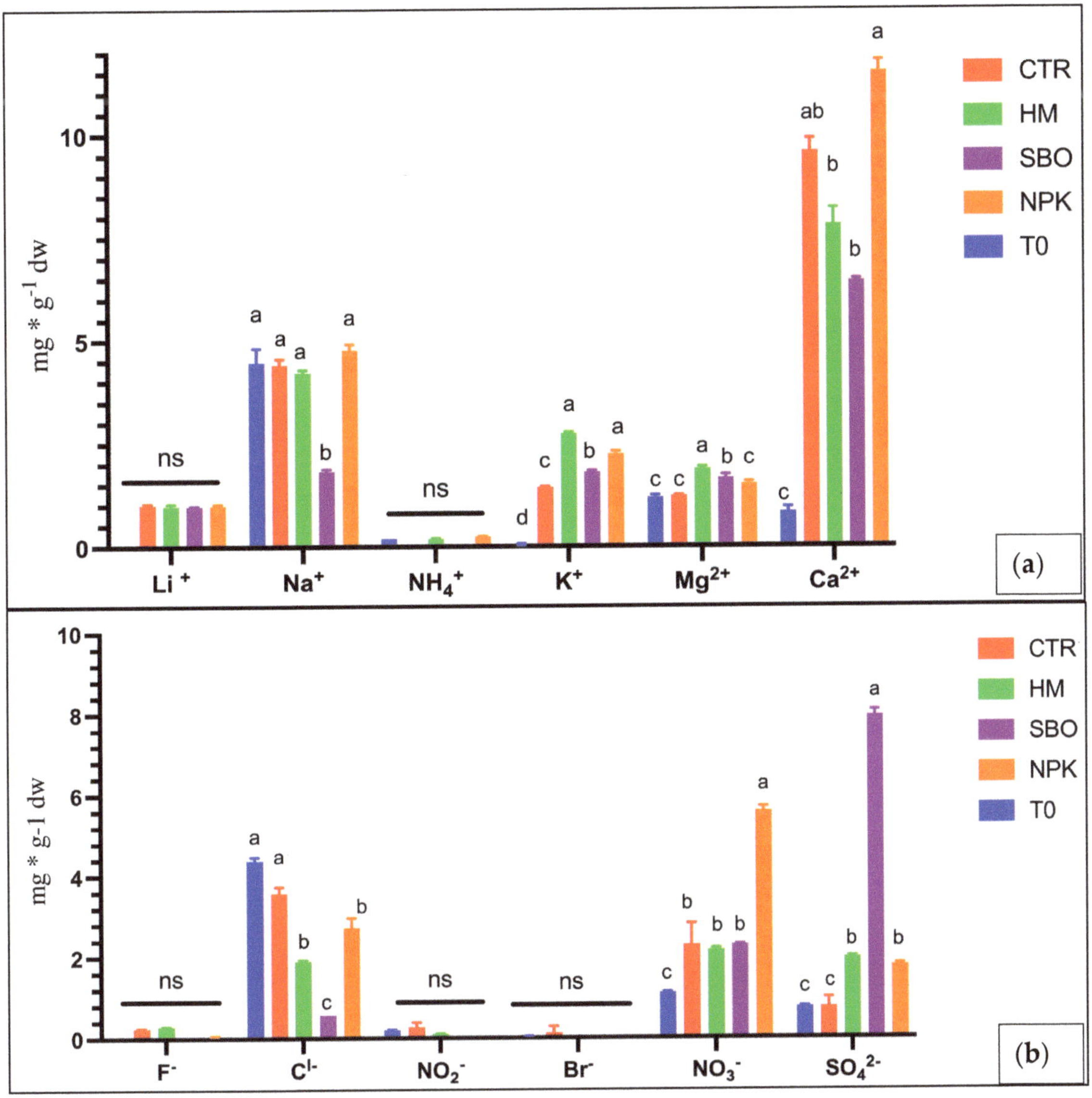

Figure 8. Cations (**a**) and anions (**b**) (mg * g^{-1} dw) in soil under the broccoli–pepper crop rotation. The second crop rotation cycle, from October 2023 to July 2024, with soil samples collected after the pepper harvest. The treatments include T0 (initial soil), CTR (unfertilized control), HM (horse manure), SBO (sulfur bentonite plus orange waste), and NPK (mineral fertilizer). Different letters indicate statistically significant differences among the treatments according to Tukey's test ($p < 0.05$). n.s. = not significant ($p > 0.05$).

To integrate the multivariate effects of treatments over time, a principal component analysis (PCA) was conducted using chemical, biochemical, and ionic soil parameters measured across both crop cycles under the broccoli–pepper rotation. The biplot (Figure 9) illustrates the distribution of treatments and active variables along the first two principal components, which together explained 62.80% of the total variance (F1: 41.38%; F2: 21.43%).

Samples from the second cycle (T02, CTR2, SBO2, HM2, NPK2) displayed a clear separation from first-cycle samples, with SBO2 and HM2 positioned on the far right of F1, closely associated with MBC, OM, DHA, SO_4^{2-}, and CEC, indicating enhanced microbial and nutrient status under organic management. First-cycle treatments, such as SBO1 and HM1, were also positively associated with fertility-related variables like EC, TN, and WC. Conversely, T02 and CTR1/CTR2 were located on the negative side of F1, clustering with Cl^-, NO_2^-, Na^+, and Mg^{2+}, suggesting lower biological activity and possible ionic accumulation. NPK1 and NPK2 were grouped along the upper right quadrant, aligning with NO_3^-, NH_4^+, and CAT, reflecting a mineral nitrogen-driven soil profile. This PCA confirms the distinct and progressive enrichment of soil biological and chemical properties under organic amendments, while distinguishing the mineral- and non-treated plots along contrasting multivariate trajectories.

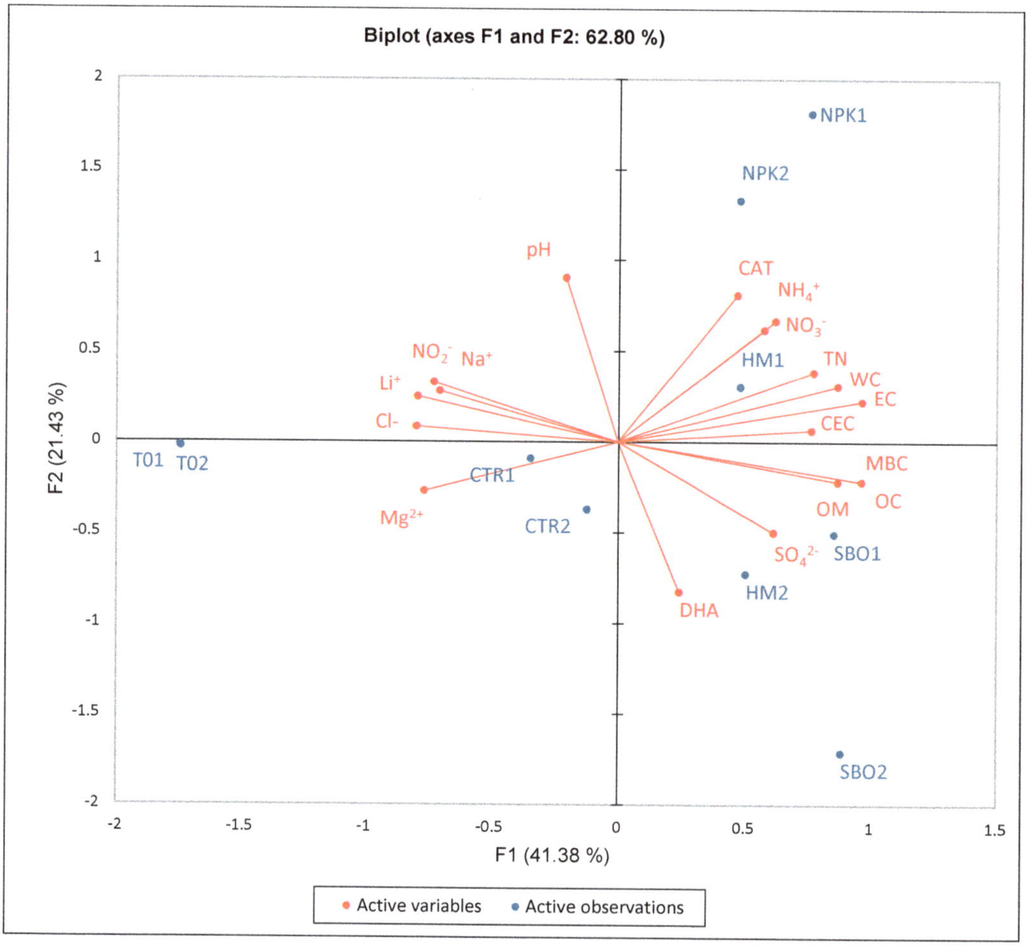

Figure 9. Principal component analysis (PCA) of chemical, biochemical, and ionic soil parameters under the broccoli–pepper crop rotation. The biplot represents soil samples collected after the first cycle (T01, CTR1, NPK1, HM1, SBO1) and the second cycle (T02, CTR2, NPK2, HM2, SBO2).

To assess whether the combined effects of fertilization and crop rotation influenced soil functioning, a correlation matrix was computed using all chemical and biochemical variables measured across both crop cycles (Table 6). This analysis revealed several significant relationships that clarify how different management strategies shape soil fertility dynamics. The correlation matrix highlights strong positive relationships among organic matter (OM), organic carbon (OC), total nitrogen (TN), cation exchange capacity (CEC), and microbial biomass carbon (MBC), indicating enhanced fertility with organic inputs. Enzymatic activities (FDA, DHA) correlate positively with MBC and negatively with pH and catalase (CAT), suggesting increased microbial activity under acidified conditions. Electrical conductivity (EC) and water content (WC) also show positive associations with key fertility indicators.

Table 6. Correlation matrix (Pearson (n − 1) among soil chemical, biochemical, and ionic variables measured across both crop rotation cycles and all fertilization treatments. Values in bold are different from 0 with a significance level, alpha = 0.005.

Variables	WC	pH	EC	WSP	FDA	CAT	DHA	OC	TN	C/N	OM	CEC	MBC
WC	1	−0.313	0.446	0.042	0.144	0.049	0.242	0.773	0.760	0.141	0.779	0.800	0.718
pH	−0.313	1	−0.242	−0.199	−0.730	0.615	**−0.901**	−0.506	−0.326	−0.132	−0.522	−0.064	−0.715
EC	0.446	−0.242	1	0.225	0.209	−0.106	−0.020	**0.820**	0.638	0.334	**0.810**	0.269	0.704
WSP	0.042	−0.199	0.225	1	−0.274	0.146	0.080	0.275	−0.265	**0.842**	0.159	0.116	0.062
FDA	0.144	−0.730	0.209	−0.274	1	**−0.901**	0.781	0.266	0.399	−0.384	0.345	−0.099	0.642
CAT	0.049	0.615	−0.106	0.146	**−0.901**	1	−0.632	−0.056	−0.180	0.394	−0.120	0.394	−0.416
DHA	0.242	**−0.901**	−0.020	0.080	0.781	−0.632	1	0.267	0.183	−0.035	0.291	0.097	0.573
OC	0.773	−0.506	**0.820**	0.275	0.266	−0.056	0.267	1	0.767	0.426	**0.992**	0.640	**0.892**
TN	0.760	−0.326	0.638	−0.265	0.399	−0.180	0.183	0.767	1	−0.225	**0.821**	0.490	**0.839**
C/N	0.141	−0.132	0.334	**0.842**	−0.384	0.394	−0.035	0.426	−0.225	1	0.320	0.361	0.089
OM	0.779	−0.522	**0.810**	0.159	0.345	−0.120	0.291	**0.992**	**0.821**	0.320	1	0.625	**0.922**
CEC	0.800	−0.064	0.269	0.116	−0.099	0.394	0.097	0.640	0.490	0.361	0.625	1	0.481
MBC	0.718	−0.715	0.704	0.062	0.642	−0.416	0.573	**0.892**	**0.839**	0.089	**0.922**	0.481	1

Three-way ANOVA (Table 7) revealed that fertilization had a significant effect on all soil chemical and biological variables ($p < 0.001$). Several parameters, including MBC, CAT, DHA, FDA, and OM, also exhibited significant two- and three-way interactions.

Table 7. The p-values from three-way ANOVA testing the effects of fertilization, crop rotation, and year on soil chemical and biological variables. Significance levels: * $p < 0.05$, ** $p < 0.01$, *** $p < 0.001$, n.s. = not significant ($p > 0.05$).

Variable	Fertilizer	Fertilizer × Rotation	Fertilizer × Year	Fertilizer × Year × Rotation	Rotation	Year	Year × Rotation
C/N	***	***	***	***	n.s	***	***
CAT	***	***	***	***	***	***	n.s
CEC	***	***	***	***	*	***	n.s
DHA	***	***	***	**	***	***	*
EC	***	n.s	***	***	n.s	***	***
FDA	***	***	***	***	n.s	***	n.s
MBC	***	**	**	***	***	***	***
OC	***	*	*	**	n.s	***	***
OM	***	***	**	***	n.s	***	***
TN	***	n.s	***	***	n.s	n.s	***
WC	***	***	***	***	***	***	***
WSP	***	***	***	***	***	***	***
pH	***	n.s	*	n.s	n.s	***	n.s

Main effects of the year were significant for most variables, especially for microbial and enzymatic indicators such as FDA and WC. By contrast, the rotation alone had limited effects, but its interaction with other factors (notably Year × Rotation) was significant for selected variables including OC, TN, and WC.

4. Discussion

Scientific studies have investigated the individual and combined effects of crop rotation and organic amendments on soil health and crop productivity illuminating that both practices offer significant benefits, with their effectiveness varying in respect to specific agricultural contexts and objectives. Crop rotation has been shown to enhance soil microbial biomass and diversity. A global synthesis revealed that crop rotation increased soil microbial biomass carbon (MBC) by 13.43% and microbial biomass nitrogen (MBN) by 15.84%, compared to continuous monoculture systems [34]. Additionally, Shannon's diversity index, a measure of microbial diversity, was significantly elevated by 7.68% under crop rotation practices [35]. These improvements were attributed to the varied root exudates and organic matter inputs from different crops, which create a more hospitable environment for diverse microbial communities [36]. These findings can also explain the increase in organic matter in the plots amended with the inorganic fertilizer NPK. The application of organic amendments instead directly added organic matter to the soil, thereby enhancing the soil organic carbon (SOC) stocks. Research carried on in UK arable systems, comparing diversified crop rotations and organic amendments have confirmed our results, evidencing significant SOC accumulation in the upper layer of the soil (0–30 cm) amended with organic fertilizers [37]. The increase in soil organic carbon (SOC) has been consistently associated with enhanced soil physical and chemical properties, including improved aggregate stability, increased water-holding capacity, and greater nutrient retention and availability. Furthermore, the application of organic amendments significantly stimulated soil biological activity, as evidenced by elevated enzymatic activities, such as dehydrogenase and FDA. These enzymatic processes are key indicators of microbial functioning and nutrient turnover, thereby playing a pivotal role in sustaining nutrient cycling and promoting long-term soil health and fertility [38]. Our results demonstrated that fertilization exerts a more dominant influence on soil fertility and ecosystem functioning than crop rotation alone. Specifically, the control plots subjected to crop rotation without fertilization exhibited significantly lower levels of soil organic matter, microbial biomass carbon (MBC), and enzymatic activities, indicating that the absence of nutrient input limits biological and biochemical soil functions. This trend was further supported by observations in the plots treated with mineral fertilizers (NPK), which also showed reduced concentrations of organic matter, MBC, and enzymatic activity compared to the organically amended plots. These findings highlighted the limited capacity of crop rotation by itself to sustain soil biological quality in the absence of external nutrient inputs. A principal component analysis (PCA) further reinforced this conclusion by revealing a strong positive correlation between dehydrogenase activity, MBC, and organic matter with the HM and SBO treatments, underscoring the central role of fertilization in enhancing microbial-driven soil processes. While both crop rotation and organic fertilization independently contributed to the improved soil health, they contribute to varying extents. Organic fertilization, particularly with biologically enriched amendments, tends to have a more immediate and pronounced effect on microbial activity, soil organic carbon (SOC) accumulation, and nutrient availability. By contrast, crop rotation—especially when incorporating diverse species or extended ley periods—contributes more gradually by enhancing root diversity, soil structure, and long-term nutrient cycling. For example, integrating multi-year grass–clover leys within crop rotations, combined with compost fertilization, has been shown to significantly increase SOC levels compared to either practice alone [39]. This combination leverages both the continuous input of organic residue and the structural and functional benefits of diverse cropping, ultimately leading to improved soil fertility, biological activity, and agroecosystem resilience. The changes observed in the soil properties across two crop rotation cycles (broccoli–tomato and broccoli–pepper) highlighted the interactive effects of both crop identity and fertilization strategies. While the crop species (tomato vs. pepper) significantly influenced microbial and nutrient dynamics, the type of fertilizer applied played a central role in modulating soil health indicators, such as pH, enzymatic activity, organic matter, and nutrient availability [40]. These findings reinforce the complexity of plant–soil–microbe–fertilizer interactions in agroecosystems, and their importance in shaping sustainable soil management practices [41]. The fertilizer type had a marked influence on the direction and magnitude of soil property changes. Notably, SBO and HM were more effective than NPK or the unfertilized control in enhancing microbial activity, organic carbon levels, and nutrient cycling. Organic fertilizers, such as SBO and HM, significantly reduced the soil pH, particularly with SBO, and with the effect being more pronounced in the broccoli–pepper system in the second cycle of both rotations. This acidification is likely due to the sulfur added with SBO that is well known to decrease the pH. Souri and Sayadi [42] demonstrated the efficiency of sulfur bentonite granules in decreasing the soil pH and improving the uptake of nutrient elements by the crop cultivated in calcareous soil. Additionally, the decrease in the soil pH in the presence of HM and mostly of SBO can be ascribed also to the oxidation of organic compounds, nitrification of ammonium, and excretion of organic acids by roots due to an increase in microorganisms enhanced by the high carbon input from organic amendments and by sulfur, as already demonstrated by other authors [43]. The mild acidification of the NPK fertilization is consistent with its targeted nutrient supply and minimal organic content. The increase in EC, most prominent under NPK, followed by SBO, reflected the accumulation of soluble ions released during the mineralization of organic matter (in SBO) or supplied directly via inorganic salts (in NPK). There is a difference in the soil electric conductivity also between tomato and pepper in rotation with broccoli, which may be due to the different absorption capacity and amount of soil ions between the two varieties. Although higher EC can indicate improved nutrient availability, it may also signal potential risks of salinity build up by fertilizer application [44].

All fertilization treatments, including the unfertilized control, even if at a different extent, led to an increase in the soil organic carbon (OC) and total nitrogen (TN), suggesting a cumulative contribution from plant residues and microbial biomass. However, the greatest increases were recorded under SBO and HM, due to their direct organic inputs and their role in stimulating microbial carbon use efficiency and residue incorporation [45]. By contrast, NPK showed a limited capacity to build the SOC, confirming the well-established notion that inorganic fertilization alone lacks the capacity to sustain long-term improvements in soil organic matter [46]. Interestingly, while SBO initially increased TN, a decline was observed in the second cycle of broccoli–pepper, likely due to the combination of rapid nitrogen mineralization and high uptake by pepper plants. This points to a potential mismatch between nitrogen release and crop demand, emphasizing the need for synchronized fertilization strategies. CEC increased under all treatments except for NPK, with the most significant enhancement seen in the SBO- and HM-treated soil. This is attributable to the high humified material supported by high value of C/N which contribute to greater ion retention and buffering capacity [44]. The decline in exchangeable calcium, sodium, and magnesium, particularly under organic treatments, may result from enhanced leaching or competitive uptake by plants and microorganisms. Organic amendments significantly promoted enzymatic activities (FDA and DHA), confirming their role in stimulating microbial metabolism through increased substrate availability [47]. The application of SBO, in particular, elevated water-soluble phenols (WSP), which serve as carbon-rich energy sources for microbes and can enhance the degradation of complex organic matter [48].

However, catalase (CAT) activity declined under SBO, suggesting a possible oxidative stress response or a shift in microbial community composition, possibly away from catalase-producing taxa [47]. While a decrease in catalase activity is often associated with oxidative stress and microbial imbalance, in certain contexts it may also reflect a stabilization or maturation of the soil microbial community. For example, in soil where organic matter inputs are reduced and overall microbial respiration rates decline, the generation of reactive oxygen species (ROS), like hydrogen peroxide, may also decrease. As a result, the microbial demand for antioxidant enzymes, such as catalase, diminishes, leading to lower catalase activity. In this case, the decline could indicate a lower oxidative burden in the soil environment, potentially associated with more stable redox conditions or a shift toward microbial communities that maintain equilibrium with reduced enzymatic antioxidant requirements. The contrasting effects of SBO—increasing overall microbial activity while reducing CAT—highlight the complex, compound-specific impacts of organic fertilizers on microbial ecology. The NPK application led to increased nitrate and ammonium concentrations, indicating high immediate nutrient availability. However, its effect on microbial activity, organic carbon, and CEC was limited, aligning with the findings that inorganic fertilization often supports short-term productivity but fails to improve the long-term soil health [49]. The reduction in WSP under the NPK treatment further underscores its limited support for microbial substrate pools. While the type of fertilizer shaped the baseline soil changes, the subsequent crop species modulated the degree and direction of those changes through rhizosphere-driven mechanisms. For instance, pepper, particularly under SBO, intensified the decline in TN and pH, and amplified enzymatic activities, compared to tomato. This suggests that pepper's rhizosphere environment is more dynamic, with higher exudate-driven microbial turnover, possibly due to greater dependence on AMF symbioses and root-induced nutrient solubilization [50]. The synergistic effect of organic amendments with pepper roots likely accelerated nutrient cycling processes, explaining the elevated nitrate and sulfate concentrations. Conversely, tomato's effects appeared more moderate and buffered, promoting a more stable microbial environment, with less dramatic shifts in nitrogen and pH.

The strong and consistent effect of fertilization across all variables supports the central role of nutrient input in determining soil fertility status. However, the significant interaction effects—especially those involving year and crop rotation—highlight that fertilization outcomes are not static, but depend on the temporal and biological context.

Microbial and enzymatic indicators, such as MBC, CAT, DHA, and FDA, were especially sensitive to interactions, suggesting that microbial dynamics are modulated by both management practices and the crops involved. The effect of rotation was more evident in the interaction terms than as a main factor, which suggests that crop sequence influences soil processes indirectly, by modifying the plant–microbe interactions over time.

These results emphasize the need to consider fertilization strategies in combination with crop rotation and seasonal dynamics to optimize soil health and biological functioning in diversified cropping systems. In the current context of high fossil fuel prices, the use of crop rotation combined with sulfur bentonite organo-mineral fertilizer (SBO) offers a more sustainable alternative to synthetic fertilizers such as NPK. With conventional fertilizers priced at €616/ton for nitrogen, €525/ton for phosphate, and €534/ton for potassium, SBO, priced at only €230/ton [7], provides a cost-effective and environmentally friendly option, particularly when considering its additional value in land restoration. This integrated approach supports both soil health and long-term economic viability for farmers.

5. Conclusions

The outcomes of this study demonstrate that both fertilization strategy and crop identity play pivotal roles in shaping soil chemical and biochemical properties across successive rotation cycles. Organic amendments, such as SBO and HM, significantly enhanced microbial activity, soil organic matter content, and cation exchange capacity. However, these benefits were accompanied by increased acidification and accelerated nutrient turnover—particularly when applied in conjunction with pepper cultivation, a biologically active crop. By contrast, the NPK fertilization promoted short-term nutrient availability but did not contribute substantially to long-term indicators of soil health.

These findings underscore the importance of considering the interaction between fertilizer type and crop species in the design of crop rotation systems. Organic inputs, when strategically paired with microbial-stimulating crops, like pepper, may optimize soil microbial functioning and nutrient cycling. However, this approach also presents challenges, such as the need for careful monitoring of soil pH, nitrogen availability and synchronization, and the potential buildup of salinity.

Looking forward, the key challenge will be to refine and implement fertilization regimes that are not only crop-specific but responsive to site conditions and long-term sustainability goals. Future research should focus on developing dynamic nutrient management frameworks that integrate organic and inorganic inputs, account for microbial functional diversity, and mitigate the risks associated with acidification and nutrient imbalance. Moreover, there is a pressing need to evaluate these systems under varying climatic and soil conditions to ensure their adaptability and resilience.

In summary, our findings indicate that, under the specific pedoclimatic conditions of this study, organic fertilization can independently enhance soil ecosystem functioning, regardless of the crop rotation scheme. This suggests that neither crop rotation nor organic amendment alone can be universally regarded as superior; rather, their effectiveness is highly context-dependent. A synergistic strategy that combines tailored organic inputs with carefully planned crop rotations appears to be the most effective approach for improving soil health, strengthening agroecosystem resilience, and advancing sustainable agricultural productivity.

Author Contributions: Conceptualization, A.M. (Adele Muscolo); methodology, F.M.; software, A.M. (Angela Maffia); validation, A.M. (Angela Maffia); S.B., F.M., M.O., and C.M.; investigation, S.B.; resources, A.M. (Adele Muscolo); data curation, A.M. (Angela Maffia); writing—original draft preparation, A.M. (Adele Muscolo); writing—review and editing, A.M. (Angela Maffia) and A.M. (Adele Muscolo); visualization, S.B.; supervision, A.M (Adele Muscolo); project administration, A.M. (Adele Muscolo); funding acquisition, A.M. (Adele Muscolo). All authors have read and agreed to the published version of the manuscript.

Funding: This research was financially supported by the project LIFE20 ENV/IT/000229—LIFE RecOrgFert PLUS, and funded by the LIFE Programme of the European Commission.

Data Availability Statement: The original contributions presented in this study are included in the article. Further inquiries can be directed to the corresponding author.

Conflicts of Interest: The authors declare no conflicts of interest.

References

1. Cole, M.B.; Augustin, M.A.; Robertson, M.J.; Manners, J.M. The Science of Food Security. *Npj Sci. Food* **2018**, *2*, 14. [CrossRef] [PubMed]
2. Techen, A.-K.; Helming, K. Pressures on Soil Functions from Soil Management in Germany. A Foresight Review. *Agron. Sustain. Dev.* **2017**, *37*, 64. [CrossRef]
3. Virto, I.; Imaz, M.; Fernández-Ugalde, O.; Gartzia-Bengoetxea, N.; Enrique, A.; Bescansa, P. Soil Degradation and Soil Quality in Western Europe: Current Situation and Future Perspectives. *Sustainability* **2014**, *7*, 313–365. [CrossRef]
4. Chew, K.W.; Chia, S.R.; Yen, H.-W.; Nomanbhay, S.; Ho, Y.-C.; Show, P.L. Transformation of Biomass Waste into Sustainable Organic Fertilizers. *Sustainability* **2019**, *11*, 2266. [CrossRef]
5. Muscolo, A.; Mauriello, F.; Marra, F.; Calabrò, P.S.; Russo, M.; Ciriminna, R.; Pagliaro, M. AnchoisFert: A New Organic Fertilizer from Fish Processing Waste for Sustainable Agriculture. *Glob. Chall.* **2022**, *6*. [CrossRef]
6. Maffia, A.; Marra, F.; Celano, G.; Oliva, M.; Mallamaci, C.; Hussain, M.I.; Muscolo, A. Exploring the Potential and Obstacles of Agro-Industrial Waste-Based Fertilizers. *Land* **2024**, *13*, 1166. [CrossRef]
7. Panuccio, M.R.; Marra, F.; Maffia, A.; Mallamaci, C.; Muscolo, A. Recycling of Agricultural (Orange and Olive) Bio-Wastes into Ecofriendly Fertilizers for Improving Soil and Garlic Quality. *Resour. Conserv. Recycl. Adv.* **2022**, *15*, 200083. [CrossRef]
8. Rodale Institute. Regenerative Organic Agriculture and Climate Change a Down- to-Earth Solution to Global Warming. 2014. Available online: https://rodaleinstitute.org/wp-content/uploads/rodale-white-paper.pdf (accessed on 2 March 2025).
9. Liu, C.; Plaza-Bonilla, D.; Coulter, J.A.; Kutcher, H.R.; Beckie, H.J.; Wang, L.; Floc'h, J.-B.; Hamel, C.; Siddique, K.H.M.; Li, L.; et al. Diversifying Crop Rotations Enhances Agroecosystem Services and Resilience. *Adv. Agron.* **2022**, *173*, 299–335. [CrossRef]

10. Pradhan, G.; Meena, R.S.; Kumar, S.; Lal, R. Utilizing Industrial Wastes as Compost in Wheat-Rice Production to Improve the above and below-Ground Ecosystem Services. *Agric. Ecosyst. Environ.* **2023**, *358*, 108704. [CrossRef]
11. Rai, A.; Singh, A.K.; Mishra, R.; Shahi, B.; Rai, V.K.; Kumari, N.; Kumar, V.; Gangwar, A.; Sharma, R.B.; Rajput, J.; et al. Sulphur in Soils and Plants: An Overview. *Int. Res. J. Pure Appl. Chem.* **2020**, *21*, 66–70. [CrossRef]
12. Lisowska, A.; Filipek-Mazur, B.; Kalisz, A.; Gorczyca, O.; Kowalczyk, A. Changes in Soil Sulfate Sulfur Content as an Effect of Fertilizer Granules Containing Elemental Sulfur, Halloysite and Phosphate Rock. *Agronomy* **2023**, *13*, 1410. [CrossRef]
13. Sayara, T.; Basheer-Salimia, R.; Hawamde, F.; Sánchez, A. Recycling of Organic Wastes through Composting: Process Performance and Compost Application in Agriculture. *Agronomy* **2020**, *10*, 1838. [CrossRef]
14. Maffia, A.; Scotti, R.; Wood, T.; Muscolo, A.; Lepore, A.; Acocella, E.; Celano, G. Transforming Agricultural and Sulfur Waste into Fertilizer: Assessing the Short-Term Effects on Microbial Biodiversity via a Metagenomic Approach. *Life* **2024**, *14*, 1633. [CrossRef] [PubMed]
15. Maffia, A.; Marra, F.; Canino, F.; Oliva, M.; Mallamaci, C.; Celano, G.; Muscolo, A. Comparative Study of Fertilizers in Tomato-Grown Soils: Soil Quality, Sustainability, and Carbon/Water Footprints. *Soil Syst.* **2023**, *7*, 109. [CrossRef]
16. Russo, M.; Di Sanzo, R.; Marra, F.; Carabetta, S.; Maffia, A.; Mallamaci, C.; Muscolo, A. Waste-Derived Fertilizer Acts as Biostimulant, Boosting Tomato Quality and Aroma. *Agronomy* **2023**, *13*, 2854. [CrossRef]
17. Marra, F.; Maffia, A.; Canino, F.; Greco, C.; Mallamaci, C.; Adele, M. Effects of fertilizer produced from agro-industrial wastes on the quality of two different soils. *Arch. Agron. Soil Sci.* **2023**, *69*, 3600–3618. [CrossRef]
18. Muscolo, A.; Mallamaci, C.; Settineri, G.; Calamara, G. Increasing soil and crop productivity by using agricultural wastes pelletized with elemental sulfur and bentonite. *Agron. J.* **2017**, *109*, 1900–1910. [CrossRef]
19. Muscolo, A.; Papalia, T.; Settineri, G.; Romeo, F.; Mallamaci, C. Three different methods for turning olive pomace in resource: Benefits of the end products for agricultural purpose. *Sci. Total Environ.* **2019**, *662*, 1–7. [CrossRef]
20. Muscolo, A.; Romeo, F.; Marra, F.; Mallamaci, C. Transforming agricultural, municipal and industrial pollutant wastes into fertilizers for a sustainable healthy food production. *J. Environ. Manag.* **2021**, *17*, 113771. [CrossRef]
21. Ben Said, I.; Muscolo, A.; Mezghani, I.; Chaieb, M. Reclaimed municipal wastewater for forage production. *Water Sci. Technol.* **2017**, *75*, 1784–1793.
22. Severson, R.C.; Shacklette, H.T. *Essential Elements and Soil Amendments for Plants: Sources and Use for Agriculture, 1017*; US Government Printing Office: Washington, DC, USA, 1988.
23. Bouyoucos, G.J. Hydrometer Method Improved for Making Particle Size Analyses of Soils[1]. *Agron. J.* **1962**, *54*, 464–465. [CrossRef]
24. Walkley, A.; Black, I.A. An examination of the Degtjareff method for determining soil organic matter and a proposed modification of the chromic acid titration method. *Soil Sci.* **1934**, *37*, 29–38. [CrossRef]
25. Kjeldahl, J. Neue Methode zur Bestimmung des Stickstoff in organishen Kopern. *Anal. Chem.* **1883**, *22*, 354–358.
26. Kaminsky, R.; Müller, W.H. A recommendation against the use of alkaline soil extractions in the study of allelopathy. *Plant Soil* **1978**, *49*, 641–645. [CrossRef]
27. Box, J.D. Investigation of the Folin–Ciocalteau reagent for the determination of polyphenolic substances in natural waters. *Water Res.* **1983**, *17*, 511–525. [CrossRef]
28. Hendershot, W.H.; Duquette, M. A Simple Barium Chloride Method for Determining Cation Exchange Capacity and Exchangeable Cations. *Soil Sci. Soc. Am. J.* **1986**, *50*, 605–608. [CrossRef]
29. Maffia, A.; Marra, F.; Canino, F.; Battaglia, S.; Mallamaci, C.; Oliva, M.; Muscolo, A. Humic Substances from Waste-Based Fertilizers for Improved Soil Fertility. *Agronomy* **2024**, *14*, 2657. [CrossRef]
30. Vance, E.D.; Brookes, P.C.; Jenkinson, D.S. An extraction method for measuring soil microbial biomass C. *Soil Biol. Biochem.* **1987**, *19*, 703–707. [CrossRef]
31. Adam, G.; Duncan, H. Development of a sensitive and rapid method for the measurement of total microbial activity using fluorescein diacetate (FDA) in a range of soils. *Soil Biol. Biochem.* **2001**, *33*, 943–951. [CrossRef]
32. Von Mersi, W.; Schinner, F. An improved and accurate method for determining the dehydrogenase activity of soils with iodonitrotetrazolium chloride. *Biol. Fertil. Soils* **1991**, *11*, 216–220. [CrossRef]
33. Sidari, M.; Santonoceto, C.; Anastasi, U.; Preiti, G.; Muscolo, A. Variations in four genotypes of lentil under NaCl-salinity stress. *Am. J. Agron. Biol. Sci.* **2008**, *3*, 410–416. [CrossRef]
34. Liu, Q.; Zhao, Y.; Li, T.; Chen, L.; Chen, Y.; Sui, P. Changes in Soil Microbial Biomass, Diversity, and Activity with Crop Rotation in Cropping Systems: A Global Synthesis. *Appl. Soil Ecol.* **2023**, *186*, 104815. [CrossRef]
35. Venter, Z.S.; Jacobs, K.; Hawkins, H.-J. The impact of crop rotation on soil microbial diversity: A meta-analysis. *Pedobiologia* **2016**, *59*, 215–223. [CrossRef]
36. Wang, L.; Zhao, Y.; Al-Kaisi, M.; Yang, J.; Chen, Y.; Sui, P. Effects of seven diversified crop rotations on selected soil health indicators and wheat productivity. *Agronomy* **2020**, *10*, 235. [CrossRef]

37. Zani, C.F.; Manning, D.A.C.; Abbott, G.D.; Taylor, J.A.; Cooper, J.; Lopez-Capel, E. Diversified Crop Rotations and Organic Amendments as Strategies for Increasing Soil Carbon Storage and Stabilisation in UK Arable Systems. *Front. Environ. Sci.* **2023**, *11*, 1113026. [CrossRef]
38. Zhou, J.; Xu, H.; Zhang, M.; Feng, R.; Xiao, H.; Xue, C. Impact of Organic Amendments on Black Wheat Yield, Grain Quality, and Soil Biochemical Properties. *Agronomy* **2025**, *15*, 961. [CrossRef]
39. Loaiza Puerta, V.; Pujol Pereira, E.I.; Wittwer, R.; van der Heijden, M.; Six, J. Improvement of Soil Structure through Organic Crop Management, Conservation Tillage and Grass-Clover Ley. *Soil Tillage Res.* **2018**, *180*, 1–9. [CrossRef]
40. Chen, Q.; Song, Y.; An, Y.; Lu, Y.; Zhong, G. Soil Microorganisms: Their Role in Enhancing Crop Nutrition and Health. *Diversity* **2024**, *16*, 734. [CrossRef]
41. Howe, J.A.; McDonald, M.D.; Burke, J.; Robertson, I.; Coker, H.; Gentry, T.J.; Lewis, K.L. Influence of Fertilizer and Manure Inputs on Soil Health: A Review. *Soil Secur.* **2024**, *16*, 100155. [CrossRef]
42. Souri, B.; Sayadi, Z. Efficiency of Sulfur-Bentonite Granules to Improve Uptake of Nutrient Elements by the Crop Plant Cultivated in Calcareous Soil. *Commun. Soil Sci. Plant Anal.* **2021**, *52*, 2414–2430. [CrossRef]
43. Huang, Y.; Kang, R.; Mulder, J.; Zhang, T.; Duan, L. Nitrogen Saturation, Soil Acidification, and Ecological Effects in a Subtropical Pine Forest on Acid Soil in Southwest China. *J. Geophys. Res. Biogeosci.* **2015**, *120*, 2457–2472. [CrossRef]
44. Tejada, M.; Garcia, C.; Gonzalez, J.L.; Hernandez, M.T. Use of Organic Amendment as a Strategy for Saline Soil Remediation: Influence on the Physical, Chemical and Biological Properties of Soil. *Soil Biol. Biochem.* **2006**, *38*, 1413–1421. [CrossRef]
45. Lal, R. Soil Carbon Sequestration to Mitigate Climate Change. *Geoderma* **2004**, *123*, 1–22. [CrossRef]
46. Titirmare, N.S.; Ranshur, N.J.; Patil, A.H.; Patil, S.R.; Margal, P.B. Effect of Inorganic Fertilizers and Organic Manures on Physical Properties of Soil: A Review. *Int. J. Plant Soil Sci.* **2023**, *35*, 1015–1023. [CrossRef]
47. Gao, Y.; Sun, S.; Xing, F.; Mu, X.; Bai, Y. Nitrogen Addition Interacted with Salinity-Alkalinity to Modify Plant Diversity, Microbial PLFAs and Soil Coupled Elements: A 5-Year Experiment. *Appl. Soil Ecol.* **2019**, *137*, 78–86. [CrossRef]
48. Badri, D.V.; Vivanco, J.M. Regulation and Function of Root Exudates. *Plant Cell Environ.* **2009**, *32*, 666–681. [CrossRef]
49. Maeder, P.; Fliessbach, A.; Dubois, D.; Gunst, L.; Fried, P.; Niggli, U. Soil Fertility and Biodiversity in Organic Farming. *Science* **2002**, *296*, 1694–1697. [CrossRef]
50. Philippot, L.; Raaijmakers, J.M.; Lemanceau, P.; van der Putten, W.H. Going Back to the Roots: The Microbial Ecology of the Rhizosphere. *Nat. Rev. Microbiol.* **2013**, *11*, 789–799. [CrossRef]

land

MDPI

Article

Estimating the Economic Cost of Land Degradation and Desertification in Morocco

Anas Laamouri [1,*] and Abdellatif Khattabi [2,*]

1 National Agency for Water and Forests, Rabat 10000, Morocco
2 National School of Forestry Engineering, Salé 11000, Morocco
* Correspondence: anaslaamouri@gmail.com (A.L.); ab_khattabi@yahoo.com (A.K.)

Abstract: Desertification affects over 90% of Moroccan territory, leading to soil degradation that reduces agricultural productivity, diminishes biodiversity, and alters environmental functions. This study estimates the total economic cost of desertification in Morocco using a zonal approach based on regional sensitivity. The methodology includes two stages: quantifying productivity losses from water and wind erosion, salinization, overgrazing, silting of dams, carbon storage loss, and land-use changes; and monetizing impacts using methods such as productivity change, replacement cost, and the social cost of carbon. The total cost is estimated at USD 2.1 billion per year, with 78.02% from agricultural and grazing land productivity losses, 2.95% from dam silting, 18.47% from carbon storage loss, and 0.56% from land-use changes. These findings underscore the urgency of public policies, including land use planning, sustainable agriculture, irrigation modernization, and community engagement. Drawing on successful initiatives in the MENA region and globally, Morocco can mitigate desertification's impacts and foster sustainable development.

Keywords: desertification; soil degradation; economic cost; land productivity; carbon storage loss

1. Introduction

Desertification, characterized by the degradation of land productivity, is a pressing environmental and developmental challenge exacerbated by anthropogenic activities, climate variability, and biodiversity changes [1–3]. This phenomenon affects over 40% of the global population, with more than 90% residing in developing countries, particularly in arid, semi-arid, and dry sub-humid regions [4]. Drylands, occupying 40% of the Earth's terrestrial surface, are the most vulnerable, with Africa accounting for 37% of global dryland areas, followed by Asia (33%) and Australia (14%) [5].

In Morocco, desertification is a critical issue, impacting over 90% of the country's territory, particularly in arid and semi-arid zones. This process is driven by fragile soils, climate conditions, and unsustainable exploitation of natural resources by rural populations [6,7]. To combat desertification, Morocco adopted a National Action Program to Combat Desertification (PANLCD) in 2001, emphasizing territorial integration, participatory approaches, and synergy between sectoral programs [8–10].

Land degradation imposes significant global economic costs, estimated at over USD 10 trillion annually, due to productivity losses, biodiversity decline, and water pollution [11]. While sustainable management projects, such as Mali's restoration initiative, have demonstrated high returns on investment (e.g., USD 12 for every USD invested), the economic burden remains disproportionately high in developing countries, where rural livelihoods are heavily dependent on natural resources. In Morocco, the economic cost of environmental degradation was estimated at USD 3.41 billion (3.52% of GDP) in 2014 [12].

Despite extensive research on global land degradation, few studies isolate desertification as a distinct phenomenon or address the economic costs associated with its impacts. Existing studies often overlook regional variations in these costs, particularly in heterogeneous landscapes like Morocco. This study seeks to fill this gap by answering the question: "How are the economic costs of desertification impacts distributed across Morocco's homogeneous zones, and what are the variations in these costs?"

To address this question, the study aims to:

- Assess land degradation and its economic impacts across Morocco's homogeneous zones.
- Quantify productivity losses resulting from water and wind erosion, salinization, overgrazing, and changes in land use.
- Develop actionable recommendations for sustainable land management tailored to the specific characteristics of each homogeneous zone.

By leveraging spatial analysis tools such as InVEST 3.10.2 software and economic evaluation methodologies—including productivity change and replacement cost—this research highlights the importance of incorporating desertification costs into decision-making. It also provides a framework for tailoring strategies to mitigate desertification and promote sustainable development in Morocco's diverse regions.

2. Literature Review

2.1. Global Perspectives on Desertification and Degradation

Land degradation and desertification have become critical global challenges, threatening ecosystems, economies, and livelihoods. The increasing reliance on satellite imagery and geospatial analysis has enhanced our understanding of these issues on a regional and global scale. Studies such as those by [13,14] have underscored the alarming rates of soil erosion under current land management practices and projected their worsening under climate change scenarios. These studies highlight the urgent need for targeted mitigation and adaptation strategies. Additionally, the authors of [15] provided a framework for linking land degradation to economic costs, emphasizing the global implications of failing to address desertification.

Other studies have shed light on regional disparities in desertification drivers and impacts. For example, ref. [16] demonstrated how varying socio-economic conditions influence land degradation trends across the MENA region. In sub-Saharan Africa, ref. [15] highlighted the role of poverty and weak governance in accelerating land degradation while simultaneously showcasing successful interventions through community-driven restoration projects. These global and regional perspectives underscore the complexity of desertification and the need for region-specific solutions.

2.2. Advances in Assessing Land Degradation

Technological advancements have played a pivotal role in advancing the methodologies used to assess land degradation. For instance, the integration of remote sensing data with field observations, as demonstrated by [15], has improved the accuracy of identifying degraded lands. Similarly, ref. [17] explored how hydrological modeling combined with socio-economic data can provide deeper insights into the cascading effects of degradation on water resources and agricultural productivity. Such interdisciplinary approaches have set a new benchmark in understanding the multifaceted nature of land degradation.

The use of spatially explicit models has also become increasingly prominent. The Universal Soil Loss Equation (USLE) remains a cornerstone for soil erosion assessments, but its integration with advanced tools like the InVEST Sediment Delivery Ratio (SDR) model has allowed for greater precision in quantifying erosion and its economic impacts [18,19]. Meanwhile, the application of machine learning techniques in land use classification, as seen in studies by [20], has improved the reliability of degradation assessments, particularly in heterogeneous landscapes.

Beyond technical tools, frameworks for understanding the socio-economic dimensions of desertification have gained traction. Ref. [21] emphasized the need to integrate socio-economic variables into desertification studies, arguing that land degradation cannot be fully understood without considering human activities, institutional weaknesses, and governance structures. Similarly, the role of community participation in combating desertification has been highlighted in studies like those of [22], who demonstrated the effectiveness of participatory approaches in reversing degradation trends in vulnerable regions.

2.3. Economic Impacts and Policy Implications

The economic consequences of land degradation are profound, with far-reaching impacts on national economies and local livelihoods. Ref. [12] estimated that land degradation costs some countries as much as 9% of their GDP annually. In the MENA region, ref. [12] found that Morocco's economic losses due to environmental degradation amounted to 3.52% of GDP, highlighting the urgency of addressing these issues. Studies have emphasized the need for robust policy frameworks that integrate land restoration with socio-economic development. For instance, ref. [15] suggested that incentivizing sustainable land management practices through subsidies and community programs can yield significant economic and environmental benefits.

2.4. Relevance to the Case Study

This study builds upon the methodologies discussed, applying them to Morocco's unique context. Unlike global studies that provide generalized findings, this research employs a zonal approach to assess economic costs across Morocco's homogeneous zones. By integrating spatial tools like InVEST and incorporating recent economic valuation techniques, this study bridges critical gaps identified in the prior literature, such as those highlighted [12]. Furthermore, it contributes to the growing body of knowledge by demonstrating how localized assessments can inform national and regional policy frameworks.

2.5. Analytical Framework for the Case Study

To ensure a robust analysis, this study adopts an interdisciplinary framework that integrates biophysical and economic methodologies. Key components include:

- Spatial analysis tools: Advanced tools like the InVEST SDR model and carbon storage models were employed to quantify erosion and ecosystem service losses. These tools provide high-resolution insights into degradation patterns, enabling targeted interventions.
- Economic valuation: The economic impacts of degradation were monetized using established methods, including replacement costs for dam siltation and social costs of carbon for ecosystem losses. This approach ensures alignment with global best practices while tailoring the results to Morocco's context.
- Zonal approach: By segmenting Morocco into homogeneous zones based on climatic, geological, and land-use characteristics, this study captures the spatial variability of degradation impacts. This approach ensures that recommendations are context-specific and actionable.

3. Materials and Methods

3.1. Study Area

3.1.1. Relief and Climate

Morocco's relief (Figure 1a) has been spatialized on the basis of USGS Earth Explorer data at 30 m resolution and is highly diversified, influencing both climate and vegetation. On the one hand, there are the mountain ranges with their high peaks, such as the mountain of Toubkal at 4167 m, which is the highest point in the Anti-Atlas, Middle Atlas, and High Atlas, as well as the Rif mountains. In addition, there are the fertile Atlantic coastal plains and the arid plateaus to the east, and finally, the south and southeast regions of the country (the Sahara) are dominated by expanses of desert and Saharan dunes.

With regard to rainfall distribution, we used [23] standardized rainfall index over 40 years (1981–2020). After studying this map (Figure 1b), we can see that the Rif mountain range in the north is the country's wettest region, with annual rainfall totals in excess of 1000 mm, records annual rainfall of less than 400 mm, with values decreasing towards the south and west, particularly in the area framed by the Atlas Mountains, with stations recording values ranging from 600 mm in the north to 200 mm in the south. The south and southeast, which are the pre-desert and Saharan domains, are characterized by extreme drought, with annual accumulations of less than 100 mm.

3.1.2. Soil and Vegetation

The land cover of Morocco 2020 (Figure 1c) with a resolution of 10m was drawn up based on the [24] SENTINEL-2 Land Cover product, which was created using SENTINEL-2 satellite imagery from the European Space Agency (ESA). The Impact Observatory is the classification model used to create this product, and it identified 10 classes of the earth's surface, including vegetation, bare surfaces, crops, and urban areas.

Morocco's soils (Figure 1d) are the fruit of varied and sometimes complex natural processes, giving rise to a great diversity of soils. According to [25], Morocco has sandy soils rich in iron oxides, clay soils near the coast, and brown soils under forests, which are highly suitable for agriculture. In limestone areas, the slope and disappearance of vegetation accelerate soil degradation. There are also rough mineral soils on rocky formations and eroded soils on marl soils, as well as shallow soils on steep slopes, which are highly subject to erosion.

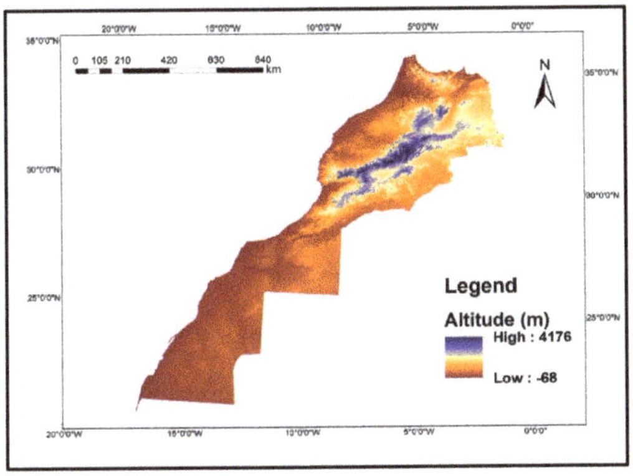

(a)

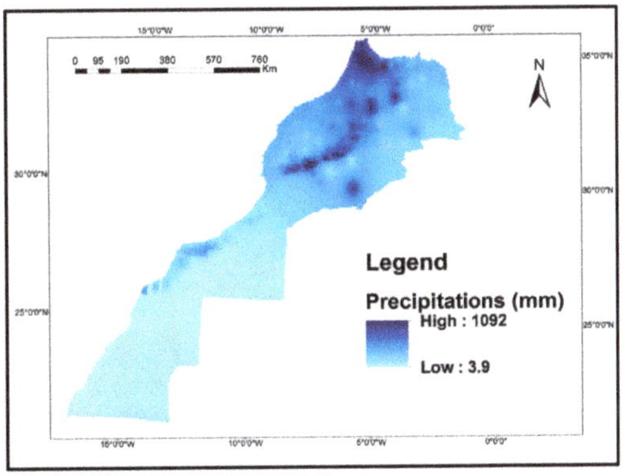

(b)

Figure 1. *Cont.*

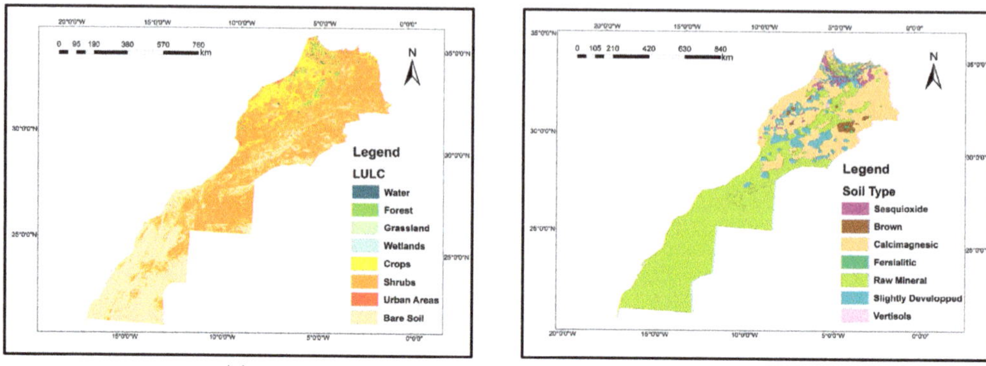

(c) (d)

Figure 1. (**a**). Relief map of Morocco. Source: [26]. (**b**). Map of mean annual precipitation in Morocco (1981–2020). Source CHIRPS data: [23]. (**c**). Impact Observatory land-use map of Morocco 2020. Source: adapted from [24]. (**d**). Typical soil map of Morocco. Source: [25].

3.2. Methodology

In order to study the phenomenon of land degradation and desertification in Morocco, we worked on eight homogeneous zones (Figure 2) identified and delimited within the framework of the PANLCD 2013 with regard to sensitivity to desertification. In fact, the delimitation of these zones was based on 3 main criteria according to [9], which are relief, climate, and soil capital.

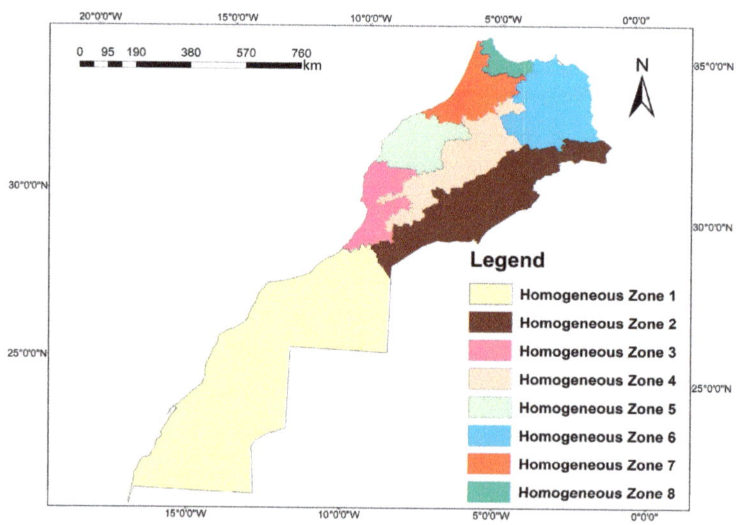

Figure 2. Map of homogeneous zones in Morocco. Source: adapted from [9].

To estimate the total economic cost of desertification in Morocco in 2020, we proceeded to monetize the impacts of desertification. The methodological approach adopted in our study does not cover all the cases likely to occur in reality but focuses on the most frequently encountered and most important cases. We have used a zonal approach to desertification sensitivity to calculate the costs of the following types of degradation: degradation of agricultural land by water and wind erosion and salinization, followed by overgrazing of rangelands, silting of dams, loss of carbon storage, and changes in land use.

To do this, our methodology is divided into 2 main stages: firstly, we quantified the various impacts of degradation based on InVEST, the PANLCD 2013 report, and secondly, we monetized these impacts using the various methods for assessing the economic cost of degradation, the totals of which are the total economic cost of desertification in Morocco in 2020.

3.2.1. Quantifying the Impacts of Desertification

A. Degradation of agricultural land

A-1 erosion

In order to assess the extent of water erosion-related land degradation in Morocco, we used the Sediment Delivery Ratio (SDR) model of the InVEST tool, which uses the universal soil loss equation of [27]: USLE = R × K × LS × C × P to estimate the potential soil loss for each pixel within the country.

The inputs to the SDR model are as follows:

- Digital terrain model raster (Figure 1a);
- Raster of the erosivity factor (R) (See Supplementary Materials Figure S1) calculated on the basis of the global rainfall erosivity map [28];
- Raster of the erodibility factor (k) (See Supplementary Materials Figure S2) determined using the harmonized soil database [13] and the erodibility factor estimation table K [29];
- Land use raster (Figure 1c);
- Biophysical table (Table 1) with two factors (—Usle_c: cover management factor—Usle_p: supporting practices factor).

Table 1. Biophysical table of sediment delivery ratio (SDR) model.

Description	Lucode	Usle_c	Usle_p	Data Source
Water	1	0.04	1	[18]
Forest	2	0.003	1	[18]
Pastures	3	0.15	0.9178	[19]
Wetlands	4	0.03	1	[18]
Cultivated Land	5	0.19	1	[18]
Shrub vegetation	6	0.05	0.9178	[19]
Built-up areas	7	0.1	1	[18]
Bare Soil	8	1	1	[18]

After running the SDR model, we obtained the total amount of potential soil loss due to water erosion in Morocco, then mapped this information by water erosive state, which according to [30] are grouped into 4 states: very low (<5 t/ha/yr), low (5–10 t/ha/yr), medium (10–20 t/ha/yr), and high (≥20 t/ha/yr). As the high level of degradation was irrecoverable, it was not included in the analysis. This map of erosive water conditions was then superimposed on the land use map to determine the surface area of agricultural land affected by light (very low, low) and moderate erosion.

A-2 Wind erosion

To measure the extent of land degradation linked to wind erosion, we used the map (Figure 3) drawn up as part of PANLCD 2013. Studying this map enabled us to calculate the surface area affected by each erosive condition, which are 35.23 million hectares for very low erosion, and 3.37, 24.17, and 7.24 million hectares of affected surface area for low, medium, and high erosion, respectively.

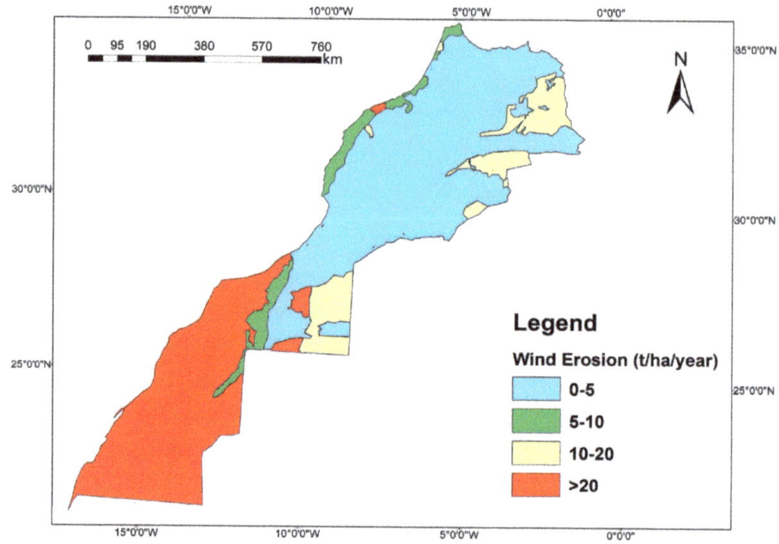

Figure 3. Wind erosion states in Morocco. Source: adapted from: [9].

A-3 Salinization

In order to quantify salinization-related degradation, we used the percentage of soil area degraded by salinization, as calculated within the framework of the PANLCD and listed in Table 2. We then calculated the area of soil affected by salinization for each homogeneous zone.

Table 2. Percentage of agricultural land area affected by salinization.

Homogeneous Zone	% Area of Agricultural Land Affected by Salinization
1	1
2	1
3	5
4	1
5	1
6	1
7	1
8	1

Source: [9].

B. Overgrazing of rangelands

To quantify the degradation of rangelands, a two-step method was used.

Firstly, we calculated the overall forage production of natural rangelands (forest, alfa grass cover, and productive rangelands) for each homogeneous zone. Based on the areas drawn from the land use map (Figure 3) and specific data from [31,32], the estimate of productive rangelands was set at 12.5 forage units/ha. The needs of livestock satisfied by stabling, representing 40% of their total consumption, were taken into account [6].

The results were calculated from data provided by the HCP in several reports [32–35] in Livestock Units (LU) according to the ratios of [6], and forage requirements in forage units per LU were set at 1200 FU/ha. Overconsumption was finally calculated by subtracting local production from the estimated offtake for each zone.

C. Siltation of dams

To quantify ex situ land degradation resulting in dam silting, we initially calculated the amount of sediment exported (in tons per year) by each major Moroccan watershed, using the InVEST (SDR) model and the same parameters as those employed to quantify water erosion (in situ), in addition to shapefiles representing Moroccan watersheds (Figure 4) and then identified the main dams located downstream of each watershed. To quantify dam siltation by homogeneous zone, the watershed map was superimposed on the homogeneous zone map.

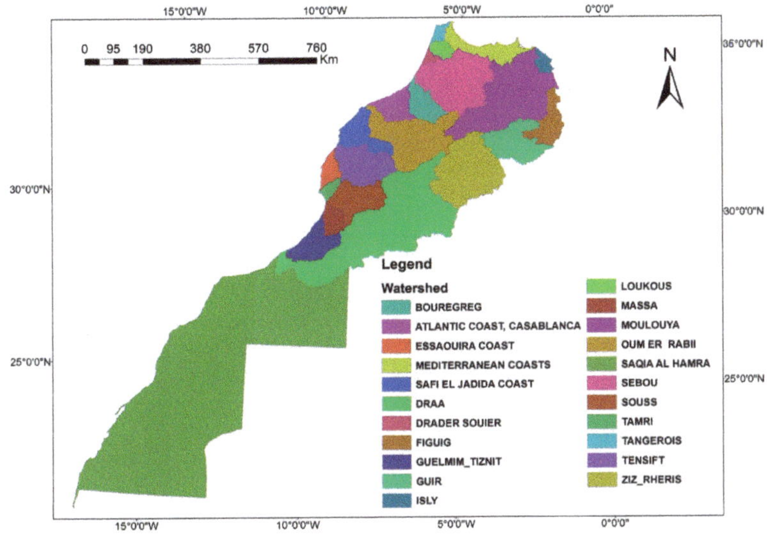

Figure 4. Watersheds of Morocco. Source: adapted from: [36].

D. Loss of carbon storage

In order to quantify the carbon storage lost annually, we used InVEST's Carbon Storage and Sequestration model, applied between 2000 and 2020. We then calculated the difference in the quantity of carbon between these two dates to determine the quantity of degraded carbon storage in Morocco, and finally to identify its annual loss in Morocco. To do this, we need two inputs: the land cover rasters for the years 2000 and 2020 (See Supplementary Materials Figures S3 and S4), which will be drawn up on the basis of the CAS EARTH [37] product at 30m resolution, using pixel-based image classification and object-based image classification, and offering 10 classes, 8 of which are suitable for Morocco, namely crop, forest, grassland, shrub vegetation, wetland, water body, urban area, and bare soil. We also use a biophysical table (Table 3) comprising above-ground biomass carbon density, below-ground biomass, soil carbon, and dead matter carbon.

Table 3. Biophysical table of carbon storage and sequestration.

LULC_Desc	Lucode	C_above	C_below	C_soil	C_dead	Data Source
Cultivated Land	1	44.43	29.3	10.45	0	[20]
Forest	2	132.4	26.14	15.67	0	[20]
Pastures	3	31.2	1.1	18.67	0	[20]
Shrub vegetation	4	10	3	25	0	[19]
Wetlands	5	82	1.3	20	0	[18]
Water bodies	6	0	0	0	0	[18]
Built-up areas	7	3	0.6	13.5	0	[20]
Bare Soil	8	3.5	0.35	16.5	0	[18]

E. Land use trends

In order to understand the evolution of land use, we used the MCD12Q1.061 MODIS product, which offers 500m resolution with an IGBP classification of 17 classes according to [38], of which only 14 are available in Morocco. Our study was based on multi-date mapping between 2001 and 2020, with change analysis carried out using the Google Earth Engine, which enabled us to obtain the change matrix for transitions between the main land uses, which we will reduce to 7 classes (Table 4). In order to attenuate the annual variations recorded due to climatic conditions, notably precipitation, we calculated a five-year moving average of the changes over a 16-year period from 2005 to 2020. Based on the moving averages, we calculated the average annual change for each land use class.

Table 4. Classes used to assess land use evolution 2001–2020.

Integration of Classes to Assess Evolution Between 2001 and 2020	Classes MCD12Q1.061
Bare soil or sparse vegetation	Bare soil or sparse vegetation
Cultivated land	A mosaic of cultivated land and natural vegetation Cultivated land
Forests	Deciduous forests Evergreen deciduous forests Mixed forests Evergreen forests
Pastures	Pastures
Shrub vegetation	Dense shrublands Open shrublands Savannahs Wooded savannahs
Built-up areas	Urbanized and built-up areas
Wetlands	Wetlands

3.2.2. Estimating the Economic Cost of Degradation

A. Degradation of agricultural land

A-1 Erosion

The economic cost of agricultural land erosion due to water and wind erosion in 2020 was estimated on the basis of the loss in value of agricultural production caused by a fall in productivity. For each homogeneous zone, we used four cropping systems: cereal, legume, sugar, and oilseed, based on [32] data by province. We then calculated a weighted average of prices in USD per quintal: 27.83 for cereals, 89.36 for pulses, 72.66 for oilseeds, and 13.38 for sugars to estimate cereal productivity in each homogeneous zone.

In short, the economic cost of lost productivity is calculated by multiplying lost agricultural production (PAP) by the price of cereals in 2020. This estimate was made for each homogeneous zone using the following formula: PAP (homogeneous zone) = lost agricultural productivity × area of land affected by different erosive conditions

A-2 Salinization

To calculate the cost of salinization-induced land degradation, we multiplied the affected land area for each homogeneous zone by 390.5 USD/ha, which is the economic cost induced by losses in agricultural productivity due to salinization according to [12]

B. Overgrazing of rangelands

The cost associated with rangeland degradation is equivalent to the quantity of forage units over-consumed, multiplied by the unit price of a forage unit. Each forage unit represents one kilogram of barley. According to [32], the unit price of a kilogram of barley is between USD 0.26 and 0.83.

C. Siltation of dams

To calculate the cost of land degradation ex situ, we first transformed the quantity of sediment exported from the t into m^3, knowing that the weight by volume of dam silt is 1.8 t/m^3 [39], then used the replacement cost method. This approach was supported by reference to the cost of developing water resources, assessed on the basis of the cost of mobilizing water from another dam intended to replace the lost storage capacity, which consisted of multiplying this quantity by the average cost of developing the mobilized water, varying between 0.21 and 0.63 USD/m^3 according to data from [40].

D. Loss of carbon storage

In order to estimate the economic cost of the loss of carbon storage, our analysis is based on the social valuation of a ton of sequestered carbon, representing the social damage avoided by preventing the release of CO_2. According to [12], the social price of carbon ranges from USD 40 to 80. To monetize the quantity of degraded carbon obtained by the InVEST model, we multiply this quantity by the extreme values of this range, then average the two.

E. Land use trends

To estimate the cost of the average annual change in land use on the total economic value (Table 5), 4 classes will be retained for reasons of significant importance.

Table 5. Total economic value of ecosystems (TEV).

Land Use	VET (USD/Year)	Data Source
Forest	103	[12]
Wetlands	9778	[41]
Cultivated land	482	*
Pastures	132	[12]

* Value estimated by the authors based on available agricultural productivity data for Morocco (2022).

4. Results

4.1. Degradation of Agricultural Land

4.1.1. Water Erosion

Running the SDR model of the InVEST tool generated a map of land degradation (Figure 5) resulting from water erosion in Morocco. The classification of the latter by erosive state shows us that Morocco is affected by light erosion (very low and low) on sixty percent of its surface, nine percent is affected by moderate erosion, and one-third is affected by strong erosion.

On the one hand, the surface area of agricultural land affected by the various erosive conditions was obtained by superimposing the land use map and the map of erosive water conditions in Morocco. Extracting the results by homogeneous zone enabled us to calculate the surface area of agricultural land degraded by water erosion by homogeneous zone.

In addition, the calculation of lost productivity in Q/ha (quintal per hectare) per homogeneous zone according to light and medium erosive conditions and its product by the agricultural areas (ha) affected by these two conditions provided us with the average production (Q) lost per homogeneous zone. Lastly, we estimated the cost of lost cropland production caused by water erosion (Table 6), which results in an annual loss of USD 263.1 million.

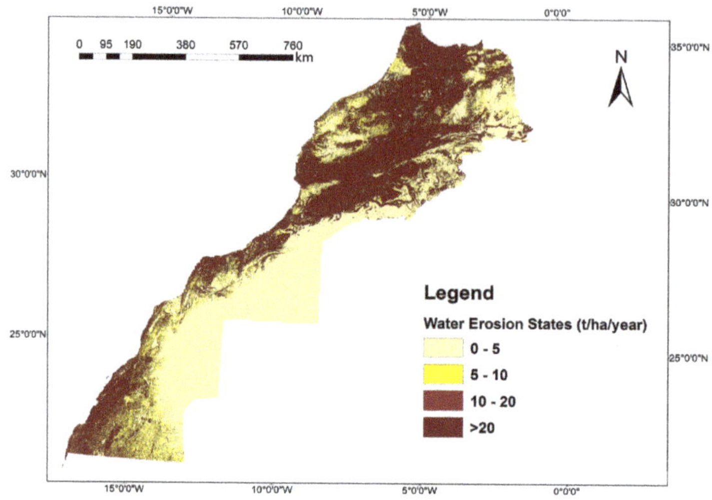

Figure 5. Soil losses in Morocco in 2020.

Table 6. Annual costs of loss of cropland production caused by water erosion (×1 million USD).

Homogeneous Zone	Average Productivity (Q/ha)	Lost Production (1000 Q/Year) Due to Light and Medium Erosion	Cost (×1 Million USD /Year)
1	4.4	4	0.1
2	20.1	71.6	2
3	6.9	265.5	7.4
4	38.4	3255.8	91.2
5	9.9	1978.4	55.4
6	23	767.54	21.5
7	27.2	3011.2	84.4
8	11	32.3	0.9
Total		**9386.6**	**263.1**

4.1.2. Wind Erosion

Following the same approach adopted to monetize the impact of water erosion, and based on the wind erosion conditions (Figure 3), we calculated the cost of lost cropland production caused by wind erosion, which results in an annual loss of USD 317.6 million (Table 7).

Table 7. Annual costs of loss of cropland production caused by wind erosion (×1 million USD).

Homogeneous Zone	Average Productivity (Q/ha)	Lost Production (1000 Q/year) Due to Light and Medium Erosion	Cost (×1 Million USD /year)
1	4.4	5.4	0.15
2	20.1	61.2	1.7
3	6.9	239.8	6.7
4	38.4	2981.9	83.6
5	9.9	1934.3	54.2
6	23	850.1	23.8
7	27.2	4988.3	139.8
8	11	269.2	7.5
Total		**11,330.67**	**317.6**

4.1.3. Salinization

The cost of agricultural land degradation associated with salinization in Morocco is USD 25.03 million (Table 8).

Table 8. Annual costs of loss of cropland production due to salinization (×1 million USD).

Homogeneous Zone	Affected Farmland (ha)	Economic Cost (USD Million)
1	119	0.04
2	603.2	0.23
3	17,284	6.75
4	7764.2	3.03
5	13,451	5.25
6	3760.5	1.47
7	18,550	7.25
8	2568.7	1.00
Total	**64,100.6**	**25.03**

4.2. Degradation of Rangelands Through Overgrazing

Total forage production from forests and productive rangelands was calculated by homogeneous zone, with Homogeneous Zone 4 standing out as the most productive, closely followed by Homogeneous Zone 6, while Homogeneous Zone 5 had the lowest production. Calculation of forage harvesting by livestock by homogeneous zone revealed that Homogeneous Zone 7 had the greatest pressure from livestock, followed by Homogeneous Zone 5, while Homogeneous Zone 8 had the lowest annual forage harvesting. The annual cost of pasture degradation due to overgrazing is estimated at USD 1.03 billion (Table 9).

Table 9. Annual costs of pasture degradation due to overgrazing (×1 million USD).

Homogeneous Zone	Annual Production (10^6 UF)	Annual Withdrawal (10^6 UF)	Annual Overconsumption (10^6 UF)	Economic Cost (Million USD)
1	177.8	310.8	132.4	37.5
2	143.8	218.2	74.4	21
3	332.6	744	411.3	116.6
4	486.2	1012.4	526.1	149.1
5	27.2	1126.5	1099.2	311.6
6	373.7	889.2	515.5	146.1
7	104	1377.5	1273.5	361
8	62.9	215.2	152.3	43.1
Total	**1708.5**	**5893.8**	**4184.9**	**1037.1**

In order to spatialize the intensity of the pastoral load (Figure 6), we calculated the index of pastoral pressure by homogeneous zone, which is the ratio between the actual load and the equilibrium load. Given that these homogeneous zones are very extensive, the calculation of these indices was reduced to the provincial level of the country, and analysis of the results shows that Homogeneous Zone 5 suffers the greatest pressure, followed by Homogeneous Zone 7, while Homogeneous Zone 2 endures the least intensity.

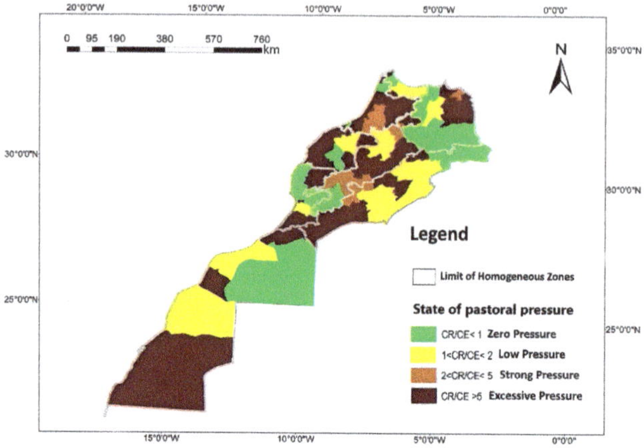

Figure 6. Degree of overgrazing by province in Morocco.

4.3. Dam Silting

The economic cost induced by dam silting is an annual value of 61.6 million USD (Table 10), established on the basis of the economic costs of dam silting by catchment area (Appendix A.3).

Table 10. Annual costs incurred by siltation of dams (×1 million USD).

Homogeneous Zone	Economic Cost (Million USD)
1	7.1
2	16.7
3	12.8
4	7.6
5	6.4
6	5.5
7	3.3
8	2
Total	**61.6**

4.4. Degradation of Carbon Storage

Implementation of the InVEST carbon storage and sequestration model has given us the quantity of carbon stored between 2000 and 2020 (Appendix A.1) in Morocco and by homogeneous zone. The carbon storage lost in Morocco between 2000 and 2020 is 12.48 10^7 t, representing a loss of USD 7.77 billion over a 20-year period, i.e., an average annual loss of 6.24 10^6 t and an average annual economic cost of 390 (million USD/year) for the Kingdom (Table 11).

Table 11. Annual costs incurred by loss of carbon storage between 2000 and 2020 (×1 billion USD).

Homogeneous Zone	Percentage Variation (%)	CO_2 Sequestered Between 2000 and 2020 (10^7 t)	Economic Cost (in Billions of USD)
1	−2.96%	15.4	9.58
2	−3.81%	1.76	1.10
3	−43.96%	2.28	1.42
4	−5.85%	26.32	16.38

Table 11. *Cont.*

Homogeneous Zone	Percentage Variation (%)	CO_2 Sequestered Between 2000 and 2020 (10^7 t)	Economic Cost (in Billions of USD)
5	19.47%	3.51	2.18
6	−13.81%	11.64	7.25
7	4.40%	8.28	5.15
8	−2.96%	2.63	1.64
Total		**12.48**	**7.77**

4.5. Land Use Trends

Evolutionary trends, using 2001 as the reference year (value = 1), highlight the non-linear nature of land use transformations. Some land occupations show an overall upward trend, others a downward trend, while some evolve in an increasing manner during one period and regressively during another (Figure 7). The annual change matrix obtained enabled us to ascertain the average evolution of each land use, namely a decline in cultivated land (19,084 ha), grassland (12,857 ha), and wetlands (87 ha), and a net increase in forest (796 ha), representing an annual lost cost of USD 11.55 million for land use change.

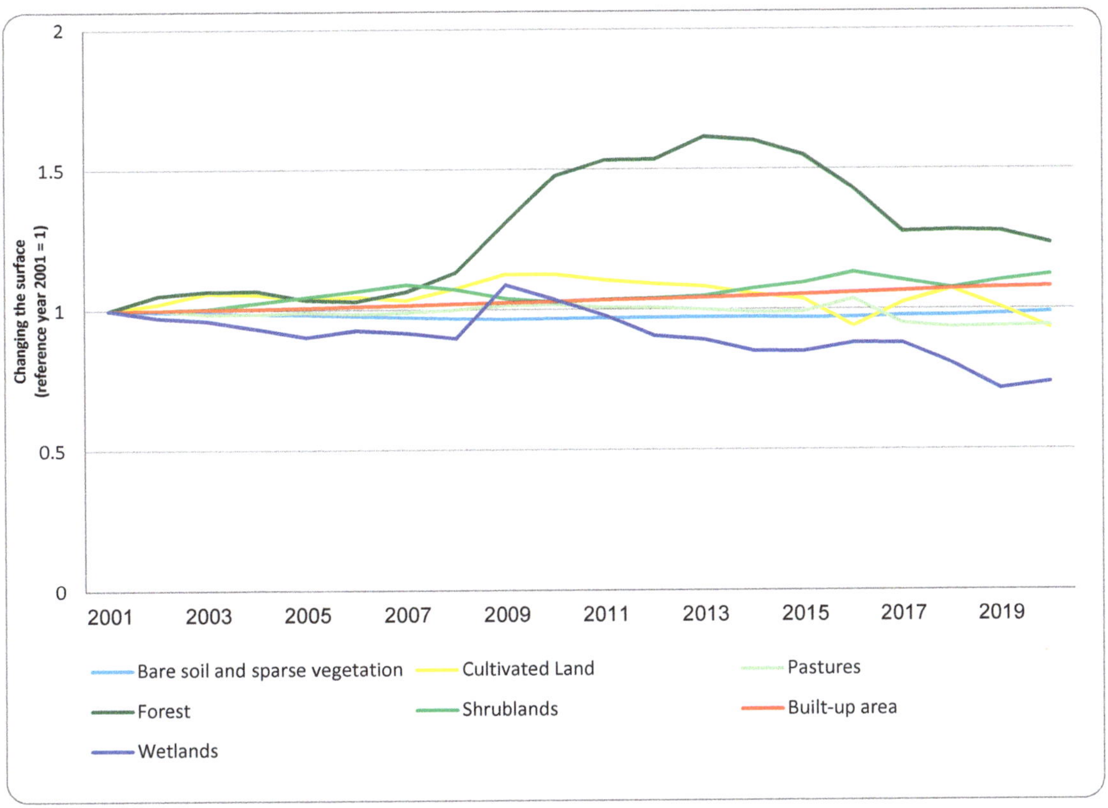

Figure 7. Surface change in land use in Morocco (2001–2020).

4.6. Total Annual Economic Cost of Desertification in Morocco

The total annual economic cost of desertification in 2020 is estimated at USD 2.1 billion (Table 12), representing 1.77% of Morocco's GDP, with degradation of agricultural land accounting for 28.75% of the total cost, and degradation of rangelands through overgrazing, degradation of carbon storage, siltation of dams and changes in land use accounting for 49.27%, 18.47%, 2.95%, and 0.56%, respectively (Figure 8).

Table 12. Total economic cost of desertification in Morocco (billions of USD).

Degradation	Economic Cost (Billion USD)
Degradation of agricultural land	0.6
Degradation of rangelands through overgrazing	1.04
Degradation of carbon storage	0.39
Siltation of dams	0.06
Land use changes	0.01
Total annual economic cost of desertification in Morocco 2020	**2.1**

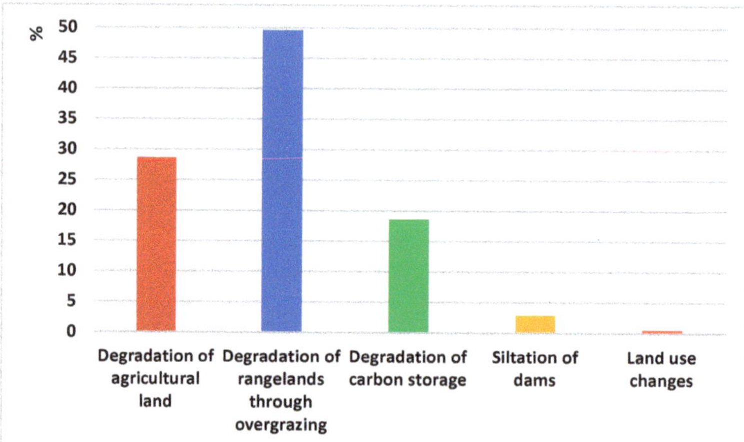

Figure 8. Contribution of each degradation (%) to the total cost of desertification.

5. Discussion

Determining the economic cost associated with desertification is intrinsically complex, and its relevance is inextricably linked to the accessibility of relevant knowledge and information. Thus, the material losses considered in this study reflect only a limited fraction of the impacts of desertification, which are likely to be underestimated. The conceptual approach adopted to assess the cost of desertification-related degradation focuses on the most prevalent and significant cases without claiming to encompass all possible situations.

To estimate the costs of desertification and land degradation, "in situ" effects such as water and wind erosion, overgrazing, salinization, carbon stock degradation, and land use change have been addressed, as well as "ex situ" effects, including water loss in dams due to transport of surface soil horizons and siltation. The data needed to estimate these degradations are the land area and its occupation, the degree of degradation, and the consequent variation in agricultural and pastoral productivity. The prices of agricultural and forestry products, the social cost of carbon, and the total economic value of ecosystems make it possible to estimate the economic cost of desertification in monetary terms.

The total economic cost of desertification is estimated at USD 2.1 billion, or 1.77% of Morocco's GDP in 2020. Pasture degradation due to overgrazing accounts for the majority of this cost (49.27%), while overconsumption of available forage resources by livestock is estimated at 4.1 billion forage units per year. Homogeneous Zone 5 is the most affected, with pressure corresponding to 25% of overconsumption. The loss of productivity of agricultural land through water and wind erosion and salinization results in a loss of more than 20 million quintals of agricultural production and accounts for 28.75% of the total cost of desertification. Wind erosion causes the most losses, and homogeneous Homogeneous Zone 7 is the most degraded. With regard to the degradation of carbon storage, more than 6 million tC is lost annually and its cost represents 18.47% of the total cost, with Homogeneous Zone 4 losing the most storage. The effects of ex situ water erosion, manifested by the silting-up of dams, account for 2.95% of the total cost, with homogeneous Zone 2 suffering the most damage. Land use change accounts for the smallest share of the total economic cost of desertification, at 0.56%. Analysis of change over a 20-year period shows that there has been an annual decline in cultivated land (19,084 ha), grassland (12,857 ha), and wetlands (87 ha), and a net increase in forest (796 ha). Evaluating the cost of degradation by homogeneous zone, the seventh homogeneous zone is the most affected by degradation with a cost exceeding USD 580 million, followed by the sixth with a cost exceeding USD 420 million, while the least degraded homogeneous zones are the first (USD 47.25 million) and second (USD 42 million).

It is essential to continue efforts to better understand the causes and dynamics of soil degradation in Morocco. This requires the development of more in-depth analyses to obtain more accurate and comprehensive estimates. In addition, carrying out cost/benefit studies could help to identify interventions that have the greatest positive impact on the environment while remaining economically viable. To guarantee sustainable management, it would be advisable to set up a rigorous monitoring and evaluation system based on observatories that are well distributed across the homogeneous zones identified. These efforts should not be limited to the physical and biological aspects of combating degradation but should also include social and economic dimensions, with the active participation of local communities. Finally, it is imperative to promote sustainable agricultural and land management practices adapted to the specific characteristics of each region to preserve soils and improve their resilience in the face of climatic and anthropogenic pressures.

5.1. Comparison with Earlier Studies

This study's findings align with and expand upon those of [12], who estimated that environmental degradation costs Morocco 3.52% of its GDP. While their research provided a comprehensive assessment of environmental degradation, including urban and industrial factors, this study focuses specifically on desertification-related land degradation. The economic cost of desertification in this study, estimated at USD 2.1 billion (1.77% of GDP), reflects a narrower scope but provides greater detail in terms of zonal analysis and specific drivers such as overgrazing and wind erosion.

Additionally, ref. [12] highlighted the significant role of land degradation in rural poverty, a finding corroborated here through the identification of homogeneous zones 5 and 7 as areas bearing the highest economic burden. Unlike their broader approach, this study incorporates newer datasets and advanced tools, such as the InVEST SDR model, to achieve spatially explicit insights. By using a zonal approach, it advances prior findings by revealing spatial variability in degradation impacts and the economic consequences for specific zones.

Moreover, while [12] emphasized the broader implications of environmental degradation on Morocco's economy, this study underscores the specific mechanisms of desertification and provides actionable recommendations for land management tailored to each zone. This targeted approach offers a framework for prioritizing interventions and allocating resources efficiently, thereby advancing earlier findings.

5.2. Public Policy Recommendations and Comparative Analysis

To address desertification, Morocco should strengthen land use planning, promote ecosystem restoration through reforestation and soil conservation, and modernize irrigation systems to combat salinization. Sustainable agricultural practices such as no-till farming and crop rotation, coupled with community engagement, are vital for long-term success.

Morocco can learn from regional examples like Tunisia's Land Degradation Neutrality targets and Jordan's water harvesting techniques, as well as global initiatives like China's Loess Plateau restoration and India's watershed programs. These highlight the value of integrating traditional knowledge with modern solutions. Morocco's PANLCD provides a strong foundation, but scaling up region-specific interventions and monitoring systems will enhance its effectiveness.

6. Conclusions

The findings of this study reveal that land degradation in Morocco is a significant and ongoing issue, affecting more than 90% of its territory and generating significant economic costs. This study estimates the total cost of land degradation at USD 2.1 billion per year, representing 1.77% of Morocco's GDP in 2020. Among the various forms of degradation, rangeland overgrazing contributes the most (49.27%), followed by the loss of agricultural land productivity (28.75%), carbon storage degradation (18.47%), dam siltation (2.95%), and land use changes (0.56%).

The findings reveal that certain homogeneous zones, particularly Homogeneous Zone 7 and Homogeneous Zone 6, bear disproportionate economic burdens due to high pressures on natural resources. In contrast, homogeneous zones 1 and 2 are the least affected, reflecting significant spatial variability in degradation impacts. Despite the robustness of the methodologies employed, this study acknowledges several limitations, including the lack of field validation for certain parameters (wind and water erosion) and the use of temporally non-harmonized data. These limitations highlight the need for updated climatic and environmental datasets and the establishment of monitoring systems tailored to regional characteristics.

To mitigate the impacts of desertification, targeted efforts must be implemented, such as:

- Strengthening sustainable land management practices for agricultural and rangeland areas.
- Promoting ecosystem restoration in the most sensitive homogeneous zones.
- Implementing a monitoring and evaluation system based on region-specific indicators.
- Engaging local communities in conservation initiatives to ensure long-term sustainability.

In conclusion, this study underscores the importance of addressing desertification as a national priority, not only due to its environmental implications but also its economic consequences. The results provide a solid foundation for guiding policies and strategies to mitigate desertification effects while promoting sustainable development.

Supplementary Materials: The following supporting information can be downloaded at: https://www.mdpi.com/article/10.3390/land14040837/s1.

Author Contributions: Conceptualization, A.K.; Methodology, A.L.; Validation, A.K.; Formal analysis, A.L.; Investigation, A.L.; Data curation, A.L.; Writing—original draft, A.L.; Writing—review & editing, A.K.; Visualization, A.L.; Supervision, A.K. All authors have read and agreed to the published version of the manuscript.

Funding: This research received no external funding.

Data Availability Statement: The original contributions presented in this study are included in the article/Supplementary Material. Further inquiries can be directed to the corresponding authors.

Conflicts of Interest: The authors declare no conflict of interest.

Appendix A

Appendix A.1. Figures of Carbon Storage in Morocco in 2000 and 2020

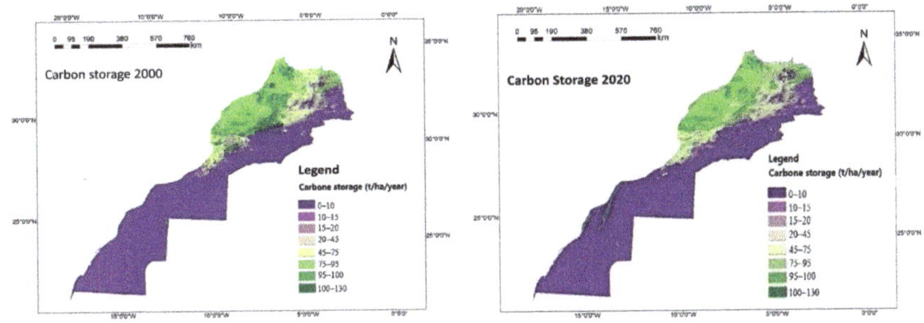

Appendix A.2. Data on Forage and Crop Production by Homogeneous Zone

Appendix A.2.1. Farmland Production Season (2018–2019)

Homogeneous Zone / Production in (10^3 Q)	1	2	3	4	5	6	7	8
Cereal crops	52.7	1156.18	1220.21	9110.66	6000.3	3331.59	24,038.2	2723.57
Growing legumes		18.2	43.58	151.14	153.3	143.33	1307.4	35.65
Sugar crops				2159.8	1270.8	3779.6	11,705.2	
Oilseed crops				9.8	1.6		1109.4	

Source: [32].

Appendix A.2.2. Area (in ha) of Forest Species by Homogeneous Zone in 2019

Homogeneous Zone / Area of Species (in ha)	1	2	3	4	5	6	7	8
Fir								4856
Alfa		520,935	1815	276,881.8		1,734,387		
Juniper		34,565	202,895	439,038.8	170	101,578.2	735	1938
Pines			2082	42,372		30,164.3		11,645
Thuja	5903		139,143.4	98,347.2	7740	151,436.9	78,866	51,365
Cedar				39,205		16,555.5	51	14,996
Argan tree	28,058		552,915.7	266,193.9	2316	132		
Zeen Oak				49		796	5657	7015
Cork oak			29	5181	645	16,015	201,525	71,910
Holm oak		3347	173,317	789,124.5	14,440	271,883	66,347	69,912
Other	689,652	116,911.3	59,238.9	31,378.3	43	11,499.5	29,761	3187.7
Total	723,613	675,760	1,131,436	1,987,772	25,354	2,334,448	382,942	236,824

Source: [32].

Appendix A.2.3. Livestock Numbers (in Thousands of Head) 2019 by Homogeneous Zone

Homogeneous Zone	Goats	Sheep	Cattle	Camels
1	774.57	447.5	18.36	170.37
2	465.993	876.972	38.28	15.44
3	1048.3	2837.09	317.82	1.32
4	1482.53	3324.12	322.11	
5	268.3	3892.5	737.6	
6	624.96	2927.16	170.16	
7	344.5	3746.6	1108.8	
8	361.24	373.74	192.4	

Source: [32–35].

Appendix A.3. Economic Cost of Dam Silting by Catchment Area

Watersheds	Storage Capacity (Millions m^3)	Exported Sediment (10^6 t/Year)	Lost Storage Capacity (Millions m^3)	Replacement Costs (USD)	
				Cost 0.21 USD/m^3 (in Millions of USD/Year)	Cost 0.63 USD/m^3 (In Millions of USD/Year)
Bouregrag, Atlantic Coastal Casa	1082.30	5.16	3.44	0.72	2.17
Guelmim Tiznit	63.30	2.69	1.79	0.38	1.13
Loukkos, Tangiers, and Mediterranean coasts	1716.56	14.64	9.76	2.05	6.15
Moulouya, isley, Figuig	703.00	12.58	8.39	1.76	5.28
OUM Er Rabii, Coastal Safi El Jadida	4952.40	44.63	29.75	6.25	18.74
Saqia el-Hamra	16.30	6.43	4.29	0.90	2.70
Sebou, Drader Souier	5552.66	22.91	15.27	3.21	9.62
Souss-Massa					
Ziz_Rheris, guir, Draa, Tamri	1853.84	121.02	80.68	16.95	50.82
Tensift, Coastal Essaouira	79.80	33.76	22.51	4.73	14.18

References

1. Cornet, A. Desertification: An environmental problem, a development problem. *Summ. Agro Mus. Conf.* **2002**, *29*, 2–6.
2. Khanamani, A.; Fathizad, H.; Karimi, H.; Shojaei, S. Assessing desertification by using soil indices. *Arabian J. Geosci.* **2017**, *10*, 1–10. [CrossRef]
3. Xu, D.; Wang, Y.; Wang, Z. Linking priority areas and land restoration options to support desertification control in northern. *China Ecol. Indic.* **2022**, *9*, 1–4. [CrossRef]
4. UNCCD. *World Territorial Outlook: Summary for Decision-Makers*; United Nations Convention to Combat Desertification: Bonn, Germany, 2022.
5. Bonnet, B.; Chotte, J.L.; Hiernaux, P.; Ickowicz, A.; Loireau, M. *Désertification et Changement Climatique, un Même Combat?* Quae: Paris, France, 2024; pp. 14–16.
6. Laouina, A. *Prospective Maroc 2030. Gestion Durable des Ressources Naturelles et de la Biodiversité au Maroc*; Haut-Commissariat au Plan: Rabat, Morocco, 2006; pp. 109–110.
7. Ghanam, A. Desertification and environmental issues in Morocco. *J. Environ. Manag.* **2003**, *68*, 237–251.

8. United Nations. *Annual Report of the Chairperson on the Activities of the Commission Covering the Period January to December 2012*; United Nations: New York, NY, USA, 2012; pp. 73–78.
9. HCEFLCD. *Le Programme d'Action National de Lutte Contre la Désertification: Actualisation et Adaptation aux Spécificités Zonales*; Haut-Commissariat aux Eaux et Forêts et à la Lutte Contre la Désertification: Rabat, Morocco, 2013.
10. El Rhaffouli, K. The "Access and Benefit Sharing" mechanism in Morocco. *Forêt Méditerranéenne* **2015**, *36*, 143–144.
11. Economics of Land Degradation Initiative. *Land Degradation Costs 10.6 Trillion USD per Year, Says New Report*; SEI: Stockholm, Sweden, 2024.
12. World Bank. Report of the High Level Committee on Carbon Pricing. In *The Cost of Environmental Degradation in Morocco*; Chapter 6 "Forêts": Environment and Natural Resources Global Practice Discussion Paper; Croitoru, L., Khattabi, A., Sarraf, M., Eds.; World Bank Group Report No. 105633-MA; World Bank: New York, NY, USA, 2017; pp. 65–74.
13. FAO; ISRIC. *JRC: Harmonized World Soil Database*; Version 1.2; FAO: Rome, Italy; IIASA: Laxenburg, Austria, 2012. Available online: https://webarchive.iiasa.ac.at/ (accessed on 9 January 2022).
14. Feng, X.; Fu, B.; Lu, N.; Zeng, Y.; Wu, B. How ecological restoration alters ecosystem services: An analysis of long-term effects in China's Loess Plateau. *Sci. Adv.* **2020**, *6*, eaax8408.
15. Nkonya, E.; Mirzabaev, A.; von Braun, J. *Economics of Land Degradation and Improvement: A Global Assessment for Sustainable Development*; Springer International Publishing: Cham, Switzerland, 2016.
16. Verstraete, M.M.; Scholes, R.J.; Stafford Smith, M. Climate and desertification: Looking at an old problem through new lenses. *Front. Ecol. Environ.* **2009**, *7*, 421–428. [CrossRef]
17. Lu, N.; Fu, B.; Feng, X.; Zeng, Y.; Liu, Y. Hydrological and socio-economic effects of ecological restoration in China's Loess Plateau. *J. Hydrol.* **2018**, *556*, 1135–1144.
18. Kusi, K.K.; Khattabi, A.; Mhammdi, N. Integrated assessment of ecosystem services in response to land use change and management activities in Morocco. *Arabian J. Geosci.* **2021**, *14*, 418. [CrossRef]
19. Marques, S.M.; Campos, F.S.; David, J.; Cabral, P. Modelling sediment retention services and soil erosion changes in Portugal: A spatio-temporal approach. *ISPRS Int. J. Geo-Inf.* **2021**, *10*, 6. [CrossRef]
20. Maanan, M.; Maanan, M.; Karim, M.; Ait Kacem, H.; Ajrhough, S.; Rueff, H.; Snoussi, M.; Rhinane, H. Modelling the potential impacts of land use/cover change on terrestrial carbon stocks in north-west Morocco. *Int. J. Sustain. Dev. World Ecol.* **2019**, *26*, 560–570. [CrossRef]
21. Geist, H.J.; Lambin, E.F. Dynamic causal patterns of desertification. *BioScience* **2004**, *54*, 817–829. [CrossRef]
22. Reynolds, J.F.; Smith, D.M.S.; Lambin, E.F.; Turner, B.L.; Mortimore, M.; Batterbury, S.P.; Downing, T.E.; Dowlatabadi, H.; Fernández, R.J.; Herrick, J.E.; et al. Global desertification: Building a science for dryland development. *Science* **2007**, *316*, 847–851. [CrossRef] [PubMed]
23. UCSB Climate Hazards Center. CHIRPS: Climate Hazards Group InfraRed Precipitation with Station Data. Available online: https://www.chc.ucsb.edu/data/chirps (accessed on 4 May 2022).
24. ESRI. Esri Releases New 2020 Global Land Cover Map. Available online: https://www.esri.com/about/newsroom/announcements/esri-releases-new-2020-global-land-cover-map/ (accessed on 24 December 2021).
25. Sabir, M.; Sagno, R.; Tchintchin, Q.; Zaher, H.; Benjelloun, H. Evaluation of Organic Carbon Stocks in Moroccan Soils. Carbone des Sols. 2020. Available online: https://papers.ssrn.com/sol3/Delivery.cfm/e7cb7935-ee5f-4680-9be4-f8a2ac1d5db3-MECA.pdf?abstractid=4576754&mirid=1 (accessed on 26 December 2021).
26. USGS. Earth Explorer: Digital Elevation Data. Available online: https://earthexplorer.usgs.gov/ (accessed on 1 December 2021).
27. Wischmeier, W.H.; Smith, D.D. *Predicting Rainfall Erosion Losses: A Guide to Conservation Planning*; No. 537; U.S. Department of Agriculture, Science and Education Administration: Washington, DC, USA, 1978.
28. ESDAC. Global Rainfall Erosivity. Available online: https://esdac.jrc.ec.europa.eu/content/global-rainfall-erosivity (accessed on 29 December 2021).
29. Roose, E.J. *Land Husbandry: Components and Strategy*; FAO Soils Bulletin 70; Food and Agriculture Organization of the United Nations: Rome, Italy, 1996; p. 147.
30. Bridges, E.M.; Oldeman, L.R. Global assessment of human-induced soil degradation. *Arid Soil Res. Rehabil.* **1999**, *13*, 319–325. [CrossRef]
31. A.E.F.C.S. *Programme Forestier National Phase I*; Situation du secteur problématique et défis; A.E.F.C.S.: Rabat, Maroc, 1998.
32. HCP. *Annuaire Statistique du Maroc, Année 2020*; Haut-Commissariat au Plan: Rabat, Maroc, 2020; pp. 59, 62–64, 73–74, 83–84.
33. HCP. *Monographie de la Région Eddakhla-Oued Eddahab*; Haut-Commissariat au Plan: Rabat, Morocco, 2018.
34. HCP. *Région Souss Massa: En Chiffres 2020*; Haut-Commissariat au Plan: Rabat, Maroc, 2020.
35. HCP. *Monographie de la Région Laayoune-Sakia El Hamra*; Haut-Commissariat au Plan: Rabat, Maroc, 2020.
36. GeoAménagement. Watersheds of Morocco. Available online: https://www.geoamenagement.com/ (accessed on 17 January 2022).
37. CAS EARTH. GLC FCS30: Global Land Cover Data. Available online: https://data.casearth.cn/thematic/glc_fcs30 (accessed on 21 March 2022).

38. USGS. MODIS Land Cover Data (MCD12Q1 Version 6.1). Available online: https://lpdaac.usgs.gov/products/mcd12q1v061/ (accessed on 11 April 2022).
39. Remini, W.; Remini, B. Sedimentation in North African dams. *LARHYSS J.* **2003**, *2*, 52–54.
40. CESE. *Governance Through Integrated Water Resources Management in Morocco: Levier Fondamental de Développement Durable*; CESE: Rabat, Morocco, 2014.
41. Khattabi, A. Étude socio-économique: Activités humaines et leurs impacts sur la Merja Zerga. In Proceedings of the Séminaire sur les Zones Humides, Administration des Eaux et Forêts et de la Conservation des Sols-MedWet, Kénitra, Morocco, 24–26 September 1997.

land

MDPI

Article

Spatio-Temporal Evolution of Water-Regulating Ecosystem Services Values in Morocco's Protected Areas: A Case Study of Ifrane National Park

Oumayma Sadgui [1,*], Abdellatif Khattabi [2] and Zouhir Dichane [3]

1 Applied Economics and Social Sciences in Agriculture, Hassan II Institute of Agronomy and Veterinary Medicine, Rabat 10 101, BP, Morocco
2 National School of Forestry Engineering, Salé 11 000, BP, Morocco; ab_khattabi@yahoo.com
3 Geoengineering and Environment Laboratory, Faculty of Sciences, Moulay Ismail University, Meknes 11 201, BP, Morocco
* Correspondence: o.sadgui@iav.ac.ma; Tel.: +21-2676-9377-78

Abstract: Water-Regulating Ecosystem Services (WRES) play a crucial role in maintaining water quality and preventing soil erosion, particularly in watershed areas that are vulnerable to Land Use Land Cover Changes (LULCC) and climate change. This study focuses on the Upper Beht Watershed, the most ecologically significant basin of the Ifrane National Park (INP). The main objective is to understand how WRES values respond to the challenges posed by grasslands degradation, agricultural intensification, and urban expansion before and after the park's creation. In this research, we first analyzed historical Land Use Land Cover (LULC) data from 1992 to 2022 using Google Earth Engine platform. We then employed the Integrated Valuation of Ecosystem Services and Tradeoffs (InVEST 3.10.2) models to quantify and map the impacts of ongoing LULCC on the watershed's capacity to retain sediments and nutrients. Finally, we used the damage costs avoided method for economic assessment of WRES. Our findings demonstrate a notable improvement in the economic value of WRES following the establishment of the park, reaching USD 10,000 per year. In contrast, prior to its creation, this service experienced a decline of USD −7000 per year. This positive trend can be attributed to the expansion of forest cover in areas prioritized for reforestation and conservation interventions. The study highlights the critical importance of continuous WRES monitoring, providing park managers with robust data to advocate for sustained conservation efforts and increased investment in restoration initiatives within protected areas. Moreover, the findings can be used to raise awareness among local communities and encourage their active engagement in sustainable development initiatives.

Keywords: water-regulating ecosystem services; LULCC; sediment retention; nutrient retention; economic assessment; Ifrane national park

1. Introduction

Ecosystem services are the benefits humans derive from natural ecosystems, ranging from the provision of resources to the regulation of environmental processes. These services are indispensable for human well-being [1]. The recognition of their importance gained prominence with Costanza's 1997 study, which introduced economic valuation as a method to integrate ecosystem services into decision-making processes [2]. Among them, water regulation ecosystem services (WRES) play a crucial role in maintaining hydrological balance by ensuring water quality, mitigating soil erosion, and reducing flooding [3,4]. Ecosystems such as forests, grasslands, and wetlands are key providers of these services [5,6]. Given their critical function, the monetization of WRES has become increasingly necessary to support their inclusion in policy-making and environmental planning [7].

In Morocco, WRES are particularly significant in regions such as the Middle Atlas, where forested watersheds contribute to hydrological stability, agricultural productivity, and urban water supply [8–10]. One of the most ecologically and hydrologically significant basins in this region is the Upper Beht Watershed, which encompasses extensive forests dominated by Atlas Cedar (Cedrus Atlantica). This watershed plays a key role in securing the country's mobilizable water resources and sustaining downstream communities [11,12]. However, the capacity of this region to provide WRES has been threatened by LULCC, including grassland degradation, agricultural intensification, and urban expansion [11,13]. In response to these challenges, the Ifrane National Park (INP) was established in 2004 to conserve biodiversity and maintain essential ecosystem services. The park integrates various land-use categories (state-owned, collective, and private lands) and promotes conservation through participatory approaches. Efforts have focused on reforestation and restoration initiatives, wildlife conservation, ecotourism development, and eco-development programs [14].

While conservation actions within protected areas have been widely studied worldwide, research on their impact on WRES remains limited [15–19]. In particular, there is a lack of quantitative assessments of WRES within Morocco's national parks, leaving a significant knowledge gap regarding the biophysical and economic implications of conservation efforts amid increasing pressures from LULCC and climate change [11,20–25]. This study aims to bridge these gaps by providing a localized and detailed assessment of WRES in the Upper Beht Watershed. Specifically, it seeks to achieve the following: (1) Evaluate the impact of Ifrane National Park's establishment on the hydrological regulation of the Upper Beht Watershed by analyzing WRES trends before and after its creation. (2) Quantify and map changes in WRES over three decades (1992–2022) using remote sensing and spatial modeling techniques. (3) Estimate the economic value of WRES using the "damage costs avoided" approach to highlight their monetary significance.

To achieve these objectives, the study employs Google Earth Engine for remote sensing analysis and integrates Sediment Delivery Ratio (SDR) and Nutrient Delivery Ratio (NDR) models within the InVEST models to quantify sediment and nutrient retention changes over time. Field verification is incorporated to enhance data accuracy. The economic valuation component provides insights into the monetary importance of WRES, reinforcing the need for their integration into conservation policies.

By assessing the evolution of WRES values in relation to LULCC and conservation efforts, this study underscores the role of restoration activities in enhancing ecosystem resilience. Furthermore, it highlights how economic valuation can inform sustainable conservation planning and policymaking, paving the way for innovative financing mechanisms that align ecological preservation with socio-economic objectives. The findings of this research will provide valuable data for park managers, policymakers, and conservation stakeholders to support ongoing WRES monitoring and strengthen water resource management strategies within protected areas.

2. Materials and Methods

2.1. Study Area

The study area encompasses the Upper Beht Watershed, the most ecologically and hydrologically significant basin within Ifrane National Park, situated in the western part of the central Middle Atlas region in Morocco (Figure 1).

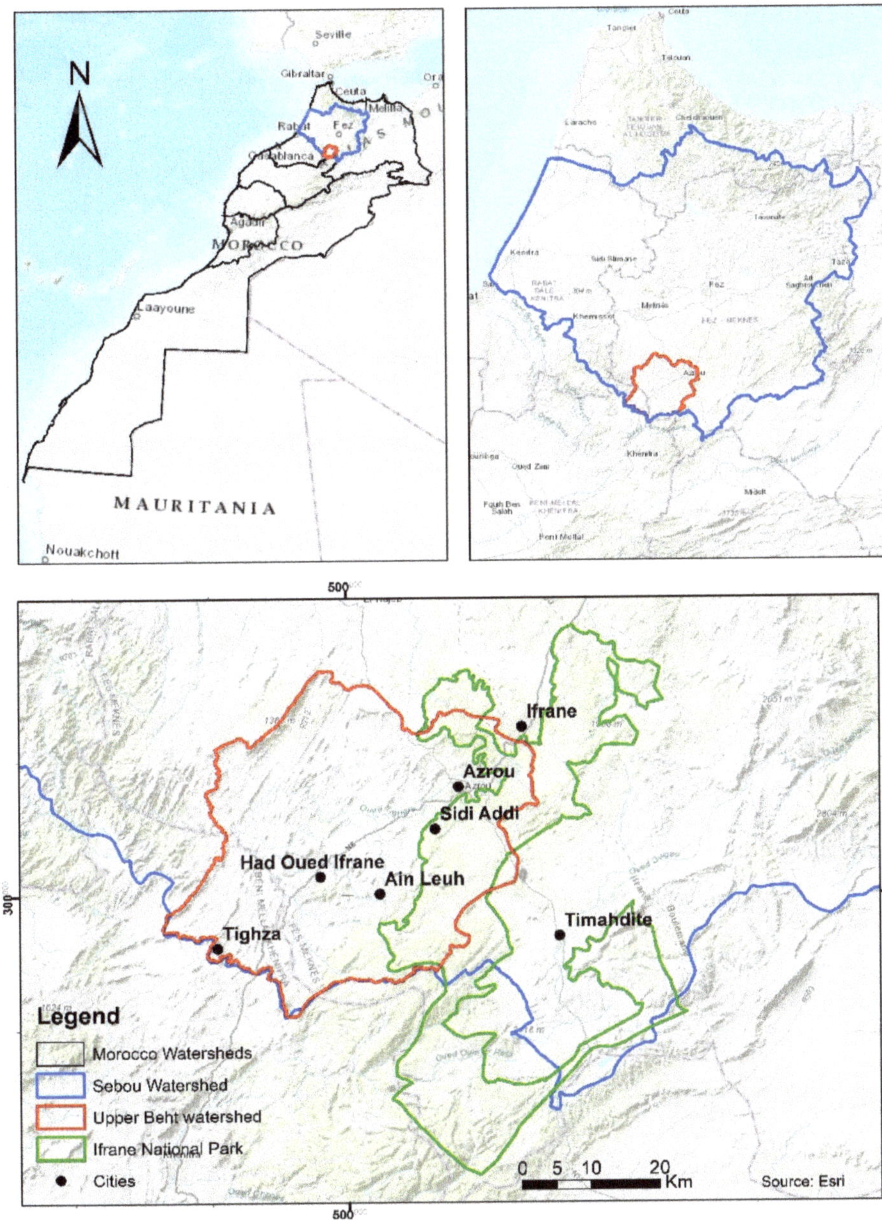

Figure 1. Map of the geographical location of the study area.

The Upper Beht Watershed is a major hydrological unit within the Sebou Watershed, receiving inflows from the R'dom river before merging with the Sebou river in the Gharb plain. It plays a critical role in water regulation and supply, acting as a primary left-bank tributary of the Sebou river. The watershed contributes significantly to water availability for downstream reservoirs and irrigated agriculture, with a total irrigated area of 29,000 ha. Its hydrological function includes precipitation capture, seasonal flow regulation, and sediment transport mitigation, which are essential for maintaining regional water resources [26].

The climate is Mediterranean with strong Atlantic influences, characterized by an annual precipitation average of 700 mm, varying from 1000 mm in the highlands to 300 mm in the Saïs plain. Winters are cold, with snowfall in the Middle Atlas, while summers are hot and dry. Annual temperatures range between 15 and 20 °C, with winter lows of 3–7 °C and summer highs of 34–36 °C [26].

The geology of the watershed is diverse, with three dominant soil types: volcanic, calcareous, and dolomitic. These substrates influence soil fertility, water retention capacity, and erosion dynamics [26].

The vegetation cover varies with altitude and land use. The Ifrane national park, established to conserve the region's biodiversity and ecosystems, hosts extensive forests, particularly in the upstream areas, while lower elevations are dominated by matorral vegetation and seasonal agriculture. The park supports 22% of Morocco's vascular plant species, with an endemism rate of 25%, notably represented by the Atlas Cedar (Cedrus Atlantica). It is also a key habitat for 33% of the country's mammalian species, including the endangered barbary macaque (Macaca Sylvanus). Beyond conservation, the INP promotes sustainable resource management and supports local development through eco-tourism and agroecology [27].

2.2. Mapping Historical LULCC

Satellite data for the years 1992, 2002, 2012, and 2022 were analyzed to assess LULCC within the INP over three decades. These datasets, obtained from the Google Earth Engine (GEE) platform, were derived from the Landsat 5, Landsat7, Landsat 8, and Landsat 9 satellite series. The analysis incorporated all available spectral bands, as well as the Normalized Difference Vegetation Index (NDVI), to enable a detailed examination of spatial and temporal LULCC.

Preprocessing steps, including radiometric and atmospheric corrections, were applied to ensure consistency and comparability across the different time periods [28,29]. LULC classification was performed using the Random Forest (RF) machine learning algorithm, selected for its reliability and high accuracy in remote sensing studies [30,31]. RF leverages ensemble learning by aggregating predictions from multiple decision trees, reducing overfitting and increasing model stability. Its inherent feature selection capabilities are particularly useful for identifying significant variables, such as spectral bands or vegetation indices, critical for accurate classification. RF has been shown to perform well in various LU/LC studies, including integration with advanced techniques like Markov chain models and multi-layer perceptrons for change prediction, achieving high accuracy and interpretability [32]. Combining RF with multi-sensor data, such as optical and SAR images, has further enhanced classification accuracy, achieving over 96% accuracy in some studies [33]. While RF generally excels, challenges remain in distinguishing spectrally similar classes and achieving consistent precision and recall measures, especially in heterogeneous landscapes [34]. Despite these limitations, RF's balance of simplicity, accuracy, and computational efficiency makes it a preferred choice for large-scale LULC classification in GEE.

In this study, supervised classification categorized LULC into six classes: forests, shrubs, crops, built-up areas, water, and bare soil. The selection of study years, encompassing both pre- and post-establishment periods of the park (created in 2004), provided insights into LULCC before and after the park's creation. Validation of classification results was conducted using confusion matrices and the Kappa index, demonstrating satisfactory accuracy levels for all time periods analyzed.

2.3. Quantifying Water-Regulating Ecosystem Services Using InVEST

After analyzing historical LULCC, we quantify WRES using the InVEST models, in particular the Sediment Delivery Ratio (SDR) and Nutrient Delivery Ratio (NDR) models [17–20]. Annual soil loss and the sediment delivery ratio, defined as the proportion of soil loss reaching the stream, are computed by the SDR model, and it is assumed that sediment is transported from the source to the stream and then to the watershed outlet. The Revised Universal Soil Loss Equation (RUSLE) is employed by the model to estimate annual soil loss. The spatial movement of nutrient masses is explained by the NDR model through a simple mass balance methodology, in which LULC and loading rates are considered to determine nutrient loads [35]. The input rasters for the InVEST models are presented in Table 1.

Table 1. InVEST models data source.

Models Inputs	InVEST Models	Source of Inputs
LULC	All InVEST models	Obtained using Google Earth engine platform
Biophysical tables (*)	All InVEST models	From literature [36,37]
Rasters of precipitation	"Nutrient Delivery Ratio"	Obtained from the website of CHIRPS [38]
Digital elevation model	"Sediment Delivery Ratio" "Nutrient Delivery Ratio"	From the website of Earth Science Data Systems (ESDS) [39]
Erosivity raster (R Factor)	"Sediment Delivery Ratio"	Calculated from annual and monthly precipitation averages over a 30-year period (1992–2022) obtained from the website of CHIRPS, using the formula of Rango and Arnoldus (1987) [38,40]
Soil erodibility raster (K factor)		Obtained by attributing the corresponding k factor values [36] to the lithologic facies of the study area

(*) Appendix A.

2.4. Economic Assessment of Water-Regulating Ecosystem Services

In this study, a revealed preference economic valuation method is employed, specifically the damage costs avoided method, which quantifies expenses that would have been incurred in the absence of a specific environmental function [41–45]. Damages related to the absence of Water-Regulating Ecosystem Services are primarily associated with potential degradation affecting agricultural lands and rangelands.

Loss in agricultural yields is obtained using the relationship of Den Biggelaar et al. (2004) [46] (Equation (1)).

$$r = EwP^{1.224} \times 0.0114 \quad (1)$$

where

r: Relative decrease in yield due to erosion (%).

EwP: Erosion rate (t/ha/year).

Then, the relative decrease in yield due to erosion was multiplied by the average crop yield to determine the decline in yield in t/ha (2).

$$\Delta R = r * \text{Average crop yield} \quad (2)$$

where

ΔR: Relative decrease in yield due to erosion (t/ha);

r: Relative decrease in yield due to erosion (%).

Forage yield losses are determined using Table 2, which provides the percentage of forage productivity losses for each erosion class [47].

Table 2. Correspondence between forage productivity loss and level of soil degradation.

Erosion Classes (t/ha/Year)	Loss of Forage Productivity (%)
0–5	2.5
5–25	25
>25	45

After quantifying the damage, its economic cost was estimated using market price. For agricultural losses, the market price of cereals was applied, given that more than 58% of the cultivated land in the study area is allocated to cereal production [48]. For rangeland damages, the valuation was based on the price of fodder barley, as one forage unit is conventionally equated to one kilogram of barley. This assessment was conducted for multiple reference years (1992, 2002, 2012, and 2022), with the corresponding market prices of cereals and barley for each year applied to ensure accurate and temporally consistent cost estimation [49].

3. Results

3.1. Historical LULCC Within the INP Watershed

The analysis of LULCC revealed a significant increase in forest cover between 2012 and 2022, the decade following the park's establishment. Prior to the park's creation, the region experienced a gradual decline in forested areas due to various anthropogenic activities, including deforestation and land conversion for agricultural purposes. However, the implementation of conservation measures and sustainable management practices within the park has led to a remarkable recovery and expansion of the forest ecosystem.

In contrast, shrubs have continued to decline even after the park's establishment, primarily due to persistent overgrazing in the area and the conversion of grasslands to agriculture and built-up areas. Similarly, water bodies have exhibited a continuous decrease despite conservation efforts. While climate change is a major contributing factor, the expansion of agricultural activities has further exacerbated the issue by increasing pressure on water resources.

Table 3 illustrates the evolution of each LULC class within the INP watershed.

Table 3. Evolution of LULC classes within the INP watershed (1992/2022).

LU Class	Surface in 1992 (Ha)	Surface in 2002 (Ha)	Surface in 2012 (Ha)	Surface in 2022 (Ha)
Crops	6999	11,762	12,456	18,502
Bare soil	47,150	46,535	54,074	58,242
Water	89	97	96	49
Built-up	14	79	106	976
Forest	19,564	17,592	16,058	20,476
Shrubs	95,365	93,116	86,391	70,936
Total	169,181	169,181	169,181	169,181

Figure 2 presents the historical LULCC within the INP watershed.

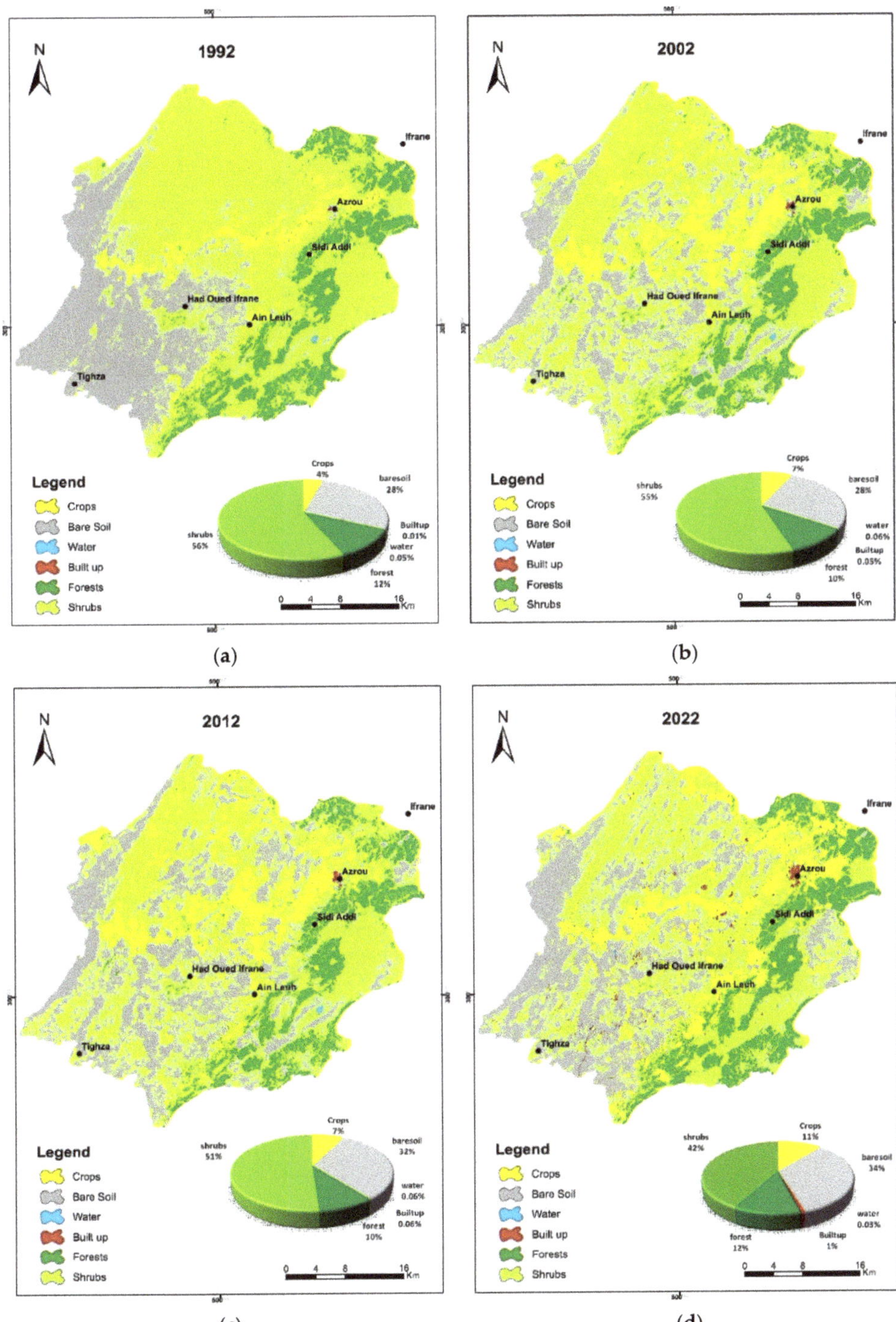

Figure 2. LULCC within the INP watershed: (**a**,**b**) represent the period before park establishment, while (**c**,**d**) illustrate the period after park establishment.

For the different years analyzed, we achieved a Kappa coefficient value equal to or greater than 90% (Table 4), indicating a near-perfect agreement between the reference classification and the automated classification performed using Google Earth Engine. This high level of concordance signifies that the model accurately replicated the reference data.

Table 4. Kappa coefficient for the classifications.

	1992	2002	2012	2022
Kappa coefficient	89%	91.3%	93.8%	98.14%

The classification accuracy results for the years 1992, 2002, 2012, and 2022 show a clear trend of improvement in the model's performance over time, as indicated by the increasing overall accuracy and the decreasing disagreement metrics. The overall accuracy increased progressively, starting from 87.5% in 1992 and reaching an impressive 98.96% in 2022. This steady improvement suggests significant advancements in the classification model's ability to align with reference data. The high accuracy achieved in 2022 highlights the impact of improved input data quality and enhanced feature extraction or modeling techniques over the study period.

The quantity disagreement, which reflects errors related to mismatches in class proportions, exhibited an initial increase from 2.78% in 1992 to 6.45% in 2002. This could be attributed to limitations in the availability or representativeness of the training data during that period. However, this metric showed a consistent decline thereafter, reducing to 4.26% in 2012 and further to 1.04% in 2022. The observed reduction in quantity disagreement over time indicates that the classification model has progressively become better at predicting the correct number of pixels for each class, aligning more closely with the reference data distributions.

Similarly, allocation disagreement, which measures spatial misallocations of classes, was notably high in 1992 (9.72%) and 2002 (11.29%), indicating significant challenges in correctly assigning class locations during the earlier years. However, this metric saw a dramatic reduction to 2.13% in 2012 and was completely eliminated in 2022 (0.00%). The absence of allocation disagreement in 2022 demonstrates a perfect spatial alignment between the predicted and reference data, suggesting substantial advancements in the classification model's spatial accuracy. These improvements may be attributed to better-distributed training data, higher-resolution satellite imagery, or the adoption of more robust classification algorithms.

These results demonstrate a significant improvement in the accuracy of the LULC classification over the study period. The refinement of classification techniques, coupled with advancements in data quality, has led to a more precise delineation of LULCC. Notably, the reduction in classification errors and inconsistencies highlights the effectiveness of the methodological approach employed. These improvements are particularly relevant for assessing LULCC and their implications for WRES within the study area.

3.2. *Water-Regulating Ecosystem Services Evolution*

Our results indicate a notable reduction in soil erosion within the INP watershed following the establishment of the park (2012/2022) (Figure 3). This decline can be attributed to the significant expansion of forest cover, which has enhanced soil stability and reduced erosion vulnerability. These findings underscore the critical role of forest restoration and protection within protected areas in mitigating soil degradation.

However, the observed reduction in erosion could have been more substantial if not for the simultaneous expansion of agricultural activities. The increase in agricultural lands has heightened soil sensitivity to erosion, partially offsetting the benefits provided by forest cover recovery.

Figure 4 illustrates the spatial evolution of the biophysical quantity of sediment loss within the INP watershed, highlighting variations across the study period.

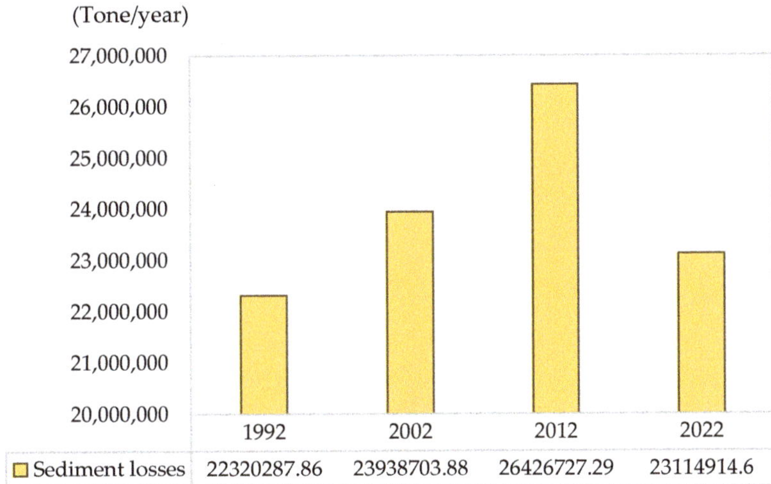

Figure 3. Sediment losses within the INP watershed in tone per year.

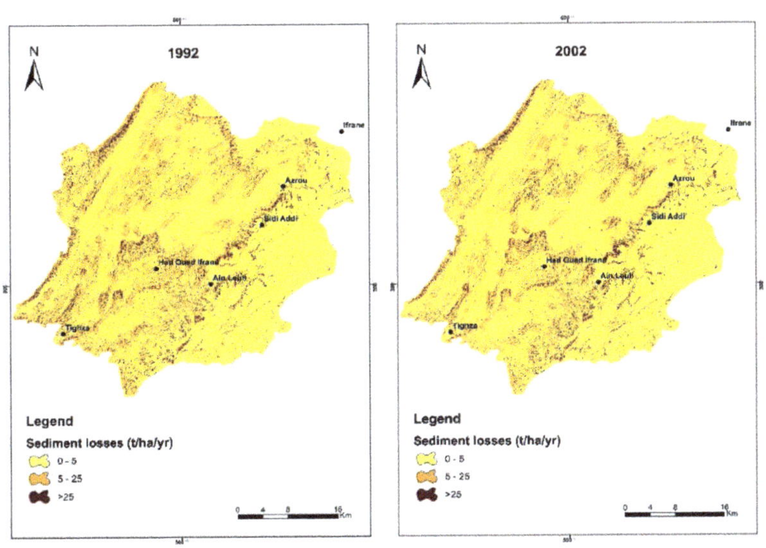

(**a**) (**b**)

Figure 4. *Cont.*

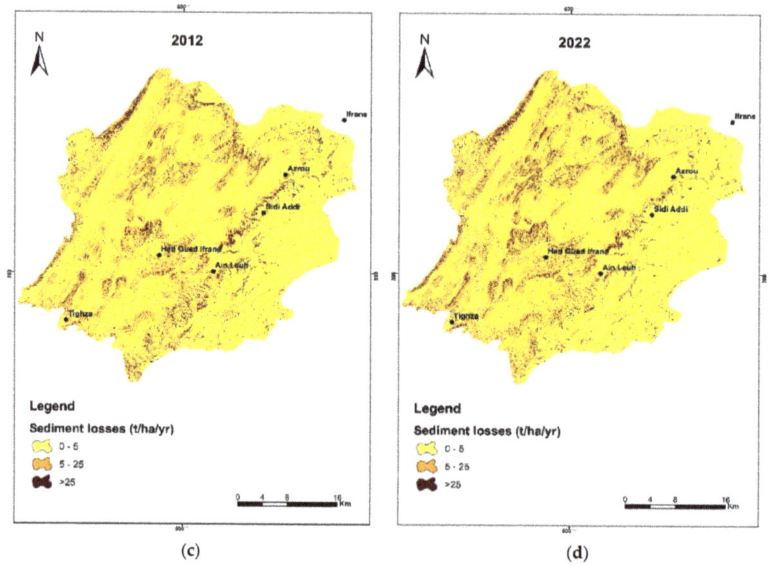

Figure 4. Sediment losses maps within the INP watershed: (**a**,**b**) represent the period before park establishment, while (**c**,**d**) illustrate the period after park establishment.

Our results also indicate a significant reduction in nutrient export, specifically nitrogen and phosphorus, within the INP watershed following the establishment of the park (2012/2022) (Figure 5). This decline can similarly be attributed to the expansion of forest cover, which demonstrates a higher efficiency in nutrient retention compared to other LULC types. These findings further highlight the pivotal role of forest restoration and protection in enhancing ecosystem services.

However, this reduction could have been significantly greater if not for the concurrent expansion of intensive agricultural activities and the substantial decrease in grasslands. These grasslands, which play a vital role in nutrient retention and erosion control, have either been degraded into bare soil due to overgrazing or converted into agricultural fields.

Figures 6 and 7 presents the spatial evolution of the biophysical quantity of nutrients exported within the INP watershed, highlighting variations from 1992 to 2022.

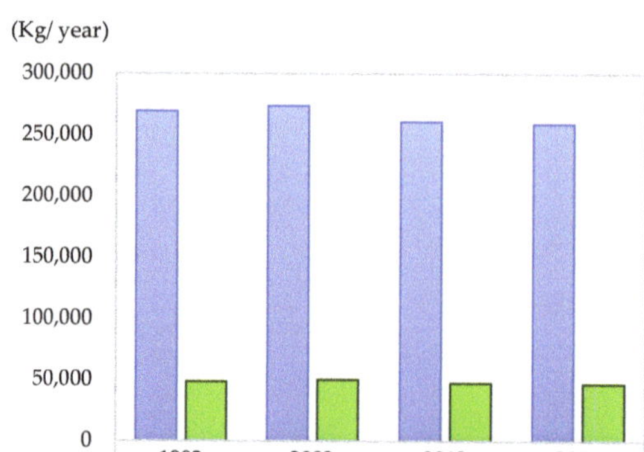

	1992	2002	2012	2022
N export	269114.5	273609.7	260858.8	258971.3
P export	48026.4	49816.3	47167.6	46546.3

Figure 5. Nitrogen and phosphorus export in the INP watershed in kg per year.

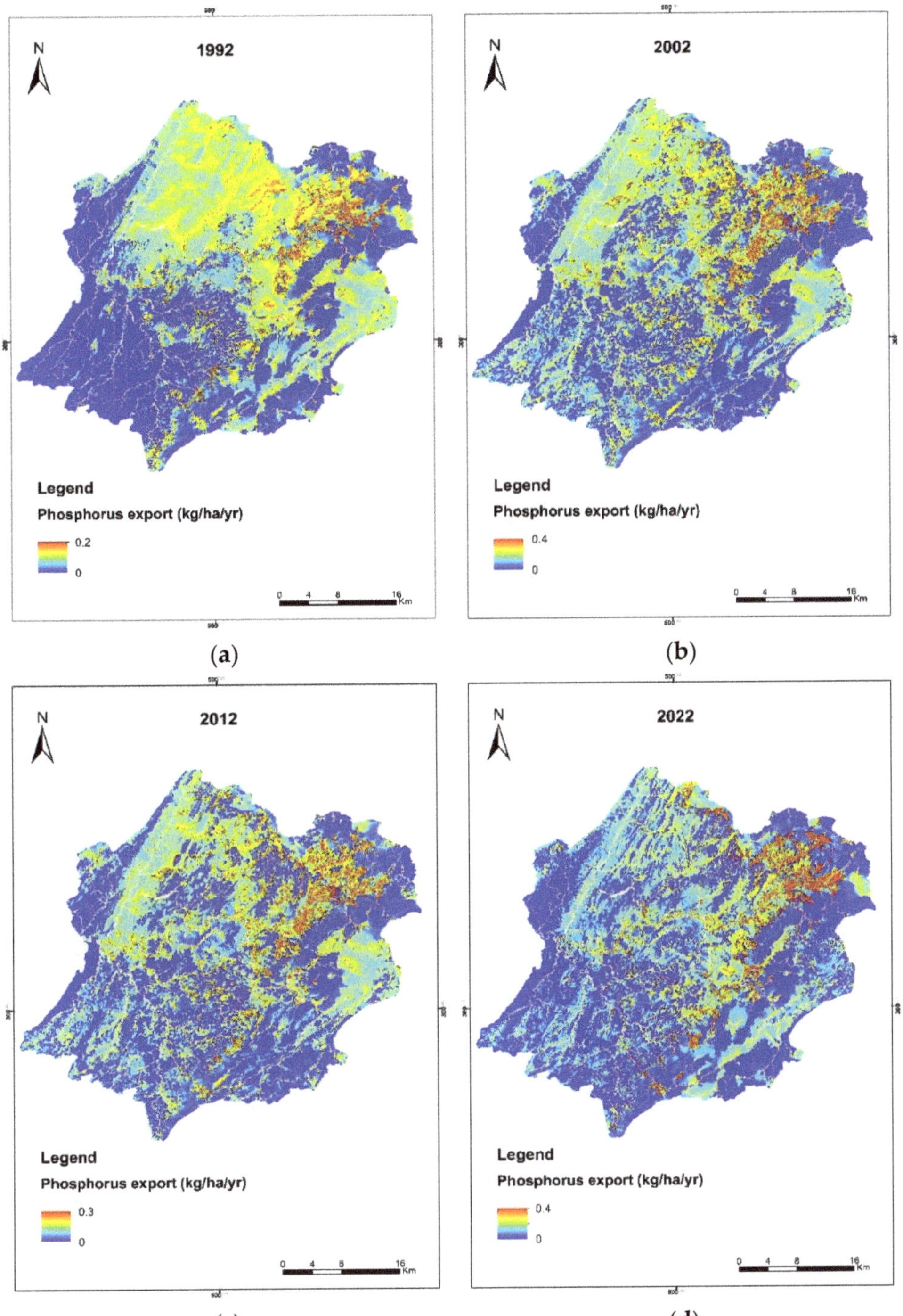

Figure 6. Phosphorus export maps within the INP watershed: (**a**,**b**) represent the period before park establishment, while (**c**,**d**) illustrate the period after park establishment.

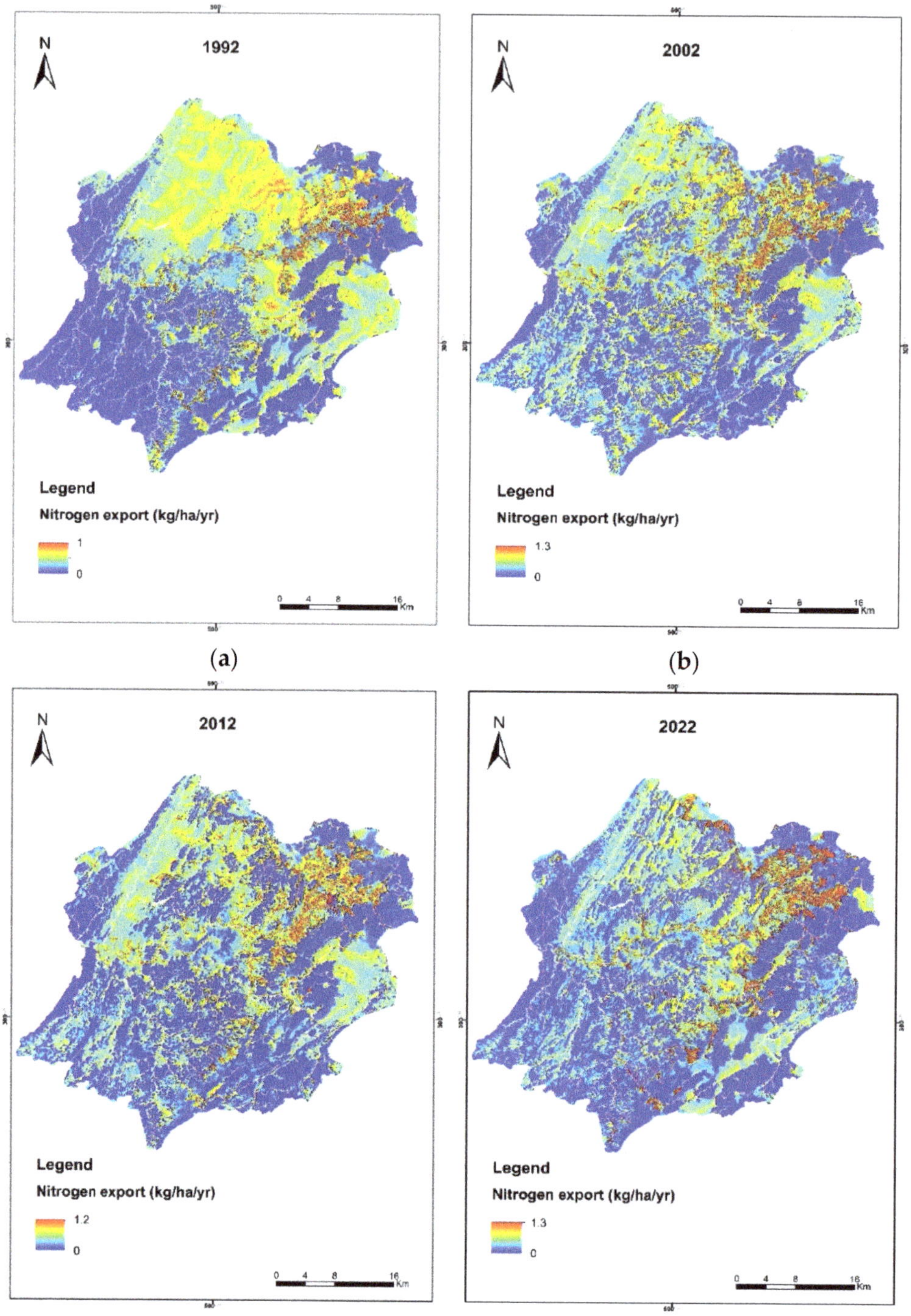

Figure 7. Nitrogen export maps within the INP watershed: (**a**,**b**) represent the period before park establishment, while (**c**,**d**) illustrate the period after park establishment.

3.3. Economic Assessment of Water-Regulating Ecosystem Services

The results show an increase in agricultural yield losses cost (Table 5), driven by the expansion of agricultural areas, even after the park's establishment, leading to greater erosion related damages. Conversely, forage yield losses cost has declined (Table 6), primarily due to a significant reduction in grazing land, which has been converted to agriculture or built-up areas. These findings highlight that, despite efforts to restore forest ecosystems and enhance WRES, the expansion of agriculture continues to undermine these gains, resulting in a low economic value, which would have been much higher.

Table 5. Losses in agricultural yields as a function of soil losses.

Soil Loss Class (t/ha/Year)	Erosion Rate Ewp (t/ha/Year)	Decrease in Yield: r (%)	Decline in Yield ΔR = rxAverage Yield (2 t/Ha) (t/ha)	Annual Cost 1992/2002 (USD/Year)	Annual Cost 2002/2012 (USD/Year)	Annual Cost 2012/2022 (USD/Year)
0–5	2.5	0.04	0.07	9255.9	1427	2219.8
5–25	15	0.31	0.06	−81.6	177	501.6
>25	25	0.59	0.12	−6.8	8.6	46.9
			Total	**9167.6**	**1612.7**	**22,741.4**

Table 6. Losses in forage yield as a function of soil losses.

Soil Losses (t/ha)	Loss of Forage Productivity (%)	Loss of Forage Productivity (UF/ha)	Annual Cost (USD/Year)	Annual Cost (USD/Year)	Annual Cost (USD/Year)
0–5	2.5	3.3	−902.1	−6284.8	−20,784.5
5–25	25	32.9	−1307.4	−3618.6	−10,803.3
+25	45	59.3	−104.7	−179.8	−1122.9
		Total	**−2314.2**	**−10,083.1**	**−32,710.8**

Note: Negative values in the tables indicate a reduction in the cost of lost productivity, reflecting an improvement in WRES, while positive values indicate the opposite.

The economic value of the WRES is determined based on the potential on-site losses in agricultural and forage yields that would occur in the absence of this ecosystem service. The results show that the WRES value was negative before the park's establishment. However, after the park's creation (in 2004), the WRES value became positive, estimated at USD 10,000 /Year (Figure 8).

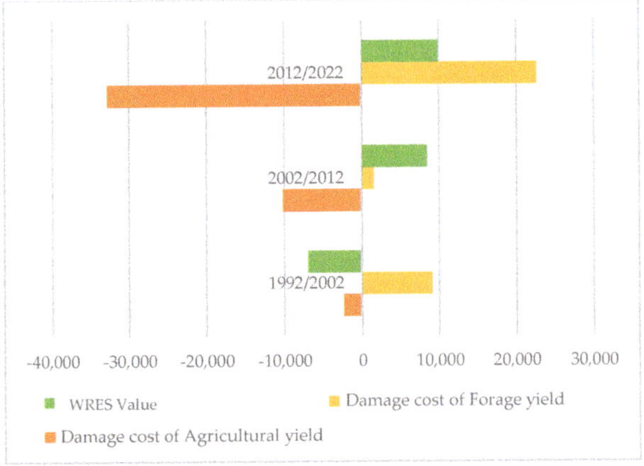

Figure 8. Economic value of Water-Regulating Ecosystem Services in USD per year.

4. Discussion

The spatiotemporal and economic assessment of Water Regulation Ecosystem Services (WRES) reveals significant trends at both spatial and economic scales. Before the establishment of Ifrane National Park in 2004, the WRES value in the Upper Beht Watershed was negative, estimated at USD −7000 per year. This deficit was primarily due to agricultural expansion and the decline of grasslands, which led to increased sediment and nutrient losses. Following the park's establishment, the WRES value turned positive, reaching USD 10,000 per year. This improvement is largely attributed to the expansion of forest cover, which enhanced sediment and nutrient retention, effectively reducing erosion-related losses and contributing to ecosystem stability. Notably, improvements in WRES were most evident in areas where forest cover had been restored.

An analysis of conservation initiatives implemented within the park reveals significant achievements. During the period 2012–2022, more than 2000 hectares were successfully reforested, and over 4000 hectares were placed under protection or designated for restoration and reforestation. These areas were managed through a participatory approach involving local communities. The number of local associations engaged in park management increased from 5 at the park's inception to 12 at present, reflecting a growing community commitment. Furthermore, there was a nearly 50% reduction in forest-related offenses during this time. These positive indicators align with the findings from remote sensing analysis, which demonstrated an increase in forest cover, thereby confirming the tangible impact of these conservation efforts [50].

However, despite this positive trend, the overall enhancement of WRES was constrained by ongoing agricultural expansion and the regression in grasslands. This shift in grasslands has not only compromised the ecological benefits provided by this areas but has also amplified sediment and nutrient runoff. Intensive agricultural practices often involve excessive use of fertilizers and irrigation, which, combined with the loss of natural vegetation, exacerbate nutrient and sediment leaching into downstream systems. As a result, while reforestation efforts have improved some ecosystem services, the negative impacts of agricultural expansion have partially offset these gains. Several studies have demonstrated that intensive agriculture and the transformation of rangelands into croplands and urban areas significantly reduce sediment and nutrient retention capacity. Agricultural expansion in grazing areas disrupts soil structure, increases surface runoff, and accelerates erosion, leading to higher sediment and nutrient losses. While cropland conversion may enhance agricultural productivity, it often comes at the expense of critical ecosystem functions and services, including WRES. The decline in grazing land, largely replaced by more economically lucrative agricultural activities, further diminishes the landscape's ability to retain sediments and nutrients, ultimately undermining conservation efforts within protected areas [51–55].

The findings of this study underscore the importance of integrated land management strategies that balance agricultural development with ecosystem conservation to maximize the benefits of soil erosion control and nutrients retention. Despite efforts to restore forest ecosystems, the continued expansion of intensive agriculture and urbanization in the study area continues to counteract these environmental gains. Implementing sustainable agricultural practices and the preservation of grazing lands are critical to optimize ecosystem services such as WRES and limit negative environmental impacts while supporting local economic activities [56–67]. To this end, we recommend the following actions:

- Enhancing ecosystem service monitoring: Investing in advanced tools for continuous mapping and assessment of ecosystem service dynamics within protected areas will enable data-driven decision-making and adaptive management strategies to address ecosystem changes effectively.

- Integrating WRES valuation into policy and decision-making: Incorporating WRES into conservation policies can strengthen environmental initiatives. For example, developing and implementing Payment for Ecosystem Services (PES) mechanisms can provide financial incentives for land users, encouraging sustainable land management practices.
- Promoting sustainable practices among local communities: Encouraging agroforestry, agroecology, soil and water conservation techniques, water-efficient crops, and the restoration of degraded grasslands (through seeding and temporary protection) will help mitigate the environmental impacts of agriculture and overgrazing. Awareness campaigns should highlight the value of WRES to foster community engagement.
- Integrating WRES dynamics into park management planning: Future revisions of the Park Management and Development Plan (PAG) should consider the spatio-temporal evolution of WRES. This integration will inform more effective conservation strategies and contribute to the sustainable management of park resources.

Notwithstanding the promising outcomes of this research, several limitations should be acknowledged. The spatial resolution of the remote sensing data could limit the accuracy of detecting subtle changes in forest cover and grassland dynamics. Future research should incorporate higher-resolution datasets and more refined modeling assumptions to enhance the reliability of the findings. In addition, while the positive impact of reforestation is evident, the persistent expansion of intensive agriculture continues to exert negative pressures on ecosystem services. More detailed, long-term case studies are needed to fully understand these trade-offs and to evaluate the cumulative impacts of different land management practices.

Finally, this study demonstrates that integrated land management strategies, which balance ecosystem conservation with sustainable agricultural practices, can lead to substantial improvements in WRES. While reforestation and community-driven conservation efforts have yielded significant benefits, the ongoing challenges posed by agricultural expansion must be addressed to secure long-term ecosystem resilience. By leveraging advanced monitoring technologies, enhancing community participation, and embedding ecosystem service valuations into policy frameworks, it is possible to achieve a sustainable balance between economic development and environmental conservation in the Upper Beht Watershed.

Author Contributions: Conceptualization, O.S. and A.K.; methodology, O.S.; software, O.S. and Z.D.; validation, A.K.; formal analysis, O.S.; investigation, O.S.; resources, O.S.; data curation, O.S.; writing—original draft preparation, O.S.; writing—review and editing, O.S., Z.D. and A.K.; visualization, O.S.; supervision, A.K. All authors have read and agreed to the published version of the manuscript.

Funding: This research received no external funding.

Data Availability Statement: All data generated or analysed during this study are included in this manuscript.

Conflicts of Interest: The authors declare no conflicts of interest.

Appendix A Biophysical Tables of SDR and NDR Models

Table A1. Biophysical tables of SDR model.

Lucode	LULC_Desc	Usle_c	Usle_p
1	Crops	0.19	1
2	Bare soil	1	1
3	Water	0.04	1
4	Built up area	0.1	1
5	Forest	0.003	1
6	Shrubs	0.5	1

Table A2. Biophysical tables of NDR model.

lucode	LULC_desc	load_n	eff_n	load_p	eff_p	crit_len_p	crit_len_n
1	Crops	12.42	0.25	2.21	0.25	150	150
2	Bare soil	1	0.05	0.1	0.05	150	150
3	Water	0	0	0	0	150	150
4	Built up area	12.78	0.08	4.17	0.05	150	150
5	Forest	2.2	0.83	0.275	0.8	150	150
6	Shrubs	6.28	0.1	1.35	0.25	150	150

References

1. *Millennium Ecosystem Assessment*; Millennium Ecosystem Assessment: Washington, DC, USA, 2003.
2. Costanza, R.; d'Arge, R.; De Groot, R.; Farber, S.; Grasso, M.; Hannon, B.; Limburg, K.; Naeem, S.; O'Neill, R.V.; Paruelo, J.; et al. The value of the world's ecosystem services and natural capital. *Nature* **1997**, *387*, 253–260.
3. Ouakhir, H.; Fortesa, J.; Reddad, H.; García-Comendador, J.; El Khalki, Y.; El Ghachi, M.; Estrany, J. Hydrological response of two contrasting small Mediterranean Mountainous catchments in the Middle Atlas–Morocco. In Proceedings of the 10th International Conference on Geomorphology, Coimbra, Portugal, 12–16 September 2022. [CrossRef]
4. Msaddek, M.; El Garouani, A. Modeling the Hydrological Impacts of Vegetation Cover Changes in the Upper Oum Er-Rbia Watershed (Morocco). *J. Ecol. Eng.* **2021**, *22*, 167–180. [CrossRef]
5. Mekonnen, S. Review on the Role of Forest Landscapes in Watershed Hydrologic Processes. *J. Environ. Earth Sci.* **2017**, *7*, 97–104.
6. Cayuela Linares, C. *Ecohydrology of Mediterranean Headwater Catchments. The Role of Forest in the Redistribution and Isotopic Modification of Water Fluxes*; Autonomous University of Barcelona: Barcelona, Spain, 2019.
7. Brauman, K.A.; Daily, G.C.; Duarte, T.K.E.; Mooney, H.A. The nature and value of ecosystem services: An overview highlighting hydrologic services. *Annu. Rev. Environ. Resour.* **2007**, *32*, 67–98.
8. Abdelaziz, E.B.; Amyay, M.; Ech-Chahdi, K.E.O. *Recent Variations of Water Area in the Middle Atlas Lakes (Morocco): A Response to Drought Severity and Land Use Changes*; Scite LLC: Henderson, NV, USA, 2023. [CrossRef]
9. Pătru-Stupariu, I.; Hossu, C.A.; Grădinaru, S.R.; Nita, A.; Stupariu, M.-S.; Huzui-Stoiculescu, A.; Gavrilidis, A.-A. A Review of Changes in Mountain Land Use and Ecosystem Services: From Theory to Practice. *Land* **2020**, *9*, 336. [CrossRef]
10. Layati, E.; El Ghachi, M. Oued Lakhdar watershed (Morocco), monitoring land use/cover changes: Remote sensing and GIS approach. *Geol. Ecol. Landsc.* **2024**, *2024*, 2395204. [CrossRef]
11. Sadgui, O.; Khattabi, A. Economic Assessment of Hydrologic Ecosystem Services in Morocco's Protected Areas: A Case Study of Ifrane National Park. *Sustainability* **2024**, *16*, 8800. [CrossRef]
12. Ennaji, N.; Ouakhir, H.; Abahrour, M.; Spalevic, V.; Dudic, B. Impact of watershed management practices on vegetation, land use changes, and soil erosion in River Basins of the Atlas, Morocco. *Not. Bot. Horti Agrobot. Cluj-Napoca* **2024**, *52*, 13567. [CrossRef]
13. Boubekraoui, H.; Attar, Z.; Maouni, Y.; Ghallab, A.; Saidi, R.; Maouni, A. Forest Loss Drivers and Landscape Pressures in a Northern Moroccan Protected Areas' Network: Introducing a Novel Approach for Conservation Effectiveness Assessment. *Conservation* **2024**, *4*, 452–485. [CrossRef]
14. Lamhamedi, H.; Lizin, S.; Witters, N.; Malina, R.; Baguare, A. The recreational value of a peri-urban forest in Morocco. *Urban For. Urban Green.* **2021**, *65*, 127339. [CrossRef]
15. Ferraro, P.J.; Ferraro, P.J.; Hanauer, M.M.; Miteva, D.A.; Miteva, D.A.; Miteva, D.A.; Nelson, J.L.; Pattanayak, S.K.; Nolte, C.; Sims, K.R.E. Estimating the Impacts of Conservation on Ecosystem Services and Poverty by Integrating Modeling and Evaluation. *Soc. Sci. Res. Netw.* **2016**, *112*, 7420–7425. [CrossRef]

16. Zeng, Y.; Koh, L.P.; Wilcove, D.S. Gains in biodiversity conservation and ecosystem services from the expansion of the planet's protected areas. *Sci. Adv.* **2022**, *8*, eabl9885. [CrossRef]
17. Zhao, C.; Su, S.; Gong, Z.; Lv, C.; Li, N.; Luo, Q.; Zhou, X.; Li, S. Effectiveness of protected areas in the Three-river Source Region of the Tibetan Plateau for biodiversity and ecosystem services. *Ecol. Indic.* **2023**, *154*, 110861. [CrossRef]
18. Langhammer, P.F.; Bull, J.W.; Bicknell, J.E.; Oakley, J.L.; Brown, M.H.; Bruford, M.W.; Butchart, S.H.M.; Carr, J.A.; Church, D.; Cooney, R.; et al. The positive impact of conservation action. *Science* **2024**, *384*, 453–458. [CrossRef]
19. Sasanifar, S.; Alijanpour, A.; Banj Shafiei, A.; Rad, J.; Molaei, M.; Álvarez-Álvarez, P. Understanding how forest ecosystem services are affected by conservation practices and differences in elevation: A study in the Arasbaran biosphere reserve, Iran. *Ecol. Eng.* **2024**, *203*, 107230. [CrossRef]
20. Amraoui, M.; Bouabidi, L.; El Amrani, M.; Ouakhir, H.; Dudic, B.; Lukic, T.; Spalevic, V. Land use dynamics and soil conservation strategies in the el kssiba region, atlas mountains of morocco. *J. Agric. For.* **2024**, *70*, 7. [CrossRef]
21. Moutaoikil, N.; Benzougagh, B.; Mastere, M.; El Fellah, B.; Lamrani, H. The impact of soil erosion on environments: A case study of the Oued Beht Watershed (Morocco). *BIO Web Conf.* **2024**, *115*, 01006. [CrossRef]
22. Kusi, K.K.; Khattabi, A.; Mhammdi, N.; Lahssini, S. Prospective evaluation of the impact of land use change on ecosystem services in the Ourika watershed, Morocco. *Land Use Policy* **2020**, *97*, 104796. [CrossRef]
23. Kusi, K.K.; Khattabi, A.; Mhammdi, N. Integrated assessment of ecosystem services in response to land use change and management activities in Morocco. *Arab. J. Geosci.* **2021**, *16*, 126. [CrossRef]
24. Kusi, K.K.; Khattabi, A.; Mhammdi, N. Evaluating the impacts of land use and climate changes on water ecosystem services in the Souss watershed, Morocco. *Arab. J. Geosci.* **2023**, *16*, 126. [CrossRef]
25. Kusi, K.K.; Khattabi, A.; Mhammdi, N. Analyzing the impact of land use change on ecosystem service value in the main watersheds of Morocco. *Environ. Dev. Sustain.* **2023**, *25*, 2688–2715. [CrossRef]
26. Idrissi, S.; Taous, A. Morphometric and hydrographic analysis of the Beht watershed and its main tributaries using geographic information systems. *Moroc. J. Geomorphol.* **2022**, *85-88*, 103.
27. *Plan for the Development and Management of Ifrane National Park*; Internal document; National Agency of Water And Forests: Rabat, Morroco, 2007; pp. 1–180.
28. Hansen, M.C.; Loveland, T.R. A review of large area monitoring of land cover change using Landsat data. *Remote Sens. Environ.* **2012**, *122*, 66–74. [CrossRef]
29. Roy, D.P.; Wulder, M.A.; Loveland, T.R.; Woodcock, C.E.; Allen, R.G.; Anderson, M.C.; Helder, D.; Irons, J.R.; Johnson, D.M.; Kennedy, R.; et al. Landsat-8: Science and product vision for terrestrial global change research. *Remote Sens. Environ.* **2014**, *145*, 154–172. [CrossRef]
30. Belgiu, M.; Drăguţ, L. Random forest in remote sensing: A review. *ISPRS J. Photogramm. Remote Sens.* **2016**, *114*, 24–31. [CrossRef]
31. Lambin, E.F.; Turner, B.L.; Geist, H.J.; Agbola, S.B.; Angelsen, A.; Bruce, J.W.; Coomes, O.T.; Dirzo, R.; Fischer, G.; Folke, C.; et al. The causes of land-use and land-cover change: Moving beyond the myths. *Glob. Environ. Change* **2001**, *11*, 261–269. [CrossRef]
32. Tahraoui, A.; Kheddam, R. LULC Change Detection Using Combined Machine and Deep Learning Classifiers. In Proceedings of the 2024 IEEE 7th International Conference on Advanced Technologies, Signal and Image Processing (ATSIP), Sousse, Tunisia, 11–13 July 2024; pp. 403–408. [CrossRef]
33. Al-Ruzouq, R.; Shanableh, A.; Gibril, M.B.A.; Kalantar, B. Multi-scale correlation-based feature selection and random forest classification for LULC mapping from the integration of SAR and optical Sentinel images. In *Remote Sensing Technologies and Applications in Urban Environments IV*; SPIE: Cergy, France, 2019; Volume 11157, pp. 58–70. [CrossRef]
34. Lin, C.; Doyog, N.D. Challenges of retrieving LULC information in rural-forest mosaic landscapes using random forest technique. *Forests* **2023**, *14*, 816. [CrossRef]
35. Naturel Capital Project. Invest User Guide_Invest Documentation. Available online: http://releases.naturalcapitalproject.org/invest-userguide/latest/index.html (accessed on 15 November 2023).
36. Achiban, H.; Taous, A.; El-Khantoury, I.; el Mderssa, M.; et Amechrouq, A. Quantification of soil loss in various lithological areas of the western Middle Atlas Central: Application to the Ras-Elma, Tamelalet and Sebab watershed (Tigrigra watershed, Morocco). In *E3S Web of Conferences*; EDP Sciences: Les Ulis, France, 2018; Volume 37, p. 04003. 8p.
37. Benez-Secanho, F.J.; Dwivedi, P. Does quantification of ecosystem services depend upon scale (resolution and extent)? A case study using the InVEST nutrient delivery ratio model in Georgia, United States. *Environments* **2019**, *6*, 52. [CrossRef]
38. CHIRPS Data: Rainfall Estimates from Rain Gauge and Satellite Observations, Climate Hazards Center. Available online: https://www.chc.ucsb.edu/data/chirps (accessed on 3 September 2024).
39. Earth Science Data Systems (ESDS) Program of the National Aeronautics and Space Administration of the United States of America. Available online: https://earthdata.nasa.gov (accessed on 15 August 2021).
40. Rango, A.; Arnoldus, H.M.J. Aménagement des bassins versants. *Cah. Tech. FAO* **1987**, *36*, 1–11.

41. Romanazzi, G.R.; Palmisano, G.O.; Cioffi, M.; Leronni, V.; Toromani, E.; Koto, R.; De Boni, A.; Acciani, C.; Roma, R. A Cost–Benefit Analysis for the Economic Evaluation of Ecosystem Services Lost Due to Erosion in a Mediterranean River Basin. *Land* **2024**, *13*, 1512. [CrossRef]
42. Pires-Marques, É.; Chaves, C.; Pinto, L.M.C. Biophysical and monetary quantification of ecosystem services in a mountain region: The case of avoided soil erosion. *Environ. Dev. Sustain.* **2021**, *23*, 11382–11405. [CrossRef]
43. Panagos, P.; Matthews, F.; Patault, E.; De Michele, C.; Quaranta, E.; Bezak, N.; Kaffas, K.; Patro, E.R.; Auel, C.; Schleiss, A.J.; et al. Understanding the cost of soil erosion: An assessment of the sediment removal costs from the reservoirs of the European union. *J. Clean. Prod.* **2023**, *434*, 140183. [CrossRef]
44. Kling, C.L.; Herriges, J. *Revealed Preference Approaches to Environmental Valuation*; Routledge: London, UK, 2008.
45. Tijani, A.; Hamadi, F. Quantifying and Accounting for Environmental Costs by the Avoidance Cost's Method: The Case of a Tunisian Firm. *J. Account. Audit. Res. Pr.* **2013**, *2013*, 491993. [CrossRef]
46. Biggelaar, C.D.; Lal, R.; Wiebe, K.; Breneman, V. The global impact of soil erosion on productivity. I. Absolute and relative erosion-induced yield losses. *Adv. Agron.* **2004**, *81*, 1–48.
47. Jorio, A. Le Coût de la Dégradation de l'Environnement au Maroc. Chapitre 5 "Sols". Environment and Natural Resources Global Practice Discussion Paper. *World Bank Group Rep.* **2017**, *105633-MA*, 49–64.
48. DPA. *Agricultural Yield in Ifrane Province*; Internal document; Provincial Direction of Agriculture of Ifrane: Ifrane, Morocco, 2024.
49. ONICL. *Cereals and Forage Market Prices from 2002 to 2024*; Internal document; The National Interprofessional Office for Cereals and Legumes: Rabat, Morocco, 2024.
50. ANEF. *The Evolution of Reforested Areas, Areas Under Protection, Forest Offenses, and Sylvopastoral Management Associations from 2012 to 2022*; Internal document; National Agency of water and forests: Rabat, Morocco, 2022.
51. Imran, M.; Haider, F. Forest ecosystem services of water-related filtration and regulation, a multi-source assessment and economic valuation in Mangla watershed. *Water Supply* **2024**, *24*, 3680–3696. [CrossRef]
52. Bisui, S.; Roy, S.; Sengupta, D.; Bhunia, G.S.; Shit, P.K. Assessment of ecosystem services values in response to land use/land cover change in tropical forest. In *Forest Resources Resilience and Conflicts*; Elsevier: Amsterdam, The Netherlands, 2021. [CrossRef]
53. Adimassu, Z.; Tamene, L.; Degefie, D.T. The influence of grazing and cultivation on runoff, soil erosion, and soil nutrient export in the central highlands of Ethiopia. *Ecol. Process.* **2020**, *9*, 23. [CrossRef]
54. Telak, L.J.; Bogunovic, I.; Rodrigo-Comino, J. Land Management Impacts on Soil Water Erosion and Loss of Nutrients. *Proceedings* **2019**, *30*, 35. [CrossRef]
55. Imbert, A.B.; Betancourt, I.F.; Jiménez, T.L.; Arencibia, M.C.; Quintana, J.R.F.; Rojas, R.S.; Lobaina, A.B.; Duran, A.C. Agricultural Practices to Mitigate Soil Degradation and Increase Carbon Capture. *Agrisost* **2017**, *23*, 24–31.
56. Bogunovic, I.; Fernández, M.P.; Kisic, I.; Marimón, M.B. Agriculture and grazing environments. In *Advances in Chemical Pollution, Environmental Management and Protection*; Elsevier: Amsterdam, The Netherlands, 2019. [CrossRef]
57. Uddin, M.M.; Haque, S.M.; Alam, M.S. Soil degradation processes under agriculture and the practices to reverse the degradation processes for environmental sustainability. *Int. J. Agric. Environ. Res.* **2017**, *3*, 4045–4063.
58. Girijaveni, V.; Reddy, K.S.; Prasad, J.; Singh, V.; Kumar, C. Regaining the Essential Ecosystem Services in Degraded Lands. In *Agroecological Approaches for Sustainable Soil Management*; Wiley: Hoboken, NJ, USA, 2023. [CrossRef]
59. Singh, B.; Kaunert, C.; Jermsittiparsert, K. Environment-Biodiversity Protection and SDG 15 (Life on Land). Practice, progress, and proficiency in sustainability. In *Maintaining a Sustainable World in the Nexus of Environmental Science and AI*; IGI Global: Hershey, PA, USA, 2024. [CrossRef]
60. Sanjay, M.A.; Rai, S.; Patil, A.A.; Nengparmoi, T.; Devi, K.B.; Dora, H.S.V.; Sharma, Y. Environmental Sustainability through Soil Conservation: An Imperative for Future Generations. *Int. J. Environ. Clim. Change* **2023**, *13*, 1700–1707. [CrossRef]
61. Dubovitski, A.A. Improving Agricultural Land Management As A Tool For Promoting Sustainable Development. In *The European Proceedings of Social and Behavioural Sciences*; European Publisher: London, UK, 2022. [CrossRef]
62. Sidemo Holm, W. Effective Conservation of Biodiversity and Ecosystem Services in Agricultural Landscapes. Ph.D. Thesis, Lund University, Lund, Sweden, 2021.
63. Velasco-Muñoz, J.F.; Aznar-Sánchez, J.A.; López-Felices, B.; García-Arca, D. Sustainable land use and management. In *Sustainable Resource Management*; Elsevier: Amsterdam, The Netherlands, 2021. [CrossRef]
64. Oloo, J. Contribution of Payment for Ecosystem Services in Addressing Adaptation to Climate Change. *Int. J. Environ. Clim. Change* **2024**, *14*, 735–747. [CrossRef]
65. Izquierdo-Tort, S.; Jayachandran, S.; Saavedra, S. Redesigning payments for ecosystem services to increase cost-effectiveness. *Nat. Commun.* **2024**, *15*, 9252. [CrossRef]

66. Altobelli, F.; Vargas, R.; Corti, G.; Dazzi, C.; Montanarella, L.; Monteleone, A.; Caon, L.; Piazza, M.G.; Calzolari, C.; Munafò, M.; et al. Improving soil and water conservation and ecosystem services by sustainable soil management practices: From a global to an italian soil partnership. *Ital. J. Agron.* **2020**, *15*, 293–298. [CrossRef]
67. Guariguata, M.R.; Atmadja, S.; Baral, H.; Boissiere, M.; Brady, M.A.; Chomba, S.; Cronkleton, P.; Djoudi, H.; Duchelle, A.E.; Duguma, L.A.; et al. *Forest and Landscape Restoration*; Center for International Forestry Research (CIFOR) and World Agroforestry Centre (ICRAF): Bogor, Indonesia, 2021. [CrossRef]

land

MDPI

Article

Digital Soil Mapping of Soil Macronutrients (N, P, K) in Emilia-Romagna (NE Italy): A Regional Baseline for the EU Soil Monitoring Law

Fabrizio Ungaro [1,*], Paola Tarocco [2] and Alessandra Aprea [2]

1 National Research Council, Institute of BioEconomy (CNR-IBE), Via Madonna del Piano 10, 50019 Sesto Fiorentino, Italy
2 Geology, Soil and Seismic Risk Area, Regione Emilia-Romagna, Viale A. Moro 52, 40127 Bologna, Italy; paola.tarocco@regione.emilia-romagna.it (P.T.); alessandra.aprea@regione.emilia-romagna.it (A.A.)
* Correspondence: fabrizio.ungaro@cnr.it

Abstract: Assessing soil fertility is a complex task as it is determined by natural and anthropogenic factors, including specific agronomic interventions (e.g., fertilization and crop rotation) and broader soil management (e.g., tillage and drainage). For agricultural management, soil represents a primary production factor whose chemical–physical characteristics and macro-elements content must be known. This work presents the maps of three macronutrients, i.e., N, K, and P, in the topsoils (0–30 cm layer) of the Emilia-Romagna (21,710.1 km^2) region in NE Italy. The maps and associated uncertainty at 100 m resolution were obtained via digital soil mapping (DSM) resorting to Quantile Random Forests using topsoil data from the regional soil database (N = 34,750). As Emilia-Romagna is characterized by two distinct major landforms, i.e., the intensively cultivated alluvial plain and the extensively managed mountain range of the Northern Apennines, each representing nearly half of the region, two distinct sets of numerical and categorical covariates were used as predictors for the DSM estimation of each macronutrient. Results highlight an average N content of approximately 1.57 ± 0.83 (standard deviation) g kg^{-1} in the alluvial plain and of 1.63 ± 0.49 g kg^{-1} in the Apennines. For exchangeable potassium (K), concentrations were 275.90 ± 92.6 mg kg^{-1} and 210.2 ± 86.3 mg kg^{-1} in the plain and Apennines, respectively. A stark contrast was observed for available phosphorus (P), with mean values of 40.4 ± 11.0 mg kg^{-1} in the alluvial plain, dropping to 15.2 ± 6.1 mg kg^{-1} in the Apennines. Such results provide useful information for assessing the fertility of regional soils and provide a reference baseline for soil quality monitoring. The resulting macronutrient maps were eventually compared with those based on the Land Use and Cover Area frame Survey (LUCAS), which represents the reference baselines at the EU scale.

Keywords: soil macronutrients; digital soil mapping; Quantile Random Forest; Emilia Romagna; LUCAS; EU Soil Monitoring Law

1. Introduction

According to a recent assessment [1], a large portion of European soils are unhealthy, with 60–70% showing signs of degradation due to unsustainable management. Nutrient imbalances affect approximately 74% of agricultural land in the EU [2], with significant consequences on soil-based ecosystem services, including agricultural productivity, carbon sequestration, and water quality [3–5]. Effective nutrient management is therefore essential to maintain soil health, production potential, and environmental quality [6–8].

The status of soil nutrients across Europe is highly heterogeneous, being influenced by parent material, climate, land use, and management history, leading to significant regional disparities [9–11]. To address this, the proposed EU Soil Monitoring Law (SML) [12] aims to establish a harmonized framework for monitoring soil health, with topsoil nitrogen (N) and phosphorus (P) concentrations among its key indicators. As first measurements should be done within three years of the directive's entry into force, a reference baseline providing the overall status of soil nutrients would be highly desirable as this would also allow to better assess soil degradation costs [13]. This legislative drive creates then an urgent need for accurate, spatially explicit baseline data on soil nutrients at a management-relevant scale.

Digital soil mapping (DSM) has emerged as a critical tool for quantifying the spatial distribution of soil properties [14,15]. By coupling soil observation data with environmental covariates through statistical or machine learning models, DSM can produce continuous maps of soil characteristics, possibly leading to improved pedological understanding [16]. At the continental scale, the Land Use/Cover Area frame statistical Survey (LUCAS) topsoil dataset has been used with Gaussian Process Regression to create baseline maps for N, P, and K at 250 m resolution [17]. While these maps highlight major drivers of nutrient distribution at a broad scale, their relatively low sample density (~1 sample per 200 km^2) and the coarse resolution of some covariates limit their applicability for regional-scale policy implementation and land management. However, the implementation of EU environmental regulations targeting water, soil, and environmental quality [18–20], aiming to reduce emissions from agriculture, still requires knowledge gaps to be filled at the national and regional scales to manage the impacts of nutrient losses on terrestrial and aquatic ecosystems on human health [21,22].

In Italy, the implementation of EU environmental directives like the Nitrates Directive (91/676/EEC) is delegated to administrative regions (NUTS2 level according to the EU Nomenclature of Territorial Units for Statistics) [23]. This directive defines the discipline on the agronomic use of livestock effluents, vegetation waters from oil mills, and waste waters from agricultural and agri-food companies [18]. In Emilia-Romagna (NE Italy), a highly productive agricultural region, specific regulations (e.g., Regional Regulation No. 2/2024) require differentiated agricultural practices based on local soil conditions, including nutrient status. This regulatory landscape requires a level of spatial detail and accuracy that EU-scale approaches like LUCAS may fail to provide [17].

Consequently, a critical research gap exists between the availability of continental-scale soil nutrient assessments and the precise data needs for effective regional policy and management. There is a pressing need for high-resolution, regional baseline maps to assess the practical applicability of continental models and to provide a reliable foundation for regional decision-making

The goal of this work is to fill this knowledge gap by providing a high-resolution (100 m) assessment of topsoil macronutrients (N, P, and K) in Emilia-Romagna using a robust DSM approach based on a large regional soil database (N = 34,750). Our specific objectives are as follows:

1. Produce and validate maps of N, P, and K concentrations and their associated uncertainty using Quantile Random Forests.
2. Identify the environmental covariates that most strongly influence the spatial patterns of each nutrient.
3. Critically compare our regional maps with the existing LUCAS-based continental maps to quantify discrepancies and evaluate the implications for regional soil health assessment within the framework of the proposed EU Soil Monitoring Law.

2. Materials and Methods

2.1. Study Area: Climate, Soils, and Land Uses

Emilia-Romagna in NE Italy (lat 43°5′ N–45°8′ N; long 9°20′ E–12°40′ E Greenwich, approximately) has an area of 22,509.67 km^2 and is characterized by a variety of landforms and landscapes, being the region's territory divided into two parts with almost equal extensions (Figure 1): the north-eastern one (47.8% of the total surface area) south of the Po River occupied by the Emiliano-Romagnola plain (ca. 12,032 km^2) which is delimited to the East by the Adriatic Sea, and the south-western one characterized by the presence of the Apennines range (hilly for 27.1% of the area and mountainous for 25.1%).

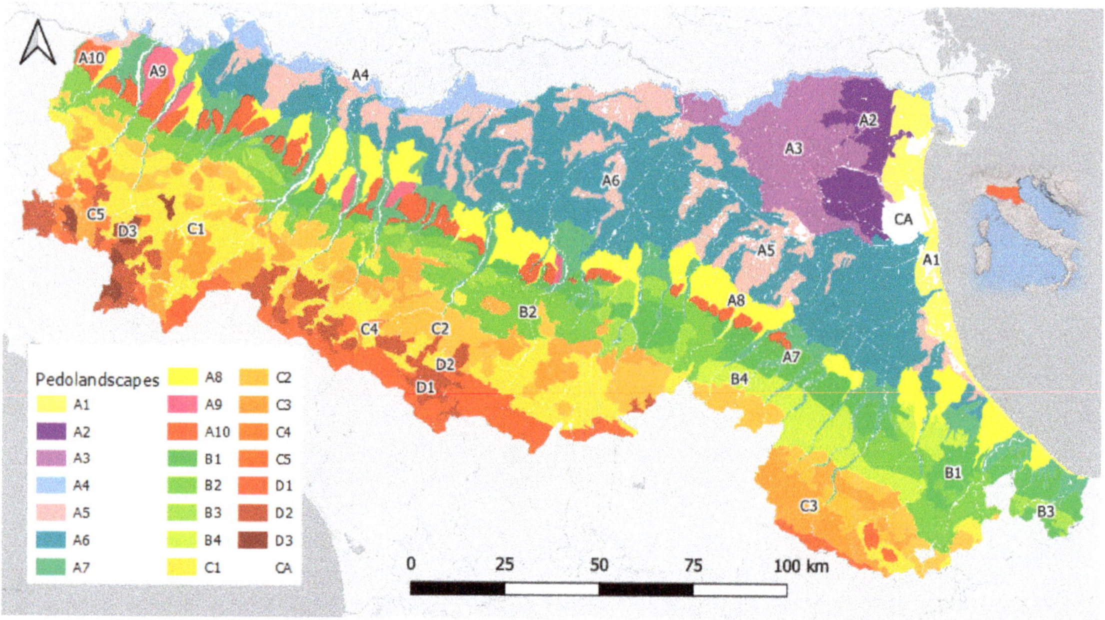

Figure 1. Study area: Pedolandscapes (soil provinces) of Emilia-Romagna. A1: Soils of the coastal plain and delta front; A2: Soils in the lower abandoned Po delta plain (Holocene); A3: Soils in the upper abandoned Po delta plain (Holocene); A4: Soils in the Po meander plain (Holocene); A5: Soils in morphologically depressed areas of the lower Apennine alluvial plain; A6: Soils of the levees and transition areas of the lower Apennine alluvial plain; A7: Soils in the fans and terraces of the upper Apennine alluvial plain (Holocene); A8: Soils in the fans and terraces of the upper Apennine alluvial plain (Holocene); A9: Soils in the terraced fans of the upper Apennine alluvial plain, located near the main river channels; A10: Soils in morphologically high areas of the ancient plain–Apennine fringe (Pleistocene); B1: Soils of the lower Apennines of Pliocene clays and sands; B2: Lower Apennines soils on unstable clays; B3: Lower Apennines soils on mudstones and sandstones; B4: Lower Apennines soils on the *Marnosa Arenacea Romagnola* (turbiditic marly sandstones); C1: Soils of the middle Apennines on unstable clays; C2: Soils of the middle Apennines on calcareous–marly flysch; C3: Soils of the middle Apennines on arenaceous–pelitic flysch; C4: Soils of the middle Apennines on gypsum and limestones; C5: Soils of the middle Apennines on ophiolitic rocks; D1: Soils of the upper Apennines on sandstones; D2: Soils of the upper Apennines on calcareous–marly flysch and mudstones: D3: Soils of the upper Apennines on ophiolites rocks; CA: water bodies.

The region's climate varies due to its diverse terrain and to the presence of the sea. The mountains are characterized by a temperate climate (Köppen-Geiger Cfb), with rainy summers followed by cold winters, while in the plain the climate is temperate subcontinental with hot summers (Köppen-Geiger Cfa). The average yearly cumulated rainfall for the reference period 1991–2015 is 927 mm, with a maximum of 1957 mm along the northwestern ridge of the Apennines and a minimum of 616 mm in the Po River delta. The average temperature for the same reference period is equal to 12.8 °C with a minimum in January (0.4 °C) and a maximum in July (27 °C) [24].

The regional soil map at the scale of 1:1,000,000 [25] identifies 22 pedolandscapes (or soil provinces): ten in the alluvial plain and four, five, and three, respectively, in the low (150–450 m a.s.l.), medium (450–900 m a.s.l.), and high (>900 m a.s.l.) Apennines (Figure 1). These pedolandscapes represent distinct geographical units where soil characteristics are relatively uniform, as they develop under a consistent combination of climate, vegetation, and parent material.

The Apennines (highest elevation in Emilia-Romagna 2165 m a.s.l.) are formed by sediments deposited in four different Meso-Cenozoic paleogeographic domains: (i) the Ligurian Domain, containing a tectonic mix and olistostromes with a high clay content; (ii) the Epi-Ligurian Domain, represented by thick series of calcareous or arenaceous turbidites, clay breccias and olistostromes; (iii) the Sub-Ligurian Domain, characterized by pelitic and evaporitic deposits (mostly gypsum), marine clays, and alternations of marine conglomerates and sands; and (iv) the Tuscan–Umbrian Domain, characterized by strongly cemented turbiditic marly sandstones. The variety of parent materials and terrain morphologies occurring in the Apennines results in twelve distinct pedolandscapes (Figure 1).

The plain area of the region (lowest elevation −8 m b.s.l.) is made up of Pleistocene–Holocene deposits from two different sources: the Po river–delta system, with a W-E orientation, and the Apennine river systems, with a dominant SW-NE orientation. The occurrence of two depositional systems resulted in an extremely complex geomorphological pattern of coastal, deltaic, fluvial, and terraced alluvial deposits, which resulted in ten different pedolandscapes (Figure 1). A description of the dominant soil types occurring in each pedolandscape unit, including their classification according to the USDA Soil Taxonomy (12th Ed., 2014), is provided in the Supplementary Materials (Table S1) along with the distribution of the FAO-WRB Major Soil Groups in the pedolandscapes of Emilia-Romagna (Figure S1).

Figure 2 illustrates the dominant land uses in Emilia-Romagna [26], highlighting the strong difference between the highly intensively cultivated plain and the extensively managed mountain rangelands and forests, which characterize 24 agricultural districts: nine in the alluvial plain, eight in the low Apennines (150–450 m a.s.l.), and eight in the medium (450–900 m a.s.l.) and high (>900 m a.s.l.) Apennines.

In the Apennines, oaks (*Quercus* L.), hornbeams (*Carpinus* L.), and chestnuts (*Castanea sativa* Mill.) dominate the broadleaf woodlands which represent more than 60% of the area. Croplands, grasslands, and permanent crops cover 22, 7, and 3% of the hilly and mountain areas, respectively.

In the plain, deep soils sustain intensive agricultural productions that, depending on local climatic conditions, range from typical permanent crops, cereals, and industrial crops in the east to more temperate climate productions such as pastures, cereals, and pig and dairy farming in the west. Orchards are prominent in agricultural districts 19, 16, and 25 located in the Romagna and Ferrara plains, while vineyards primarily grow along the Apennines foothills in the high plain and are also found in the central part of the Modena and Reggio Emilia plains (agricultural districts 10 and 7). The plain permanent meadows are typical of the Parma and Reggio Emilia districts (4 and 7 in Figure 2, respectively), being linked to the local traditional dairy production of the *Parmigiano Reggiano* cheese.

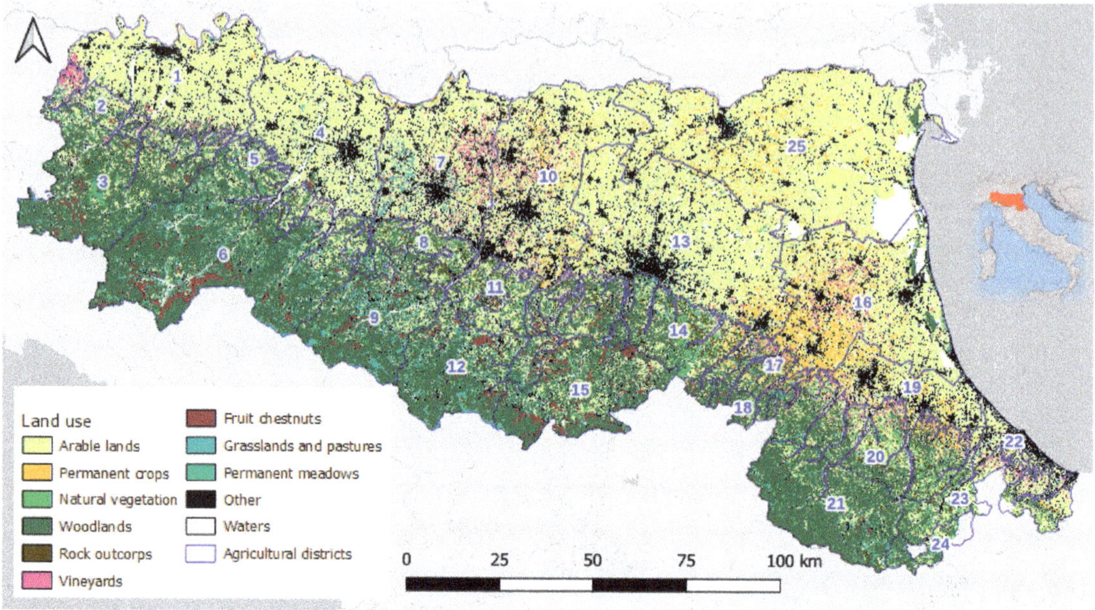

Figure 2. Dominant land use and agricultural districts (1–25) of Emilia-Romagna. Districts 1 to 14 are in Emilia; districts 16–24 are in Romagna. Plain districts: 1, 4, 7, 10, 13, 16, 19, 22, and 25; Hills districts: 2, 5, 8, 11, 14, 17, 20, and 23; and Mountain districts: 3, 6, 9, 12, 15, 18, 21, and 24. Administrative provinces: Piacenza (1, 2, and 3), Parma (4, 5, and 6), Reggio-Emilia (7, 8, and 9), Modena (10, 11, and 12), Bologna (13, 14, and 15), Ravenna (16, 17, and 18), Forlì-Cesena (19, 20, and 21), and Rimini (22, 23, and 24).

At the regional level, livestock farming and animal products represent a key economic and social value, accounting for approximately half of the agricultural gross marketable production, which is important for regional development and helps counteract the depopulation of mountain and hillside areas. Historically, Emilia-Romagna has been associated with high-quality products that carry origin designations, making it one of the regions with the highest livestock production in the country. As of June 2021, the region counted nearly 6000 cattle farms with over 570,000 heads (59% in the plain), ca. 2800 pig farms with nearly 1,100,000 heads (61% in the plain), and 421 poultry farms with over 19,377,000 heads (61% in the plain) [27]. In the agricultural districts of the plain, cow farms occur mainly in districts 4 and 7 (Parma and Reggio Emilia plains), pig farms in districts 10 and 13 (Modena and Bologna plains), and poultry farms in districts 16 and 19 (Ravenna and Forlì-Cesena plains).

2.2. Soil Macronutrient Data

The Geology, Soil, and Seismic Risk Service of Emilia-Romagna provided the soil macronutrient data that laid the foundation for the digital soil mapping (DSM). The overall dataset consists of 36,054 sites with analytical data for the 0–30 cm depth interval, sampled over a time span between 1974 and 2023. The analytical data in the soil database comes from three different sources: 1. Soil observations collected by the Geology, Soils, and Seismic Area (n = 4376, 12.2% of observations); 2. Samples taken as part of technical assistance activities for agriculture, owned by the Planning, Land Development, and Production Sustainability Sector of the Agriculture Directorate (n = 31,191, 86.5% of observations); and 3. Monitoring data from various sources, including the LUCAS dataset [17] (n = 487, 1.3% of observations). In this case, only the most recent analytical data were used for DSM.

Macronutrient concentrations were assessed following the Italian official analytical methods [28], which are coherent with ISO standards. In the case of nitrogen, the data refer to total content expressed in g kg^{-1}, measured with the Kjeldahl method (ISO 1871:2009) [29] or with the combustion method. The Olsen method (ISO 11263:1994) [30] was used to determine plant available phosphorus, which is expressed as mg kg^{-1} P_2O_5. Exchangeable K_2O content in soil was determined using the ammonium acetate method (ISO 22171:2023) [31] and is expressed in milligrams per kilogram (mg kg^{-1}) of soil.

In addition to the diversity of data sources, there were also differences in sampling depth which required a preliminary data harmonization over the 0–30 cm reference depth: in the case of multiple values within the 0–30 cm layer, the data were then interpolated using cubic splines [32,33] in the R environment [34] using the R package (v4.4.2) splines2 [35].

Successively, the statistical distributions of each nutrient were analyzed to check for possible outliers using Rosner's outlier test [36]. Once the critical value was identified, individual cases were evaluated to determine whether it was appropriate to eliminate the "anomalous" data from the dataset. In some cases, for example, it was decided to remove them because they were particularly anomalous and likely associated with sampling taken close to soil fertilization periods (in these cases, two or three elements often result in outliers). Only the N data referring to organic soils, which have particularly high values, were retained for DSM even though the procedure had identified them as outliers.

Table 1 summarizes the descriptive statistics of the three soil macronutrients. The overall sampling density (observation per km^2) is 1.6, which would be coherent with a 1:50,000 scale, but as the data points in the plain area of the region represent 76% of the total, and considering the differences in land use intensity and the role played by environmental drivers in the two contexts, this led to the implementation of two separate DSM procedures for the plain and for the Apennines. Figure 3 shows the classed post-plots of the three macronutrients' concentrations for the 0–30 cm reference layer.

Table 1. Descriptive statistics of N (g kg^{-1}), P (P_2O_5, mg kg^{-1}), and K (K_2O, mg kg^{-1}) concentrations for the plain and the mountainous areas of Emilia-Romagna and for the whole regional dataset. Std. Dev.: standard deviation.

Area	Variable	Num	Mean	Std. Dev.	Min.	P10	Q25	Median	Q75	P90	Max
Plain	N tot	27,512	1.46	0.58	0.05	0.98	1.10	1.40	1.70	2.07	14.60
	P_2O_5	26,680	44.11	30.30	0.50	14.00	23.00	37.00	57.00	83.00	185.00
	K_2O	26,711	292.19	151.47	3.63	127.00	184.00	264.60	368.00	485.00	1010.00
Hills and mountains	N tot	8626	1.42	0.66	0.10	0.79	1.00	1.30	1.70	2.20	13.00
	P_2O_5	8241	25.14	24.02	0.10	5.00	9.00	18.00	32.50	53.90	183.00
	K_2O	8097	264.08	152.71	3.63	106.00	155.90	230.00	337.00	465.00	1100.00
Whole region	N tot	36,138	1.45	0.60	0.05	0.90	1.10	1.36	1.70	2.10	14.60
	P_2O_5	34,921	39.63	30.04	0.10	10.00	18.00	32.00	53.00	78.00	185.00
	K_2O	34,808	285.65	152.22	1.00	122.00	177.15	257.00	361.00	480.00	1100.00

As the data were collected over a few decades, the existence of possible trends over time of the three macroelements was analyzed by dividing the data sets into 5 groups: data collected in the 1980s or earlier, those collected in the 1990s, data collected up to 2009, data collected between 2010 and 2016, and finally the most recent data from 2017 onwards. For all macronutrients, concentrations in the plain remained stable over time, except for P which decreased in the last period considered. The concentrations of N and K in the Apennine soils, along with their observed variability over time, are linked to the low sample size and to the specific areas where sampling campaigns were conducted. In the case of P, topsoil concentration in the Apennine remained substantially stable over time.

Summary results are presented in the Supplementary Table S2, along with localization of sampling points in the five different decades (Figure S2).

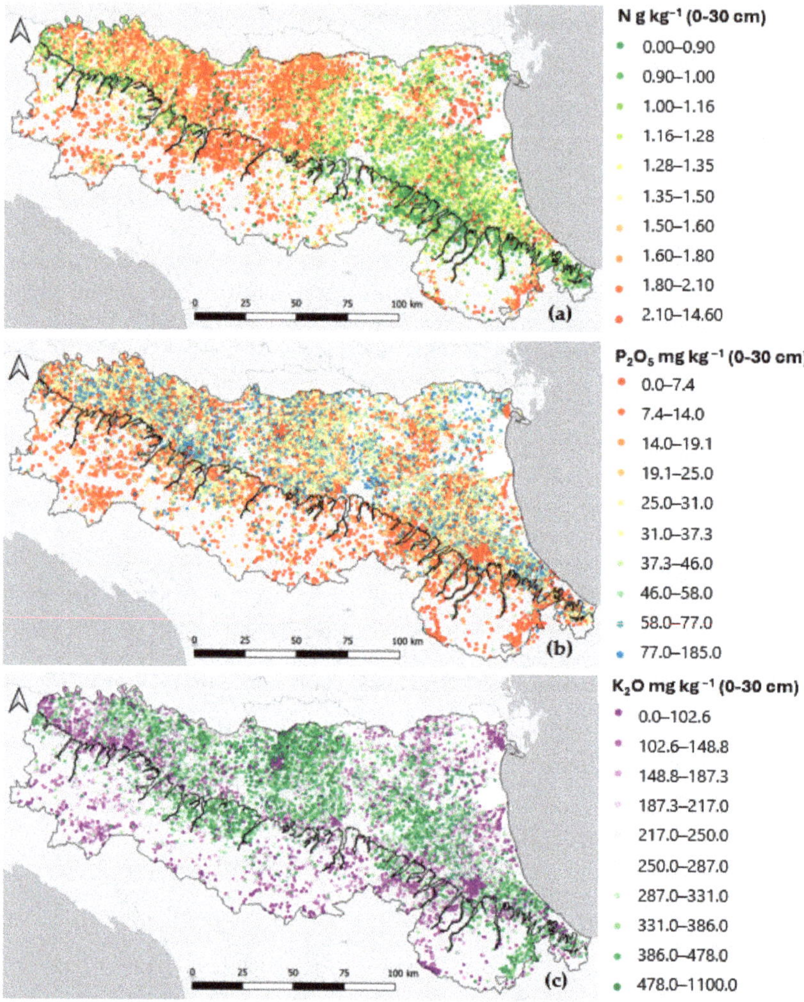

Figure 3. Classed post-plots of topsoil (0–30 cm) macronutrients' concentrations: (**a**) N (g kg^{-1}); (**b**) P_2O_5 (mg kg^{-1}); (**c**) K_2O (mg kg^{-1}). Class intervals are defined based on the deciles of the observed distributions.

2.3. Digital Soil Mapping

A digital soil mapping (DSM) approach was used to assess and map the three soil macronutrients at 100 m resolution in the two areas of the region, i.e., the plain and the Apennines. This approach relied on machine learning (ML) calibrated regression algorithms to estimate the spatial distribution of soil macronutrients using a variable number of covariates as predictors. These are typically continuous variables, such as elevation and other parameters derived from the digital elevation model (DEM), meteorological and climatic variables, and spectral and vegetation indices from remote sensing. Continuous variables are often complemented by categorical variables, such as land use, and soil map units at different scales. In both cases, independently from the original resolution, all covariates covering the entire regional territory were harmonized at 100 m resolution after being reprojected into the same reference system (EPSG:7791) [37].

The DSM models were calibrated in the R environment [34], using a DSM Workflow developed by ISRIC [38]. The workflow modelling approach is based on Quantile Random Forest (QRF) [39,40] and generates soil nutrient maps (median values) with quantified uncertainty as outputs. The workflow implements the ranger package [41], with the option quantreg to build QRF [40], estimating the cumulative probability distribution of the soil macronutrients at each location from an ensemble of 500 decision trees in the RF. Uncertainty of estimates was assessed by calculating the interquartile ranges of the values calculated for each pixel of the estimation grid. Using 75% of the available data randomly selected for calibration, model performance was evaluated with a 10-fold cross-validation, splitting the datasets into a training and a validation subset. The predictive performance of the models was eventually assessed for the calibration and validation data sets resorting to the following error metrics: mean error (ME), absolute error (AE), root mean squared error (RMSE), coefficient of determination (R^2), and index of agreement (IoA). IoA is a dimensionless index with values ranging between 0 and 1, with 0 indicating no agreement at all and 1 indicating a perfect match [42].

For the spatial prediction of topsoil macronutrients, a set of 35 covariates was used; these are listed in Table 2 along with the SCORPAN factors [14] they refer to.

Table 2. List of the covariates used in the DSM of the Emilia-Romagna plain and Apennines to estimate soil macronutrients. SCORPAN factor: C, climate; O, organisms; P, parent material; R, relief; S, soil (measured properties of the soil at a point).

Covariates	Description	SCORPAN Factor	Ref. Year	Layer Type	Variable Type	Spatial Resolution	Units
aspect	Aspect from DEM	R	2016	Raster	Num.	10 m	°
dem	Elevation	R	2016	Raster	Num.	10 m	m
geomorfo	geomorphological forms	R + P	2016	Raster	Cat.	25 m	class
mrivbf	Multi Resolution Index of Valley Bottom Flatness	R	2016	Raster	Num.	10 m	-
nort	Northness (orientation and slope)	R	2016	Raster	Num.	10 m	index
slope	Slope, from DEM	R	2016	Raster	Num.	10 m	%
twi	Topographic Wetness Index from, DEM	R	2016	Raster	Num.	10 m	$m^2\ rad^{-1}$
vdnc	Vertical dist. channel network, from DEM	R	2016	Raster	Num.	10 m	m
vdepth	Valley depth, from DEM	R	2016	Raster	Num.	10 m	m
mwmtemp	July mean temperature	C	1970–2000	raster	Num.	100 m	°C
landuse	Land use map	O	2020	Vector	Cat.	10 k	class
evi	Enhanced Vegetation Index (Modis)	O	2015–2023	Raster	Num.	250 m	-
gfc_tcov	Global forest tree canopy cover [36]	O	2019	Raster	Num.	30 m	%
ndvis5	NDVI, mean of medians sum June-September	S + O	2015–2023	Raster	Num.	30 m	-
ndvi	NDVI, mean of annual median values	S + O	2015–2023	Raster	Num.	30 m	-
ndsi	NDSI, mean of annual median values	S + O	2015–2023	Raster	Num.	30 m	-
ndwi	NDWI, mean of annual median values	S + O	2015–2023	Raster	Num.	30 m	-
nir	Sentinel2 Band 8 (Near Infrared)	S + O	2015–2023	Raster	Num.	30 m	DN 8 bit
red	Sentinel2 Band 4 (Red)	S + O	2015–2023	Raster	Num.	30 m	DN 8 bit
sosi2	SOSI, mean of annual median values	S + O	2015–2023	Raster	Num.	30 m	-

Table 2. *Cont.*

Covariates	Description	SCORPAN Factor	Ref. Year	Layer Type	Variable Type	Spatial Resolution	Units
swir	Sentinel2 Band 11 (Short wave infrared)	S + O	2015–2023	Raster	Num.	30 m	DN 8 bit
erosion	RUSLE map (actual soil erosion)	S + R + C + O	2019	Raster	Num.	20 m	Mg/ha/yr
soilrer_L2	Pedolandscapes (L2, Soil provinces)	C + O + R + P	2021	Vector	Cat.	1 M	class
soilrer_250k	Soil subsystems (L4)	S	2021	Vector	Cat.	250 k	class
soilrer_50k	Soil units (L6, only for the plain)	S	2021	Vector	Cat.	50 k	class
clay	Clay content (0–30 cm)	S	2023	Raster	Num.	100 m	%
sand	Sand content (0–30 cm)	S	2023	Raster	Num.	100 m	%
C org.	Soil organic C content (0–30 cm)	S	2023	Raster	Num.	100 m	%
pH	Soil pH (0–30 cm)	S	2023	Raster	Num.	100 m	-
avg_lulcprov	NPK mean contents per LULC class per district	S	2022	Vector	Num.	50 k–250 k	g kg^{-1}, mg kg^{-1}
avg_L2prov	NPK mean contents per pedolandscape per district	S	2022	Vector	Num.	50 k–250 k	g kg^{-1}, mg kg^{-1}

Eight covariates were derived from the 10 m resolution DEM and nine were remote sensing indices and spectral bands reflectance derived from Sentinel-2 images retrieved via Google Earth Engine [43], taking the mean value of the yearly medians for the reference time interval 2015–2023. Among the numerical soil-based covariates, four covariates were derived via DSM of basic soil properties [44], namely topsoil clay and sand contents, soil organic carbon, and pH; two provided the NPK mean concentrations for the LULC classes and for the pedolandscapes within each agricultural district; and one the RUSLE-based soil erosion loss [45]. Additionally, three categorical covariates describing the soil geography at different hierarchically linked spatial scales, namely 1:1,000,000, 1:250,000, and 1:50,000, were used [25]. Further categorical covariates used in the DSM of soil macronutrients were the land use map and the map of the geomorphological forms.

To assess the relevance of each covariate in predicting macronutrient concentrations, the workflow performed a Recursive Feature Elimination (RFE) using the caret package [46]. RFE is a feature selection technique that identifies the most relevant predictors when building a predictive model. The predictive power of each covariate was defined in terms of "node purity," which describes the homogeneity of the data within each node deriving from the partition of the data based on the values of any given covariate. The node purity is calculated as the difference in terms of the rooted mean squared error (RMSE) before and after the division performed on that specific covariate.

2.4. Postprocessing of Results and Comparison of DSM Outputs at the Regional and the EU Scale

The resulting three macronutrients' maps were postprocessed considering the 22 functionally distinct pedolandscapes based on the Emilia Romagna Soil Map at scale 1:1,000,000 [25] and the 25 agricultural districts. In addition, the regional maps were compared with the topsoil nutrient status maps for the same area based on the 2009–2012 LUCAS dataset, which in Emilia-Romagna include 117 sampling points (density ~1.0 sample per 200 km^2) where macronutrients' content was analyzed following the ISO standards [17]. To this goal the maps were resampled and reprojected to the same resolution and reference system as the LUCAS maps, i.e., 250 m and EPSG:3035, using a bilinear interpolation algorithm in R. As in LUCAS maps K and P concentrations are given as mg kg^{-1}, the regional estimates for exchangeable K and available P were multiplied by 0.830 and 0.436, respectively. Raster statistics were then computed, and macronutrient concentrations were compared in terms of agricultural districts and pedolandscapes. Figure S3 in the

Supplementary Materials shows the location of the LUCAS points in the pedolandscapes of Emilia-Romagna.

3. Results

The DSM procedure allowed the identification of the covariates' relevance in calibrating the QRF predictive models. Figure 4 shows the covariates' importance in terms of node purity; the values shown in the figure were normalized to the same 0–1 range to plot together the three macronutrients considered in the two major landforms of RER.

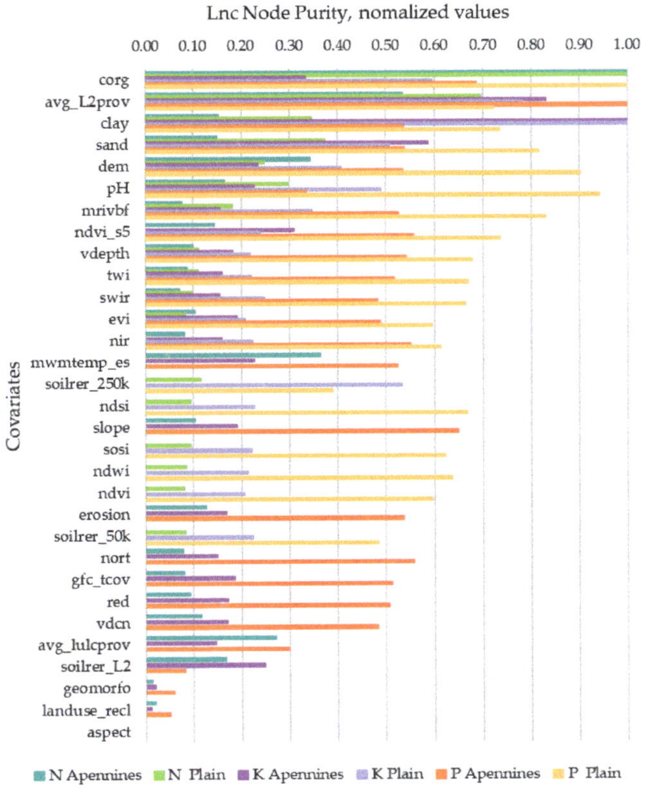

Figure 4. Importance of the covariates for predicting the levels of soil macronutrients using DSM with QRF in Emilia-Romagna.

The predictors on the Y axis of the bar plot are in decreasing order of importance based on their average ranks and can be separated into five categories: (i) climate covariates (n = 1), (ii) topography covariates (n = 9), (iii) land use land cover covariates (n = 3), (iv) soil covariates (n =10), and (v) surface reflectance covariates (n = 9).

In the case of N concentrations, soil organic matter content ranked first in terms of predictive power in both the plain and the Apennines. The average N concentration per pedolandscape, at the level of agricultural districts, ranked second, showing the same relevance in both the plain and the Apennines. Elevation ranked fourth in the Apennines but only sixth in the plain. In the plain, the normalized node purity value of soil textural fractions (sand and clay) and pH was double that observed in the Apennines; however, these factors were still among the top ten covariates by relevance in that region. In the Apennines, the average temperature of the warmest month and the N concentrations per land cover type at the district level ranked third and fourth, respectively, while showing no predictive power for estimating N concentration in soils of the plain. Among the surface reflectance covariates, only NDVI (calculated as the sum from June to September) ranked among the top ten in the Apennines, whereas three additional topography covariates—mrivbf, vdepth, and twi—demonstrated good predictive power in the plain. Still among the first ten ranking covariates for N concentration prediction were two categorical predictors related to soil geography, i.e., the pedolandscapes in the Apennines (12 classes) and the 1:250,000 soil map units in the plain (59 classes).

As for K topsoil concentration in both the plain and the Apennines, the most relevant predictor was clay content, followed, as in the case of N, by the average concentration per pedolandscape at the level of agricultural districts, again with same relevance in both the plain and the Apennines. Differences between the two major landforms occurred in the relevance of sand and C org contents, ranking third and fourth, respectively, in the Apennines and fifth and third in the plain. Elevation ranked seventh in both landforms, while the categorical covariates describing soil geography at different levels of detail ranked fourth in the plain (1:250,000 soil map units) and sixth in the Apennines (pedolandscapes). Among the ten more relevant covariates, pH ranked sixth in the plain and eighth in the Apennines, and NDVI (sum June–September) ranked fifth and tenth, respectively, in the Apennines and in the plain. As in the case of N concentration, the average temperature of the warmest month proved to be a relevant predictor only in the Apennines, where it ranked ninth, while in the plain the same rank was gained by SWIR. Two additional DEM-derived covariates were among the first ten, namely the slope in the Apennines and the mrivbf in the plain, ranking tenth and eighth, respectively.

The relevance of covariates in predicting P topsoil concentrations highlighted more differences between the two major landforms compared to what was observed in the case of N and K concentrations. Organic C content ranked first in the plain and second in the Apennines, while the average P concentration per pedolandscape at the level of agricultural districts ranked first in the Apennines but only eighth in the plain. Soil pH ranked second in the plain but only 20th in the Apennines, while among the first ten covariates sand and clay contents ranked eighth and tenth in the Apennines and fifth and sixth in the plain, respectively. Elevation ranked third in the plain but only 11th in the Apennines, where the most important DEM-derived predictors were slope (rank 3), nort (rank 5), and vdepth (rank 7). In the plain, though, three additional DEM-derived predictors were among the first ten: mrivbf ranked fourth, vdepth ranked ninth, and twi ranked tenth. Soil erosion ranked ninth in the Apennines, and P is the only macronutrient for which erosion ranks among the top ten covariates: in the case of nitrogen, it ranked 11th, and 16th in the case of potassium. Among the surface reflectance covariates, NDVI (sum June-September) ranked seventh in the plain and fourth in the Apennines where a second predictor from remote sensing, the nir reflectance (Sentinel-2 band 8), ranked sixth. It is interesting to note that in the plain, although not in the group of first ten predictors in terms of importance, a moderate predictive power was observed for nearly all the other remote sensing-based predictors (ndsi, swir, ndwi, sosi, nir, evi, and ndvi), which ranked from 11th to 17th.

Table 3 eventually reports the error metrics for the calibration and the validation datasets of the three soil macronutrients. Overall, the error metrics for the validation data sets highlighted a slightly higher precision in the prediction of soil macronutrients for the plain than for the Apennines, which was expected considering the size of the data sets. However, the DSM performance metrics for the validation data sets in both the plain and the Apennines are in most cases very good, with R^2 values between 0.87 and 0.93, IoA values $\geq$ 0.88, and mean absolute errors ranging from 0.02 to 0.21 g kg^{-1} for N, from 9.9 to 11.9 mg kg^{-1} for K, and from 1.9 to 2.5 mg kg^{-1} for P concentrations.

Table 3. Error indices for calibration (Train) and validation (Test) datasets.

Variable	Data Set	Num. Obs.	ME	AE	RMSE	IoA	R2
N g kg^{-1} Plain	Train	20,536	−0.004	0.206	0.317	0.882	0.910
	Test	6846	0.016	0.032	0.560	0.948	0.898
N g kg^{-1} Apennines	Train	6406	0.003	0.022	0.130	0.965	0.928
	Test	2136	0.033	0.067	0.241	0.945	0.897
K_2O mg kg^{-1} Plain	Train	19,941	6.455	10.331	41.840	0.959	0.908
	Test	6648	6.719	10.451	42.574	0.945	0.893
K_2O mg kg^{-1} Apennines	Train	6015	5.810	9.910	38.930	0.961	0.923
	Test	2005	7.900	11.890	48.580	0.936	0.880
P_2O_5 mg kg^{-1} Plain	Train	19,917	1.583	2.525	9.155	0.955	0.902
	Test	6639	1.439	2.395	8.504	0.943	0.900
P_2O_5 mg kg^{-1} Apennines	Train	6124	1.790	2.240	9.090	0.935	0.874
	Test	2042	1.410	1.920	8.170	0.925	0.869

3.1. Nitrogen Content

N concentrations, as discussed above, are strongly correlated with organic carbon content and, consequently, the distribution of the element across the region closely mirrors that of organic carbon content (Figure 5).

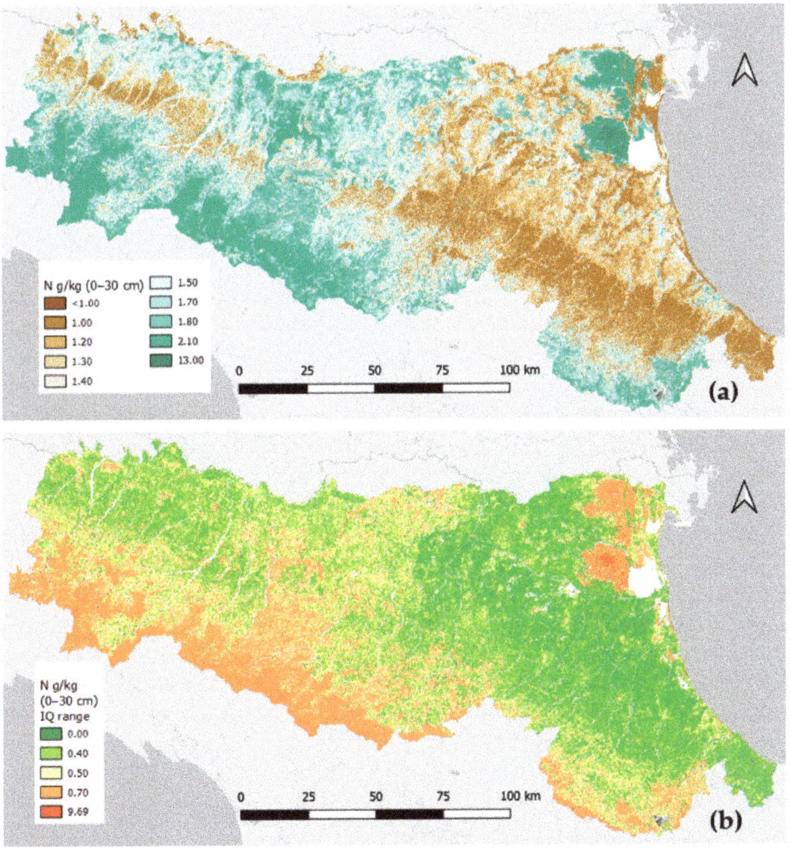

Figure 5. Median N topsoil (0–30 cm) concentrations (g kg^{-1}) estimated with QRF (**a**), and its spatial uncertainty (**b**).

At the regional level, N concentrations fall most frequently in the range 1–2 g kg^{-1}. Considering the reference soil depth 0–30 cm the Emilia-Romagna plain is characterized by a mean total N content of ca. 1.57 g kg^{-1} ± 0.83 (standard deviation). In the Apennines, again for the same depth interval, the mean total N content is ca. 1.63 ± 0.49 g kg^{-1}. Considering the entire region, a mean content of ca. 1.60 ± 0.66 g kg^{-1} is observed. Figure 5 shows the map of N concentrations (g kg^{-1}) and its spatial uncertainty summarized in terms of interquartile range.

In the plains, the areas with the highest quantities of N are in the lower delta plain (pedolandscape A2) of the Ferrara agricultural district due to the presence of soils developed on peaty deposits of formerly marshy and now reclaimed areas (average 4.3 ± 2.15 g kg^{-1}). In contrast, in the Reggio Emilia district, high N content is attributed to the presence of forage crops, including both rotational and permanent pastures, associated with livestock and dairy production (Parmigiano-Reggiano cheese district).

Pedolandscapes A9, A7, A8, and A5 are particularly rich in N (average values between 2.52 and 1.87 g kg^{-1}). The clayey soils of the floodplains (pedolandscape A5) on the right-hand side of the Taro River in the Parma district are also supplied with N above the regional average (1.86 ± 0.27 g kg^{-1}). In the rest of Emilia (central-western part of the region), moderate levels are found, especially in the floodplains' pedolandscape unit A5 (1.8 g kg^{-1} in Modena and Piacenza districts, 1.66 g kg^{-1} in Ferrara) and on terraces and alluvial fans (units A8, from 1.7 g kg^{-1} in Parma to 1.65 g kg^{-1} in Modena).

The large Emilian river natural levees (Taro, Crostolo, Secchia, Panaro) of the lower alluvial plain (pedolandscape A6) show average N contents between 1.75 g kg^{-1} and 1.61 g kg^{-1}; the situation is different in Romagna (eastern part of the region), where fruit orchards are widespread and N values vary between 1.25 g kg^{-1} and 1.19 g kg^{-1}, respectively in the Bologna and Forlì-Cesena districts. These low N levels are a consequence of changes in land use and management since the 1950s, with a sharp decline in forage crops and organic fertilization from livestock manure. In recent years, the widespread practice of grassing vineyards and orchards, as well as the reduction in tillage intensity, could help slow the decline in total N stocks. Intermediate values are found on gravelly alluvial fans (pedolandscape A9: average value 1.73 ± 1.6 g kg^{-1}), ranging from 2.53 g kg^{-1} in the Reggio Emilia to 1.51 g kg^{-1} in the Modena districts. The lowest N values are found where sandy soils prevail, namely in the coastal plain (pedolandscape A1, 1.16 ± 1.1 g kg^{-1}), and in this area the Forlì-Cesena agricultural district has the lowest value (0.83 g kg^{-1}). Low values are also found in the desaturated soils of the Apennine margin (pedolandscape A10, average 1.24 ± 0.26 g kg^{-1}), particularly in the agricultural districts of Ravenna (0.87 ± 0.13 g kg^{-1}) and Bologna (0.94 ± 0.16 g kg^{-1}).

In the lower Apennines (150–450 m a.s.l.), soils on Pliocene sands and clays (pedolandscape B1) have the lowest average concentrations, equal to ca. 1.08 ± 0.19 g kg^{-1}, with a negative trend from north-west to south-east, characterized by average values between 1.44 and 1.18 g kg^{-1} in the lower Apennines districts of Reggio-Emilia, Parma, Piacenza and Modena; with values of approximately 1.06 ± 0.18 g kg^{-1} in the lower Apennines district of Bologna; and between 1.04 and 0.96 g kg^{-1} in the lower Apennines of Romagna (districts 17, 20 and 23 Figure 2). The B2 pedolandscape on unstable clays has an average value of 1.44 g kg^{-1} ± 0.3 (the highest on the hills), with a range of average values between 1.82% ± 0.26 in the Reggio Emilia district and 1.01 ± 0.17 g kg^{-1} in the Forlì district. The mudstones and sandstones of the lower Apennines (pedolandscape B3) have an average content of 1.26 g kg^{-1} ± 0.22, with higher average values in the Reggio-Emilia district (1.46 g kg^{-1} ± 0.22) and lower in Rimini (1.09 g kg^{-1} ± 0.26). Finally, the marly-limestone formation of the Romagna lower Apennines (pedolandscape B4) is characterized by average values of ca. 1.06 g kg^{-1} ± 0.19.

The mean value of N concentration in the pedolandscapes of the middle Apennines (450–900 m a.s.l.) ranges from 1.98 ± 0.3 g kg^{-1} of the soils on ophiolitic rocks (pedolandscape C5) to 1.59 ± 0.28 g kg^{-1} of the soils on calcareous–marly flysch (C2); additionally, the soils on unstable clays (C1) and the soils on arenaceous–pelitic flysch (C3) have estimated concentration equal to ca. 1.82 ± 0.31 g kg^{-1} and 1.65 ± 0.30 g kg^{-1}, respectively, while higher average contents of 1.83 ± 0.22 g kg^{-1} are found in soils on gypsum and cavernous limestone (C4). In the C1, C2, and C3 pedolandscapes of the middle Apennines, a generally decreasing trend is observed between the Piacenza and the Bologna districts, a trend that then reverses to rise again in the Romagna provinces with a maximum in the Rimini district.

In the upper Apennines (>900 m a.s.l.), forests and meadows prevail, with higher organic matter contents than arable land, and consequently N contents also follow the same trend. Higher average values are observed in ophiolitic soils, with an average value of 2.42 ± 0.47 g kg^{-1} (pedolandscape D3), while lower average values (2.16 ± 0.58 g kg^{-1}) characterize soils derived from sandstones (D1). Intermediate values, equal to approximately 2.35 ± 0.33 g kg^{-1}, are observed in soils on calcareous–marly flysch and mudstone (D2). Here too, the Emilian districts (where, moreover, these units are more widespread) show the highest values, Parma and Piacenza in particular.

In terms of estimation uncertainties, the mean IQ range for the whole region is equal to 0.51 ± 0.34 g kg^{-1}; the estimated N concentrations in the plain and in the lower Apennine are characterized by lower mean IQ values, equal respectively to 0.41 ± 0.18 g kg^{-1} and 0.40 ± 0.09 g kg^{-1}. In the plain, the districts of Emilia have higher mean IQ range values than those of Romagna, with maximum values in Ferrara and Reggio-Emilia, respectively, at 0.57 ± 0.45 g kg^{-1} and 0.51 ± 0.12 g kg^{-1}, and minimum values in the districts of Bologna and Ravenna, at 0.29 ± 0.03 and 0.31 ± 0.07 g kg^{-1}, respectively. A very similar trend in spatial uncertainty of N concentration estimates is observed in the agricultural districts of the lower Apennine, with maximum mean IQ ranges found in Reggio-Emilia (0.54 ± 0.07 g kg^{-1}) and Modena (0.41 ± 0.05 g kg^{-1}) in Emilia, and minimum in the districts of Ravenna (0.29 ± 0.02 g kg^{-1}) and Forlì (0.32 ± 0.03 g kg^{-1}) in Romagna. In the agricultural districts of the mid- and high Apennines the mean values of the IQ range are always above the regional mean value, ranging between 0.55 ± 0.10 g kg^{-1} (Forlì district) and 0.72 ± 0.07 g kg^{-1} (Reggio-Emilia district) in the mid-Apennines and between 1.02 ± 0.18 g kg^{-1} (Modena district) and 1.69 ± 0.33 g kg^{-1} (Rimini district) in the high Apennines.

To enhance communication about uncertainty to potential stakeholders, the IQ ranges of estimated N concentration were aggregated at the municipality level and categorized as very low to very high based on the ventiles of the resulting IQ range distribution; the resulting map is presented in Figure S4 in the Supplementary Materials.

3.2. Potassium Content

Figure 6 shows the estimated K concentrations, expressed as exchangeable potassium (K_2O, mg kg^{-1}), and its spatial uncertainty in terms of interquartile range.

The clay content in this case strongly influences the resulting patterns, which closely reflect the textural characteristics of the soil parent material. For the 0–30 cm reference depth the estimated mean exchangeable K content (mg kg^{-1} K_2O) is 275.9 ± 92.6 mg kg^{-1} in the alluvial plain and of 210.2 ± 86.3 mg kg^{-1} in the Apennines. The regional estimated mean concentration is equal to 244.8 ± 95.6 mg kg^{-1}.

The mean K value in the Apennines is lower than in the plains, but in both cases, there is strong variability, evidenced by the high standard deviation values. As highlighted in Figure 6, the exchangeable K supply is high across much of the region, particularly in the plains and mid-hill areas. Lower values are found in the upper Apennines, in the foothill alluvial fans belt, and in the coarser-textured areas of the outer Po delta and of the coast.

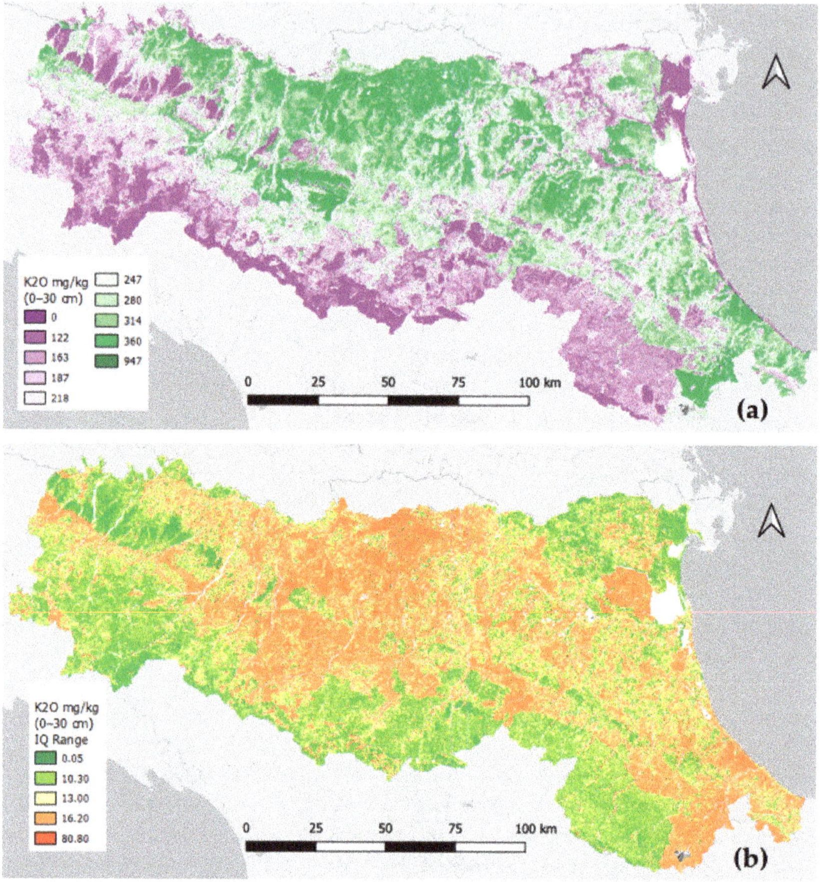

Figure 6. Median exchangeable K topsoil (0–30 cm) concentrations (K_2O mg kg^{-1}) estimated with QRF (**a**), and its spatial uncertainty (**b**).

In the plain, the areas with the highest K concentrations are on the clayey soils of the valleys of pedolandscape A5 (389.5 ± 65.0 mg kg^{-1} K_2O), with the highest values in the districts of Modena and Reggio Emilia, followed by soils developed on peaty deposits of formerly marshy areas and now reclaimed in the lower delta plain (A2) in the Ferrara district (311.4 ± 65.5 g kg^{-1} K_2O). Soils of pedolandscapes A7, A6, and A8 have concentrations above 200 mg kg^{-1} (between 262.0 and 294.0 g kg^{-1} K_2O), and these soils are predominantly medium and medium-fine textured. The soils of the remaining pedolandscapes are characterized by values below 200 mg kg^{-1}: the lowest concentrations are found on the sandy soils of the coast (A1, 142.1 ± 65.1 mg kg^{-1} K_2O) and the desaturated soils of the Apennine margin (A10, 142.5 ± 47.0 mg kg^{-1} K_2O), with the lowest values in the agricultural district of Piacenza.

As regards the lower Apennines, the soils of pedolandscape B4 (Romagnola marly–sandstone formation) have the lowest mean values (163.2 ± 30.4 mg kg^{-1} K_2O), because soils in this unit very rarely have high clay content. This unit is followed by soils on the mudstones and sandstones of the lower Apennines (B3), which have an average content of 216.7 ± 70.6 mg kg^{-1} K_2O with a high variability, and then those on Pliocene sands and clays (B1, 247.5 ± 72.9 mg kg^{-1}), although locally the soils on the sandstones (Bologna and Piacenza districts) are those with the lowest absolute values (<100 mg kg^{-1} K_2O). Unit B2 on unstable clays has a mean concentration of 329.0 ± 98.8 mg kg^{-1} K_2O, which is the highest in the hills and the second in the whole region after the soils of unit A5 on the plain, proving the high correlation between high clay values and exchangeable K contents.

The mean concentration of exchangeable K in the mid-Apennine pedolandscapes varies from 149.2 ± 29.2 mg kg^{-1} in soils on ophiolitic rocks (C5) to 200.1 ± 55.1 mg kg^{-1} in soils on calcareous–marly flysch (C2); the soils of pedolandscape C1 (196.0 ± 57.3 mg kg^{-1}) have values comparable to those of C2, but with notable differences between districts: the mean value ranges from 178.8 ± 40.5 mg kg^{-1} in Bologna to 251 ± 56.9 mg kg^{-1} in Reggio Emilia. In the case of unit C2 in the Ravenna district, topsoils have an average exchangeable K concentration of 161.9 ± 23.4 mg kg^{-1}, compared to 252.1 ± 54.6 mg kg^{-1} in Reggio Emilia.

In the case of the pedolandscapes of the upper Apennines, exchangeable K content in soils has low average values (from 111 to 168 mg kg^{-1}). The lowest values are found in soils of ophiolitic origin in the D3 unit (111.3 ± 18.1 mg kg^{-1}) and are comparable to soils derived from sandstones in the D1 unit (117.2 ± 26.7 mg kg^{-1}), while the highest values are found in soils derived from calcareous–marly flysch and mudstones (D2), which generally have medium-textured soils (168.4 ± 34.3 mg kg^{-1}). Even in this latter case, there is some variability among districts, ranging from an average value of 118.4 ± 20.8 mg kg^{-1} in Bologna to that of 190.9 ± 31.3 mg kg^{-1} in Piacenza.

The overall spatial uncertainty of the two QRF models' predictions, expressed as mean IQ range, was equal to 13.5 ± 46 mg kg^{-1} K_2O. Notwithstanding the difference in available data in the two major landforms, the predictions for the Apennines were characterized by a spatial uncertainty slightly lower than that observed for the exchangeable K predictions in the plain. The former has a mean IQ range equal to 12.8 ± 4.3 mg kg^{-1}, while for the latter it is equal to 14.2 ± 4.9 mg kg^{-1}. In the plain, the eastern districts of Emilia have higher mean IQ range values than those of Romagna, except Rimini (17.3 ± 4.6 mg kg^{-1}), with maximum values in Modena (17.8 ± 5.62 mg kg^{-1}), and minimum values in the district of Piacenza (10.3 ± 4.0 mg kg^{-1}). In the agricultural districts of the lower Apennine, the maximum mean IQ ranges are in the districts of Reggio-Emilia (17.8 ± 3.7 mg kg^{-1}) and Parma (14.4 ± 4.8 mg kg^{-1}) in Emilia, and minimum in the districts of Forlì (13.5 ± 3.8 mg kg^{-1}) and Ravenna (12.3 ± 3.3 mg kg^{-1}) in Romagna. Similar local maximum and minimum mean IQ range values are observed for the mountain districts: in Reggio-Emilia and Rimini mean estimation uncertainties were 14.7 ± 4.2 mg kg^{-1} and 16.9 ± 2.4 mg kg^{-1}, respectively, while minimum IQ ranges were observed in the districts of Ravenna (10.2 ± 1.4 mg kg^{-1}) and Forlì (10.2 ± 2.1 mg kg^{-1}) in Romagna.

As for N concentration, a classed uncertainty map at the municipality level for the estimated K concentration is given in Figure S4 in the Supplementary Materials.

3.3. Phosphorus Content

Figure 7 shows the estimated P concentrations, expressed as plant-available phosphorus (P_2O_5, mg kg^{-1}), along with its spatial uncertainty in terms of interquartile range.

The mean concentration of available phosphorus for the entire region is to 28.2 ± 15.5 mg kg^{-1} P_2O_5. However, this regional average is not representative, as values differ significantly between the plains (40.4 ± 11.0 mg kg^{-1}) and the Apennines (15.2 ± 6.1 mg kg^{-1}). Plant-available phosphorus is governed by soil properties that control its sorption and desorption, including clay mineralogy, organic matter content, soil pH, and the concentrations of exchangeable Al, Fe, and Ca [47].

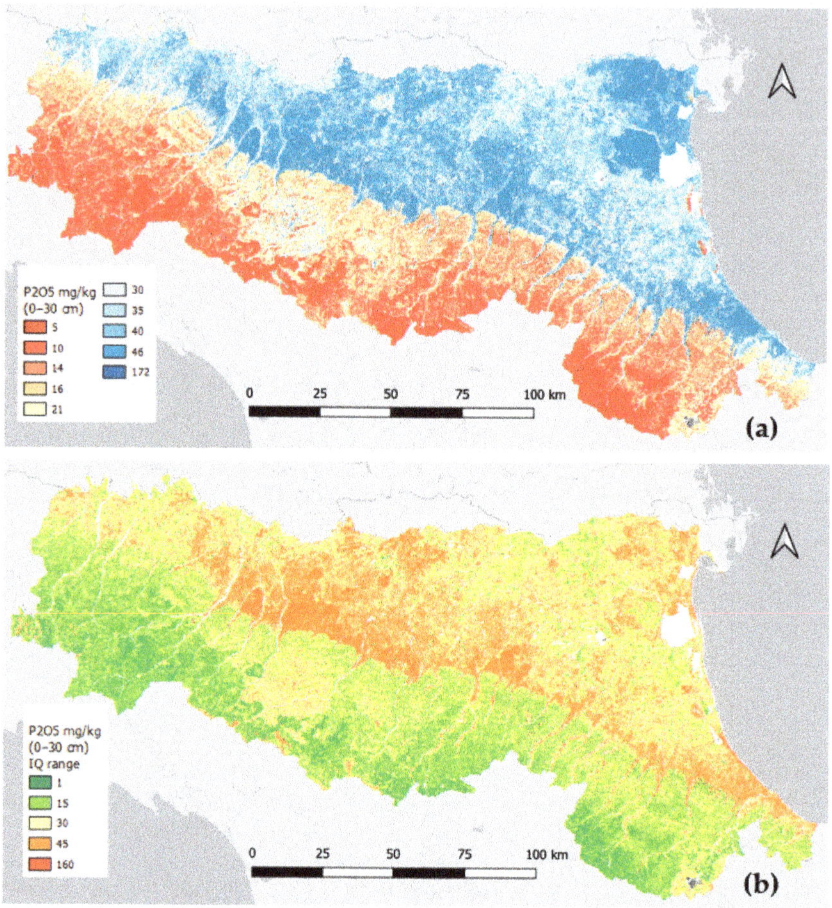

Figure 7. Median available P topsoil (0–30 cm) concentrations (P_2O_5 mg kg^{-1}) estimated with QRF (**a**), and its spatial uncertainty (**b**).

Within the plain, the highest P concentration are found in pedolandscape A2, which features a mean value of 60.6 $\pm$ 9.7 mg kg^{-1}. This contrasts sharply with the other pedolandscape units, where mean P concentrations range from 33 to 47 mg kg^{-1} P_2O_5. Notably high concentrations (above 46 mg kg^{-1} P_2O_5) also occur in units A9 and A8. The soils here are completely decarbonated, have a neutral pH (6.5–7.3), and a medium to fine texture, conditions known to increase phosphorus availability. The highest values in this group are found in the soils of unit A9, particularly in the Parma and Piacenza districts (mean concentrations up to 74.3 mg kg^{-1}), which are intensively used for horticultural crops, like tomatoes and onions.

Conversely, the soils of pedolandscape unit A10 have the second lowest mean p values in the plain (34.7 $\pm$ 12.9 mg kg^{-1}). These soils are also decarbonated but tend towards a neutral to moderately acidic pH. While phosphorus fixation in acidic soils is a known phenomenon, the predominantly weak acidity to neutrality (pH 6–7.3) of these Alfisols suggests other factors are at play. Significant spatial variation exists within this unit:

values are below average in the Bologna and Ravenna districts (24.3 and 28.6 mg kg^{-1}, respectively) but are higher in the Reggio Emilia district (50.0 mg kg^{-1}). These differences likely stem from variations in agronomic management; the practice of organic fertilization (manure and slurry) is more common in the Emilian districts, leading to higher topsoil organic carbon. Furthermore, soil erosion on the steeper slopes of the southern portions of this unit may also contribute to lower P concentrations [48].

The lowest available P mean concentration is found in the pedolandscape A4 (33.4 ± 7.1 mg kg^{-1}), with an increasing trend from Piacenza (29.7 ± 6.1 mg kg^{-1}) to Ferrara (34 ± 6.2 mg kg^{-1}). The frequent use of these soils for poplar plantations may be a contributing factor to these lower values.

Finally, similar intermediate mean values (38–40 mg kg^{-1}) are observed in pedolandscape units A3, A6, A5, A7, and A1. These soils have a highly variable texture (from sandy to clayey) but are generally calcareous and moderate alkaline. Their consistent, moderate P levels are likely sustained by continuous fertilization linked to their intensive agricultural use.

In the four pedolandscapes of the lower Apennines, available P mean concentrations vary little (ranging from 15.6 to 18.0 mg kg^{-1} P_2O_5). The lowest values are found in the Romagna and Bologna Apennines (probably due to high erosion rates and low organic matter levels), especially in units B1 and B2, while the highest values are found in the provinces of Reggio Emilia and Parma, where they range between 20 and 21 mg kg^{-1}. The highest values are found in the province of Rimini on the soils of pedolandscape B1 (mean value 21.4 ± 6.6 mg kg^{-1}), which, compared to the other provinces, is characterized by more clayey soils.

In the Middle Apennines, land use becomes an important factor influencing available P concentration, as forests occupy approximately 62% of the area. Very low values (average value 10.7 ± 4.1 mg kg^{-1} P_2O_5) are found in the acidic soils of ophiolitic origin of the C5 pedolandscape unit. These are followed by the soils of the C3 unit, which are predominantly forested and often non-calcareous (average value 11.0 ± 5.8 mg kg^{-1}) but also with some zonal variability across districts (from 9.3 ± 3.2 mg kg^{-1} in Forlì-Cesena to 21.1 ± 7.3 mg kg^{-1} in Reggio Emilia). The soils on the unstable clays of the C1 pedolandscape unit fall midway (14.1 ± 5.3 mg kg^{-1}), with the highest concentrations in the district of Reggio Emilia (22.9 ± 5.7 mg kg^{-1}). The soils of the C2 and C4 units have similar mean available P concentration (15.9 and 16.0 mg kg^{-1}, respectively) but present rather different soils: the soils on the Triassic gypsum of the pedolandscape unit C4 are mostly under forest and natural vegetation while the soils on the calcareous–marly flysch of unit C2 are characterized by different land uses, including grassland and arable land.

Finally, the mean available P concentration in the pedolandscapes of the upper Apennines are particularly low, ranging from 10.0 to 11.2 mg kg^{-1} P_2O_5. Land use again plays a role here in interpreting results: forestry prevails (84% of the area), followed by pastures and meadows, while above the upper tree line, blueberry bushes and spikenard meadows predominate. Furthermore, moderately to extremely acidic soils are prevalent, especially at higher altitudes and on arenaceous lithotypes. Where medium- or fine-textured soils prevail, surface pH values are higher (neutral to slightly acidic), as in the D1 units in the district of Modena and Rimini, with mean available P concentrations of 15.0 ± 4.0 mg kg^{-1} and 17.0 ± 2.9 mg kg^{-1} respectively.

Available P mapping uncertainty in terms of amplitude of mean IQ range over the entire region is equal to 27.0 ± 12.9 mg kg^{-1}, with the mean values for the plain (33.2 ± 12.0 mg kg^{-1}) being notably higher than those returned for the Apennines (20.2 ± 8.2 mg kg^{-1}). In the plain, lower uncertainty characterizes the agricultural district of Piacenza in the northwestern corner of the plain (27.8 ± 10.3 mg kg^{-1}), while the highest

IQ range mean values occur in the Reggio-Emilia district with 41.0 ± 12.3 mg kg^{-1} and in the Rimini district (38.3 ± 11.2 mg kg^{-1}), at the opposite southeastern corner of the plain. The province of Reggio-Emilia exhibits the highest IQ range mean value also in the Apennine districts of the hills and of the mountains, where their values are equal to 25.0 ± 7.2 mg kg^{-1} and 26.5 ± 10.1 mg kg^{-1}, respectively. The local IQ range minimum mean values are observed in the hilly district of Bologna (21.4 ± 7.9 mg kg^{-1}) and in the mountain district of Forlì (15.5 ± 6.4 mg kg^{-1}), where estimates for available P concentration have the lowest uncertainty of the region.

The classed uncertainty map for estimated P concentration at the municipality level, along with N and K concentrations, is provided in Figure S4 of the Supplementary Materials.

3.4. Comparing LUCAS and Regional Macronutrients Maps

Figure 8 shows the maps of topsoil macronutrients based on the LUCAS survey data at the EU-scale for Emilia-Romagna, along with the corresponding maps based on regional data (RER maps). The original RER maps of the macronutrients' content were resampled to match the 250 m resolution of the LUCAS maps; the classes in the legends of the six maps in Figure 8 are those of the LUCAS maps [17]. Table 4 summarizes the raster statistics for six maps, computed over the entire region and for the two major landforms. At the regional level, macronutrient contents based on LUCAS data resulted systematically above those based on the regional datasets, with a minimum average overestimation of 17.8% in the case of K concentration and a maximum of 48.1% in the case of P; the average overestimation for N concentration in the LUCAS map was equal to 26.2%. The degree of the overestimation observed for the LUCAS-based concentration maps was significantly different in the plain and in the Apennines: in the former it ranged from 7.0% for K concentration to 42.2% for P concentrations, while in the latter the corresponding figures were 29.3% and 59.9% for K and P concentrations, respectively. As for N concentrations, the LUCAS estimates were on average 13.4% and 36% higher than the RER estimates in the plain and the Apennines, respectively.

Table 4. Raster statistics for macronutrient maps at 250 m resolution based on LUCAS and RER datasets. Std. Dev.: standard deviation.

Variable	Area	Mean	Std. Dev.	Min.	Median	Max.
N g kg^{-1} LUCAS	Region	2.16	0.74	0.72	1.93	7.91
	Plain	1.81	0.47	0.72	1.70	6.40
	Apennines	2.54	0.79	0.93	2.43	7.91
N g kg^{-1} RER	Region	1.60	0.68	0.20	1.50	13.00
	Plain	1.57	0.83	0.20	1.40	13.00
	Apennines	1.62	0.48	0.22	1.60	4.40
K_2O mg kg^{-1} LUCAS	Region	247.92	69.91	31.47	248.80	796.38
	Plain	246.43	45.51	56.46	252.37	637.45
	Apennines	249.53	88.92	31.47	240.61	796.38
K_2O mg kg^{-1} RER	Region	203.86	77.88	15.77	196.63	742.85
	Plain	229.26	76.67	15.77	231.99	742.85
	Apennines	176.44	69.43	28.47	159.36	655.70
P_2O_5 mg kg^{-1} LUCAS	Region	23.88	10.33	0.00	24.60	107.14
	Plain	30.53	7.21	0.00	29.56	64.42
	Apennines	16.73	8.18	0.00	16.17	107.14
P_2O_5 mg kg^{-1} RER	Region	12.39	6.72	1.31	11.78	61.54
	Plain	17.66	4.79	2.18	17.02	61.54
	Apennines	6.71	2.61	1.31	6.29	53.82

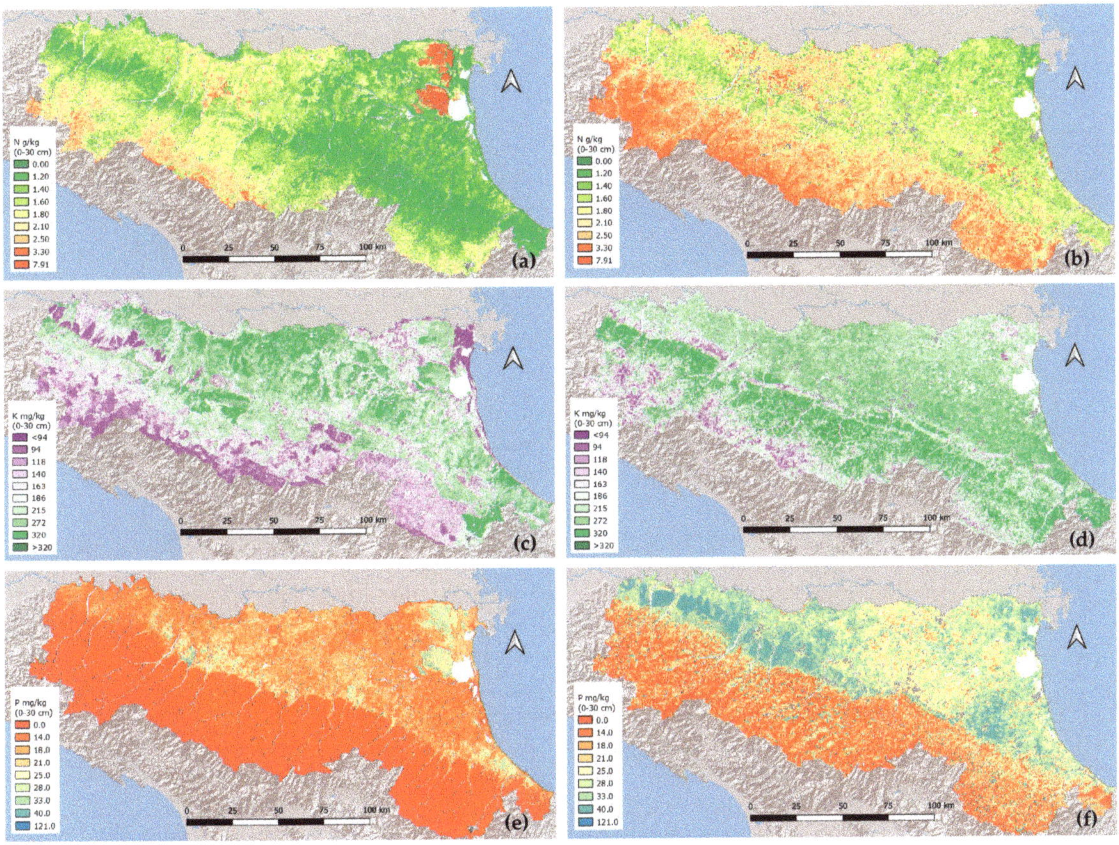

Figure 8. RER (**a**,**c**,**e**) and LUCAS (**b**,**d**,**f**) maps of topsoil (0–30 cm) nutrients concentrations in Emilia-Romagna: nitrogen (**a**,**b**) (g kg^{-1}), potassium (**c**,**d**) (mg kg^{-1}), and phosphorus (**e**,**f**) (mg kg^{-1}).

Figure 9 visually summarizes the differences between the maps derived from the two data sets by using violin plots, which illustrate the probability density of three macronutrient concentrations at various values, as well as the median and quartile values of the DSM estimates. In the case of N concentrations, RER estimates are characterized by a narrower IQ range that does not overlap with those of the LUCAS estimates and a more positively skewed distribution; in both cases, the distributions are markedly leptokurtic. The IQ ranges of the two K concentration distributions overlap, but the RER estimates exhibit a moderate bimodality which is not observed in the distribution of the LUCAS estimates, which are more markedly positively skewed and leptokurtic. A noticeable bimodality characterizes the distribution of the RER estimates of P concentrations, reflecting the observed difference in P concentrations between the plain and the Apennines. The distribution of the LUCAS P estimates also exhibits this feature, albeit with a significantly smoother bimodality. The LUCAS estimates are less positively skewed than the RER ones and slightly platykurtic.

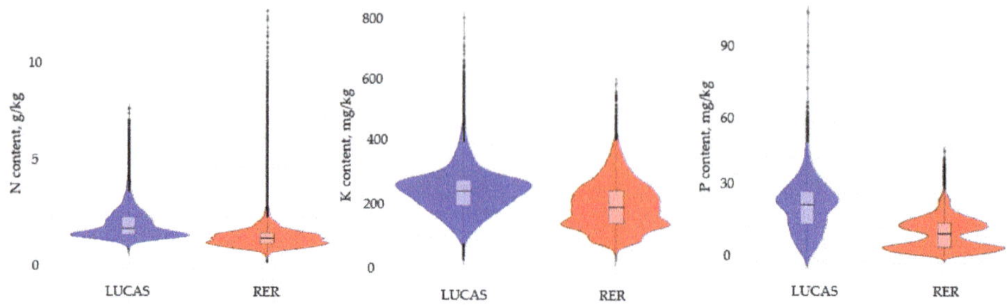

Figure 9. Violin plots displaying the spread of DSM estimates for macronutrient concentrations using the LUCAS (blue) and RER (red) datasets.

Eventually, the three macronutrient maps stemming from the LUCAS and the RER databases were compared in terms of mean concentrations in the agricultural districts and pedolandscapes of Emilia-Romagna. Figure 10 shows the radar charts of the relative difference between the LUCAS and RER DSM estimates of macronutrient concentrations, standardized over the LUCAS concentrations. In the agricultural districts of the plain, LUCAS estimates resulted systematically in higher values than the corresponding RER-based estimates, except for N concentration in the Ferrara plain (district 25), where they were 22% lower, missing the detection of large areas of reclaimed soils with peat layers characterized by the highest N concentrations of the region. On average the eastern districts of Romagna showed the largest relative differences, with values exceeding 30% in Ravenna (district 16) and Forlì (district 19). As for K concentrations in the agricultural districts of the plain, RER estimates were 4 to 8% higher than the LUCAS ones in three districts of Emilia, while in all the other sectors of the plain the opposite was observed, with larger differences in Piacenza (district 1) and Ravenna (district 16). As evident from Figure 10, in all districts, the overestimation of P topsoil concentrations based on the LUCAS database exceeded those observed for N and K, with larger relative differences in Piacenza (district 1, 60%) and Ravenna (district 16, 52%).

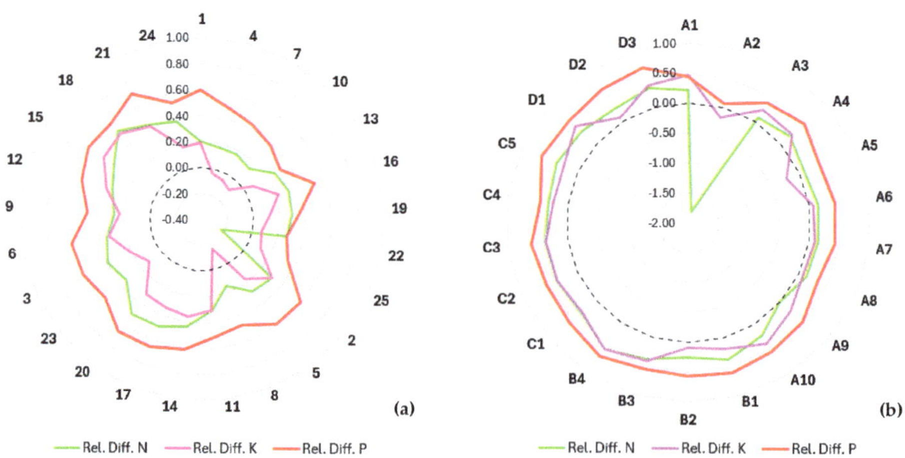

Figure 10. Radar charts of the relative differences of DSM estimates of N, K, and P concentrations based on the LUCAS and RER datasets: (**a**) agricultural districts (see Figure 2 for legend description), (**b**) pedolandscapes (see Figure 1 for legend description). The dashed line marks a relative difference equal to zero.

In the agricultural districts of the lower and mid-upper Apennines, relative differences in DSM-estimated macronutrients' concentrations followed a similar pattern, with values below or close to the average in the districts of Emilia, with a minimum in Reggio Emilia (districts 8 and 9) and a maximum in the districts of Romagna. In the lower Apennine districts, maximum relative differences were observed in Forlì (district 20), Bologna (district 14) and again in Forlì (district 20) for N (50%), K (36%), and P (67%) concentrations, respectively. In the agricultural districts of the mid-upper Apennine, maximum relative differences were observed in Ravenna (district 18) for N (53%) and K (50%), and again in Forlì (district 21) for P (70%).

In the pedolandscapes of the plain, the major difference between the LUCAS and the RER macronutrient maps was observed in the lower abandoned Po delta plain (unit A2, 512 km^2, ca. 4.5% of the plain), where the widespread occurrence of reclaimed soils with organic horizons (e.g., Histic Humaquepts, Sulfic Endoaquepts, Taphto-Histic Endoaquolls, Terric Sulfisaprists, Typic Sulfihemists, and Typic Sulfisaprists, classified according to USDA Soil Taxonomy, 12th Ed.) was not acknowledged in the LUCAS macronutrient map, which severely underestimated N concentration compared to the RER map (−181%). In the same unit, K concentration was also underestimated in the LUCAS map (−16%), while P concentration was slightly overestimated (+8%). In all the other pedolandscape units of the plain, LUCAS-based maps provided higher macronutrient concentrations, notably for P contents, with relative differences ranging from 35 (unit A8) to 54% (unit A10). In the case of N and K, relative difference ranges were between 1 (unit A9) and 24% (unit A10), and 7 (unit A6) and 46% (Unit A1), respectively. However, a notable exception regarding K content is unit A5, which consists of soils in morphologically depressed areas of the lower Apennine alluvial plain (1641 km^2, 14% of the plain) and is characterized by fine-textured soils with high to very high clay contents (e.g., Chromic Udic Haplusterts, Halic Endoaquerts, Sodic Endoaquerts, Udic Calciusterts, Ustic Endoaquerts, Vertic Calciustepts, and Vertic Endoaquepts, classified according to USDA Soil Taxonomy, 12th Ed.). In this pedolandscape, LUCAS-based K concentrations resulted on average 22% less than those observed in the corresponding unit of the RER-based map.

In the pedolandscapes of the Apennine, relative differences in DSM-estimated nutrient contents increased with elevation for P and K, while in the case of N a decreasing trend was observed. In the units of the lower Apennine the lowest relative differences occurred in the pedolandscape B2 and the maximum in B4 for all nutrients, with mean values ranging from 9 (K) to 55% (P) in B2 and from 51% (K) to 65% (P) in B4. In the pedolandscapes of the mid-Apennines, the smallest relative differences were detected in the unit C4 for all nutrients, with average values ranging from 25 to 46% for K and P concentration, respectively. Relative differences were largest in unit C3 for N (38%) and K (35%) contents, while unit C5 showed the largest differences in terms of P contents (65%). Eventually, in the upper-Apennine pedolandscapes the relative differences between the LUCAS- and RER-based maps were the largest for P contents, being above 60% in all units, with a maximum in D3 (69%) and a minimum in D1 (60%). The mean relative differences in estimated K contents varied greatly, ranging from 8% in unit D2 to 45% in unit D1. As already detected in most cases, relative differences in terms of N content were intermediate, ranging between 26% in unit D2 and 34% in unit D3.

Maps of the relative difference between the three concentration maps based on the LUCAS and RER datasets are shown in Figure S5 in the Supplementary Materials

4. Discussion

4.1. Model Performance and Covariate Interpretation

The results presented in the previous section confirm the predictive effectiveness of the DSM approach in estimating topsoil macronutrients' concentrations in Emilia-Romagna. The analysis of covariates' importance revealed that continuous soil (e.g., organic carbon, clay, sand, and pH) and categorical variables representing regional spatial trends were the most relevant predictors. This aligns with the recommendation to integrate pedological knowledge into DSM frameworks [49], which not only potentially improved estimation precision but was crucial for the interpretation of the resulting spatial patterns. The limited direct relevance of Land Use Land Cover (LULC) as a categorical predictor, notable only in the Apennines, suggests that its effects were better captured by remote sensing indices and reflectance bands [50,51]. However, while the increasing availability and accessibility of high-resolution spectral indices derived from remote sensing offers great potential in DSM applications [52], it can also result in a frequent lack of awareness of definitions and limitations, causing redundancies and inconsistencies [53]. Therefore, rather than employing an exhaustive set of predictors [54,55], we prioritized model interpretability and parsimony using a robust set of covariates tested in the region [44], with feature selection guided by Recursive Feature Elimination [56].

4.2. Root Causes of the LUCAS–RER Discrepancy and Its DSM Implications

The most significant finding of this study is the substantial and systematic discrepancy between the regional (RER) and continental (LUCAS) DSM products. We identify three primary root causes for this:

- Sampling density: the fundamental difference in observation density, LUCAS (~1.5 sample/200 km^2) versus RER (~1.6 samples/km^2), is the most critical factor. The RER dataset's high density allows it to capture local variability and nutrient cold spots and hotspots that are statistically invisible at the LUCAS sampling density. It is noteworthy that recent research highlighted that for most soil properties, macronutrients included, the differences in survey design and sampling protocols between LUCAS and Italian methods did not lead to significant differences, showing consistency among the different sampling procedures [57].
- Scale of covariates and model generalization: continental-scale models like LUCAS necessarily rely on covariates at a coarser resolution and must generalize across vastly different pedo-climatic regions. This process inherently smooths out extremes. Our regional model, using higher-resolution predictors tailored to the local context, preserves this critical fine-scale variation.
- Inability to capture specific pedolandscape units: a telling example is the failure of the LUCAS-based map to identify the high N concentrations in the organic soils of the lower Po delta plain (pedolandscape A2). This unit, covering over 500 km^2, contains distinct soil types characterized by organic horizons (e.g., Histic Humaquepts and Typic Sulfisaprists) that greatly affect nutrient levels. Continental models lack the contextual knowledge and data density to represent such specific, yet extensive, features.

This discrepancy underscores a major challenge in digital soil mapping: the loss of critical information when upscaling models or applying coarse-scale products at a regional level [58]. Our results provide concrete evidence that accuracy and relevance are significantly enhanced when DSM is conducted at a scale commensurate with the management and policy questions being addressed.

4.3. Direct Implications for the EU Soil Monitoring Law and Regional Soil Management

The discrepancy between the datasets has profound and immediate policy implications. Using the proposed SML phosphorus threshold of 50 mg kg^{-1} P_2O_5 as a benchmark, the choice of baseline data leads to drastically different outcomes (Figure 11):

- LUCAS baseline: would classify 92.8% of the plain and 25.0% of the Apennines as exceeding the admissible concentration.
- RER baseline: suggests only 16.05% of the plain and 0.11% of the Apennines are above this threshold.

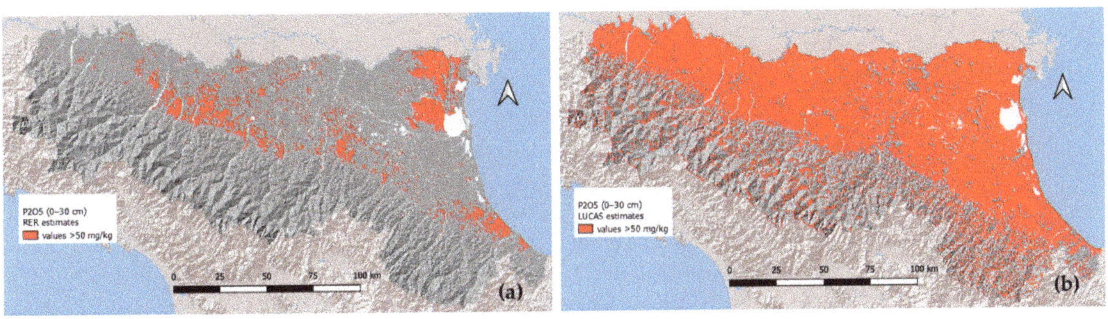

Figure 11. Values above the SML threshold of 50 mg kg^{-1} P_2O_5 as resulting from the RER (**a**) and LUCAS (**b**) estimates.

This represents a potential misclassification of over 76% of the plain's agricultural area, which would have severe consequences for farmers, land managers, and the perceived severity of phosphorus-related environmental risk in the region. While the "one-out, all-out" principle may have been removed from the final SML text, the definition of reference baselines remains a cornerstone of the legislation. Our study demonstrates that adopting a baseline based on continental-scale assessment could lead to inappropriate regulatory pressures and misdirected resources.

Therefore, the findings strongly advocate for a hybrid approach to DSM. The "top-down" paradigm of continental models, like LUCAS and SoilGrids [59], should be systematically integrated with "bottom-up," regionally coordinated efforts. Initiatives such as the FAO-Global Soil Partnership's soil organic carbon map [60] and the EU-funded EJPSOIL project [61] champion this very concept, promoting participatory, multi-scale data collection that leverages local knowledge and priorities. Our RER baseline illustrates the relevance of such integration. The future of robust soil governance under the SML lies not in choosing one scale over the other, but in creating a framework where continental models provide the broad context and regional baselines, like the one presented here, provide the essential, high-fidelity data for effective local implementation and validation at the level of the soil districts foreseen by the SML proposal [62].

4.4. Communication of Uncertainty and Study Limitations

To enhance the practical utility of our maps, we coupled concentration estimates with spatial uncertainty based on QRF prediction intervals [63]. In line with best practices for communicating with end-users [64,65], we aggregated and presented this uncertainty at the municipality level (Figure S4). This supports risk-aware decision-making in land planning and can guide future targeted sampling campaigns to enhance survey efficiency [66].

This study has two major limitations. First, due to the long time span of the data used in this study, it was not possible to assess temporal dynamics within the study area, because the data grouped by survey time (e.g., by decades) were also spatially clustered from successive surveys. This limitation is often encountered in DSM applications [49,67] as data are mostly collected and analyzed for several purposes apart from solely mapping soil properties. This limitation indeed prevented the analysis of the changes in soil nutrients' status and the assessment of the impact of land use history.

Second, the focus is exclusively on the topsoil (0–30 cm), as most data were sourced from agricultural fertilization planning. Only about 12% of the available data provide information on subsoil macronutrient concentrations, and future work will focus on extending this framework to subsoil nutrients.

5. Conclusions

This study effectively generated high-resolution (100 m), uncertainty-quantified maps of topsoil macronutrients (N, P, and K) for the Emilia-Romagna region, providing a robust baseline for soil fertility assessment and monitoring. The key conclusions are as follows:

1. The DSM approach using Quantile Random Forests proved highly effective, with models demonstrating excellent performance ($R^2 \geq 0.9$) and identifying soil organic carbon and texture as the dominant controls on macronutrient spatial patterns.
2. A critical comparison with the continental-scale LUCAS-based maps revealed significant systematic overestimations by LUCAS, particularly for phosphorus (48% at regional level), and a failure to detect important local features, such as nutrient hotspots in organic soils.
3. The root of this discrepancy lies in the extremely different sampling densities, the scale of environmental covariates, and the inability of continental models to capture specific soil–landscape relationships.
4. The practical implications are substantial: the choice of baseline data dramatically alters the assessment of soil quality against regulatory thresholds, as demonstrated for the EU Soil Monitoring Law. Relying solely on continental-scale data for regional policy implementation carries a high risk of misinformed decisions.

In conclusion, while continental-scale models like LUCAS are valuable for broad-scale assessments, they are insufficient for regional-scale land management and policy. Our results highlight one of the potential difficulties in the actual implementation of the Soil Monitoring Law resulting from the absence of soil information of adequate detail at the level of the soil districts identified by member states as required by the proposal. This eventually would lead to possible differences among and within member states in its implementation. Our work underscores the indispensable need for the integration of high-resolution, region-specific soil data to ensure the accurate and effective implementation of environmental regulations like the EU Soil Monitoring Law.

Supplementary Materials: The following supporting information can be downloaded at: https://www.mdpi.com/article/10.3390/land14112142/s1, Table S1: Dominant soils occurring in the pedolandscape units of Emilia-Romagna. Soil are classified according to Soil Taxonomy (Soil Survey Staff, 2014, Keys to Soil Taxonomy. 12th Edition, USDA-NRCS, Washington DC). The soil classifications are listed in order of prevalence. Table S2: Mean and median macronutrients' concentrations over five sampling periods. Figure S1: Distribution of the FAO-WRB Major Soil Groups in the pedolandscapes of Emilia-Romagna. Figure S2: Location of macronutrients sampling points over time. Figure S3: Location of the LUCAS points (n = 117) in the pedolandscapes of Emilia-Romagna. Figure S4: Classed uncertainty maps at municipality level for DSM macronutrients estimates: (a) Nitrogen, (b) Potassium, (c) Phosphorus. The five uncertainty classes are based on the ventiles of the distribution of the IQ range values. Figure S5. Relative differences in estimated macronutrient concentrations between the LUCAS and the RER datasets: (a) Nitrogen, (b) Potassium, (c) Phosphorus. Relative differences are calculated for each cell of the 250m estimation grid as [LUCAS -RER]/[LUCAS].

Author Contributions: Conceptualization, P.T. and F.U.; methodology, F.U. and P.T.; software, F.U. and P.T.; validation, P.T.; formal analysis, F.U., investigation, F.U., P.T. and A.A.; resources, P.T.; data curation, F.U. and P.T.; writing—original draft preparation, F.U.; writing—review and editing, P.T. and A.A.; visualization, F.U. and P.T.; supervision, F.U. and P.T.; project administration, P.T.; funding acquisition, P.T. All authors have read and agreed to the published version of the manuscript.

Funding: This research was funded by Emilia-Romagna Region (RER)—General Directorate for Environment and Land Care—Geological, Seismic and Soil Service, grant n. 558 26/04/2021, within the framework of the RER-CNR four years research agreement "Digital soil mapping applications for ecosystem services assessment and climate change strategy support and knowledge tools to support the EU Nitrates Directive".

Data Availability Statement: The 1:50,000 soil map of Emilia Romagna (Ed. 2021) is available for download at the following link: https://mappegis.regione.emilia-romagna.it/moka/ckan/suolo/Carta_Suoli_50k.zip (accessed on 9 October 2025), while the maps of soil macronutrients are downloadable from the following web pages: N: https://mappegis.regione.emilia-romagna.it/moka/ckan/suolo/Azoto_N_totale_0_30_cm_rst.zip (accessed on 9 October 2025); K: https://mappegis.regione.emilia-romagna.it/moka/ckan/suolo/Potassio_K_scambiabile_0_30_cm_rst.zip (accessed on 9 October 2025); P: https://mappegis.regione.emilia-romagna.it/moka/ckan/suolo/Fosforo_P_assimilabile_0_30_cm_rst.zip (accessed on 9 October 2025). All the maps are also available via WMS service (URL: https://servizigis.regione.emilia-romagna.it/wms/suoli (accessed on 9 October 2025)).

Acknowledgments: The authors wish to thank the CNR-IBE and RER office staff for all the administrative work and the technical support allowing the accomplishment of the research activities. The authors would like also to thank Giampaolo Sarno and Chiara Ferronato of the Sustainable Agriculture Area of the Programming, Territorial Development and Production Sustainability Sector—Emilia-Romagna Region for contributing to the discussion of the results. The Authors wish to express their gratitude to the anonymous reviewers who provided insightful comments and constructive criticisms. Their feedback was precious in significantly improving the quality, clarity, and impact of the manuscript.

Conflicts of Interest: The authors declare no conflicts of interest.

Abbreviations

The following abbreviations are used in this manuscript:

AE	Absolute Error
EC	European Commission
EPSG	European Petroleum Survey Group
EU	European Union
DEM	Digital Elevation Model
DSM	Digital Soil Mapping
IoA	Index of Agreement
IQ range	Interquartile range
ISO	International Organization for Standardization
LUCAS	Land Use/Cover Area frame statistical Survey
LULC	Land Use Land Cover class(es)
ME	Mean Error
ML	Machine Learning
MS	Member States
NDVI	Normalized Difference Vegetation Index
NDSI	Normalized Difference Soil Index
NDWI	Normalized Difference Water Index
NUTS	Nomenclature of Territorial Units for Statistics
QRF	Quantile Random Forest
RER	Regione Emilia-Romagna
RFE	Recursive Feature Elimination

RMSE	Rooted Mean Square Error
RUSLE	Revised Universal Soil Loss Equation
SML	Soil Monitoring Law
SOSI	Soil Salinity Index

References

1. Arias-Navarro, C.; Baritz, R.; Jones, A. (Eds.) *The State of Soils in Europe—Fully Evidenced, Spatially Organized Assessment of the Pressures Driving Soil Degradation*; Publications Office of the European Union: Luxembourg, 2024. [CrossRef]
2. *Soil Monitoring in Europe-Indicators and Thresholds for Soil Health Assessments*; EEA Report 08/2022; European Environment Agency: Copenhagen, Denmark, 2023. [CrossRef]
3. White, A.; Faulkner, J.W.; Conner, D.; Barbieri, L.; Adair, E.C.; Niles, M.T.; Mendez, V.E.; Twombly, C.R. Measuring the Supply of Ecosystem Services from Alternative Soil and Nutrient Management Practices: A Transdisciplinary, Field-Scale Approach. *Sustainability* **2021**, *13*, 10303. [CrossRef]
4. Hou, D. Soil health and ecosystem services. *Soil Use Manag.* **2023**, *39*, 1259–1266. [CrossRef]
5. Van Groenigen, J.W.; Van Kessel, C.; Hungate, B.A.; Oenema, O.; Powlson, D.S.; Van Groenigen, K.J. Sequestering Soil Organic Carbon: A Nitrogen Dilemma. *Environ. Sci. Technol.* **2017**, *51*, 4738–4739. [CrossRef]
6. Oenema, O.; van Liere, L.; Schoumans, O. Effects of lowering nitrogen and phosphorus surpluses in agriculture on the quality of groundwater and surface water in the Netherlands. *J. Hydrol.* **2005**, *304*, 289–301. [CrossRef]
7. Vigiak, O.; Udías, A.; Grizzetti, B.; Zanni, M.; Aloe, A.; Weiss, F.; Hristov, J.; Bisselink, B.; de Roo, A.; Pistocchi, A. Recent regional changes in nutrient fluxes of European surface waters. *Sci. Total Environ.* **2023**, *858*, 160063. [CrossRef]
8. Pandao, M.R.; Akshay, A.T.; Rupeshkumar, J.C.; Nagesh, R.N.; Dhananjay, D.S.; Sindhu, R.R. Soil Health and Nutrient Management. *Int. J. Plant Soil Sci.* **2024**, *36*, 873–883. [CrossRef]
9. Ţopa, D.-C.; Căpşună, S.; Calistru, A.-E.; Ailincăi, C. Sustainable Practices for Enhancing Soil Health and Crop Quality in Modern Agriculture: A Review. *Agriculture* **2025**, *15*, 998. [CrossRef]
10. Virto, I.; Imaz, M.J.; Fernández-Ugalde, O.; Gartzia-Bengoetxea, N.; Enrique, A.; Bescansa, P. Soil Degradation and Soil Quality in Western Europe: Current Situation and Future Perspectives. *Sustainability* **2015**, *7*, 313–365. [CrossRef]
11. Zhang, S. Heterogeneity of Soil Nutrients: A Review of Methodology, Variability and Impact Factors. *J. Environ. Earth Sci.* **2019**, *1*, 6–28. [CrossRef]
12. European Commission. Proposal for a Directive on Soil Monitoring and Resilience. 416 (Final). 2023. Available online: https://environment.ec.europa.eu/publications/proposal-directive-soil-monitoring-and-resilience_en (accessed on 5 July 2025).
13. Panagos, P.; Jones, A.; Lugato, E.; Ballabio, C. A Soil Monitoring Law for Europe. *Glob. Chall.* **2025**, *9*, 2400336. [CrossRef]
14. McBratney, A.B.; Mendonça Santos, M.L.; Minasny, B. On digital soil mapping. *Geoderma* **2003**, *117*, 3–52. [CrossRef]
15. Wadoux, A.M.C.; Minasny, B.; McBratney, A.B. Machine learning for digital soil mapping: Applications, challenges and suggested solutions. *Earth-Sci. Rev.* **2020**, *210*, 103359. [CrossRef]
16. Arrouays, D.; McBratney, A.; Bouma, J.; Libohova, Z.; Richer-de-Forges, A.C.; Morgan, C.L.S.; Roudier, P.; Poggio, L.; Mulder, V.L. Impressions of digital soil maps: The good, the not so good, and making them ever better. *Geoderma Reg.* **2020**, *20*, e00255. [CrossRef]
17. Ballabio, C.; Lugato, E.; Fernández-Ugalde, O.; Orgiazzi, A.; Jones, A.; Borrelli, P.; Montanarella, L.; Panagos, P. Mapping LUCAS topsoil chemical properties at European scale using Gaussian process regression. *Geoderma* **2019**, *355*, 113912. [CrossRef]
18. *Implementation of Council Directive 91/676/EEC Concerning the Protection of Waters Against Pollution Caused by Nitrates from Agricultural Sources*; Synthesis from year 2000 Member States Reports; EC, Office for Official Publications of the European Communities: Luxembourg, 2002; ISBN 92-894-4103-8.
19. *Common Implementation Strategy for the Water Framework Directive (2000/60/EC)*; Guidance Document No 7. Monitoring under the Water Framework Directive. Produced by Working Group 2.7-Monitoring; EC, Office for Official Publications of the European Communities: Luxembourg, 2003; ISBN 92-894-5127-0.
20. Directive 2006/118/EC of the European Parliament and of the Council of 12 December 2006 on the Protection of Groundwater Against Pollution and Deterioration. EC, 2006. Official Journal of the European Union. 27 December 2006. L372. pp. 19–31. Available online: https://eur-lex.europa.eu/eli/dir/2006/118/oj/eng (accessed on 9 October 2025).
21. Report from the Commission to the Council and the European Parliament on the Implementation of Council Directive 91/676/EEC Concerning the Protection of Waters Against Pollution Caused by Nitrates from Agricultural Sources Based on Member State Reports for the Period 2016–2019. EC, 2021, Brussels, 11.10.2021, COM(2021) 1000 Final. Available online: https://eur-lex.europa.eu/legal-content/EN/TXT/?uri=celex:52021DC1000 (accessed on 9 October 2025).
22. de Vries, W.; Schulte-Uebbing, L.; Kros, H.; Voogd, J.C.; Louwagie, G. Spatially explicit boundaries for agricultural nitrogen inputs in the European Union to meet air and water quality targets. *Sci. Total Environ.* **2021**, *786*, 147283. [CrossRef] [PubMed]

23. Regulation (EC) No 1059/2003 of the European Parliament and of the Council of 26 May 2003 on the Establishment of a Common Classification of Territorial Units for Statistics (NUTS). Available online: https://eur-lex.europa.eu/legal-content/EN/TXT/?qid=1405939838475&uri=CELEX:32003R1059 (accessed on 4 July 2025).
24. Antolini, G.; Pavan, V.; Tomozeiu, R.; Marletto, V. Atlante Idroclimatico Dell'emilia-Romagna 1961–2015, 2017; ISBN 978-88-87854-44-2. Available online: https://www.arpae.it/it/temi-ambientali/clima/rapporti-e-documenti/atlante-climatico (accessed on 3 October 2025).
25. Regione Emilia-Romagna. Carta dei Suoli Della Regione Emilia-Romagna in Scala 1: 50.000. Edizione 2021. Available online: https://geo.regione.emilia-romagna.it/gstatico/documenti/dati_pedol/carta_suoli_50k.pdf (accessed on 10 July 2025).
26. Garberi, M.L.; Lenzi, D.; Mariani, M.C.; Masi, S.; Orlandi, F.; Vigilante, E. Database Uso del Suolo di Dettaglio 2017–Documentazione Regione Emilia-Romagna, Servizio Statistica e Sistemi Informativi Geografici, 2020. Available online: https://geoportale.regione.emilia-romagna.it/approfondimenti/contenuti-allegati/documentazione-uso-del-suolo/capitolato-tecnico-uso-suolo-dettaglio-2017-ed-2021.pdf (accessed on 10 July 2025).
27. Schipani, T.; Censi, D.; Laruccia, N.; Rossi, R.; Solferini, A.; D'Aloia, M.; Michetti, M. Analisi del Sistema Agricolo, Agroindustriale e del Territorio Rurale Dell'emilia-Romagna. Regione Emilia-Romagna Servizio Programmazione e Sviluppo Locale Integrato e ART_ER S. cons. p.a. 2022. Available online: https://agricoltura.regione.emilia-romagna.it/pac-2023-2027/documenti/analisi-del-sistema/analisi-contesto.pdf/@@download/file/Analisi%20contesto.pdf (accessed on 10 July 2025).
28. Minister for Agricultural Policies. Approval of the Official Methods for Soil Chemical Analysis. Ministerial Decree of September 13, 1999, GU n. 248 del 2110.1999-S.O. n.185. Available online: https://www.certifico.com/component/attachments/download/32182 (accessed on 11 July 2025).
29. *ISO 1871:2009*; Food and Feed Products—General Guidelines for the Determination of Nitrogen by the Kjeldahl Method. International Organization for Standardization: Geneva, Switzerland, 2009.
30. *ISO 11263:1994*; Soil Quality—Determination of Phosphorus—Spectrometric Determination of Phosphorus Soluble in Sodium Hydrogen Carbonate Solution. International Organization for Standardization: Geneva, Switzerland, 1994.
31. *ISO 22171:2023*; Soil Quality—Determination of Potential Cation Exchange Capacity (CEC) and Exchangeable Cations Buffered at pH 7, Using a Molar Ammonium Acetate Solution. International Organization for Standardization: Geneva, Switzerland, 2023.
32. Bishop, T.F.A.; McBratney, A.B.; Laslett, G.M. Modelling soil attribute depth functions with equal-area quadratic smoothing splines. *Geoderma* **1999**, *91*, 27–45. [CrossRef]
33. Malone, B.P.; McBratney, A.B.; Minasny, B.; Laslett, G.M. Mapping continuous depth functions of soil carbon storage and available water capacity. *Geoderma* **2009**, *154*, 138–152. [CrossRef]
34. R Core Team. R: A Language and Environment for Statistical Computing; R Foundation for Statistical Computing. Vienna, Austria, 2021. Available online: https://www.R-project.org/ (accessed on 9 October 2025).
35. Wang, W.; Yan, J. Shape-Restricted Regression Splines with R Package splines2. *J. Data Sci.* **2021**, *19*, 498–517. [CrossRef]
36. Rosner, B. Percentage Points for a Generalized ESD Many-Outlier Procedure. *Technometrics* **1983**, *25*, 165–172. [CrossRef]
37. *EPSG:7791*; RDN2008/UTM Zone 32N. EPSG Geodetic Parameter Dataset. Geodesy Subcommittee of the IOGP Geomatics Committee: London, UK, 2016.
38. Genova, G.; Poggio, L.; Kempen, B.; Colman, B. DSM Workflow Seedling v.0.3.6. 2024, ISRIC—World Soil Information. Available online: https://git.wur.nl/isric/dsm-general/dsm.workflows/seedling/-/tree/0.3.6 (accessed on 21 August 2025).
39. Breiman, L. Random Forests. *Mach. Learn.* **2001**, *45*, 5–32. [CrossRef]
40. Meinshausen, N. Quantile regression forests. *J. Mach. Learn. Res.* **2006**, *7*, 983–999. Available online: https://www.jmlr.org/papers/volume7/meinshausen06a/meinshausen06a.pdf (accessed on 21 August 2025).
41. Wright, M.N.; Ziegler, A. Ranger: A fast implementation of random forests for high dimensional data in c++ and r. *J. Stat. Softw.* **2017**, *77*, 1–17. [CrossRef]
42. Willmott, C.J. On the validation of models. *Phys. Geogr.* **1981**, *2*, 184–194. [CrossRef]
43. Gorelick, N.; Hancher, M.; Dixon, M.; Ilyushchenko, S.; Thau, D.; Moore, R. Google Earth Engine: Planetary-scale geospatial analysis for everyone. *Remote Sens. Environ.* **2017**, *202*, 18–27. [CrossRef]
44. Ungaro, F.; Tarocco, P.; Calzolari, C. Leveraging Soil Geography for Land Use Planning: Assessing and Mapping Soil Ecosystem Services Indicators in Emilia-Romagna, NE Italy. *Geographies* **2025**, *5*, 39. [CrossRef]
45. Staffilani, F. Carta Dell'erosione Idrica Attuale Della Regione Emilia-Romagna. Note Illustrative, Edizione 2019. Available online: https://mappegis.regione.emilia-romagna.it/gstatico/documenti/dati_pedol/NOTE_ILLUSTRATIVE_EROSIONE.pdf (accessed on 25 August 2025).
46. Kuhn, M. Building Predictive Models in R Using the caret Package. *J. Stat. Softw.* **2008**, *28*, 1–26. [CrossRef]
47. Prisca, D.J.; Osumanu, H.A.; Latifah, O.; Aainaa, H. Phosphorus Transformation in Soils Following Co-Application of Charcoal and Wood Ash. *Agronomy* **2021**, *11*, 2010. [CrossRef]
48. Alewell, C.; Ringeval, B.; Ballabio, C.A.; Robinson, D.A.; Panagos, P.; Borrelli, P. Global phosphorus shortage will be aggravated by soil erosion. *Nat. Commun.* **2020**, *11*, 4546. [CrossRef]

49. Chen, S.; Arrouays, D.; Mulder, V.L.; Poggio, L.; Minasny, B.; Roudier, P.; Libohova, Z.; Lagacherie, P.; Shi, Z.; Hannam, J.; et al. Digital mapping of GlobalSoilMap soil properties at a broad scale: A review. *Geoderma* **2022**, *409*, 115567. [CrossRef]
50. Mashala, M.J.; Dube, T.; Mudereri, B.T.; Ayisi, K.K.; Ramudzuli, M.R. A Systematic Review on Advancements in Remote Sensing for Assessing and Monitoring Land Use and Land Cover Changes Impacts on Surface Water Resources in Semi-Arid Tropical Environments. *Remote Sens.* **2023**, *15*, 3926. [CrossRef]
51. Alawode, G.L.; Oluwajuwon, T.V.; Hammed, R.A.; Olasuyi, K.E.; Krasovskiy, A.; Ogundipe, O.C.; Kraxner, F. Spatiotemporal assessment of land use land cover dynamics in Mödling district, Austria, using remote sensing techniques. *Heliyon* **2025**, *11*, e43454. [CrossRef]
52. Abdulraheem, M.I.; Zhang, W.; Li, S.; Moshayedi, A.J.; Farooque, A.A.; Hu, J. Advancement of Remote Sensing for Soil Measurements and Applications: A Comprehensive Review. *Sustainability* **2023**, *15*, 15444. [CrossRef]
53. Chen, Q.; Vaudour, E.; Richer-de-Forges, A.C.; Arrouays, D. Spectral indices in remote sensing of soil: Definition, popularity, and issues. A critical overview. *Remote Sens. Environ.* **2025**, *329*, 114918. [CrossRef]
54. Nussbaum, M.; Spiess, K.; Baltensweiler, A.; Grob, U.; Keller, A.; Greiner, L.; Schaepman, M.E.; Papritz, A. Evaluation of digital soil mapping approaches with large sets of environmental covariates. *SOIL* **2018**, *4*, 1–22. [CrossRef]
55. Dash, P.K.; Ferhatoglu, C.; Miller, B.A.; Panigrahi, N.; Mishra, A. Influence of sample size and machine learning algorithms on digital soil nutrient mapping accuracy. *Environ. Monit. Assess.* **2025**, *197*, 996. [CrossRef]
56. Kasraei, B.; Schmidt, M.G.; Zhang, J.; Bulmer, C.E.; Filatow, D.S.; Arbor, A.; Pennell, T.; Heung, B. A framework for optimizing environmental covariates to support model interpretability in digital soil mapping. *Geoderma* **2024**, *445*, 116873. [CrossRef]
57. Del Duca, S.; Tondini, E.; Vitali, F.; Lumini, E.; Garlato, A.; Vinci, I.; Tagliaferri, E.; Brenna, S.; Motta, S.; Maillet, E.; et al. Comparison of LUCAS and Italian Sampling Procedures for Harmonising Physicochemical and Biological Soil Health Indicators. *Eur. J. Soil Sci.* **2025**, *76*, e70108. [CrossRef]
58. Lagacherie, P. Operational Digital Soil Mapping: Achievements, Challenges and Future Strategies to Go Beyond. *Eur. J. Soil Sci.* **2025**, *76*, e70139. [CrossRef]
59. Poggio, L.; de Sousa, L.M.; Batjes, N.H.; Heuvelink, G.B.M.; Kempen, B.; Ribeiro, E.; Rossiter, D. SoilGrids 2.0: Producing soil information for the globe with quantified spatial uncertainty. *SOIL* **2021**, *7*, 217–240. [CrossRef]
60. FAO and ITPS. *2018. Global Soil Organic Carbon Map (GSOCmap) Technical Report*; FAO and ITPS: Rome, Italy, 2018; p. 162, ISBN 978-92-5-130439-6. Available online: https://openknowledge.fao.org/handle/20.500.14283/i8891en (accessed on 2 October 2025).
61. Froger, C.; Tondini, E.; Arrouays, D.; Oorts, K.; Poeplau, C.; Wetterlind, J.; Putku, E.; Saby, N.P.A.; Fantappiè, M.; Styc, Q.; et al. Comparing LUCAS Soil and national systems: Towards a harmonized European Soil monitoring network. *Geoderma* **2024**, *449*, 117027. [CrossRef]
62. Wadoux, A.M.J.-C.; Courteille, L.; Arrouays, D.; De Carvalho Gomes, L.; Cortet, J.; Creamer, R.E.; Eberhardt, E.; Greve, M.H.; Grüneberg, E.; Harhoff, R.; et al. On soil districts. *Geoderma* **2024**, *452*, 117065. [CrossRef]
63. Kasraei, B.; Heung, B.; Saurette, D.D.; Schmidt, M.G.; Bulmer, C.E.; Bethel, W. Quantile regression as a generic approach for estimating uncertainty of digital soil maps produced from machine-learning. *Environ. Model. Softw.* **2021**, *144*, 105139. [CrossRef]
64. Vaysse, K.; Heuvelink, G.B.M.; Lagacherie, P.; Goss, M. Spatial aggregation of soil property predictions in support of local land management. *Soil Use Manag.* **2017**, *33*, 299–310. [CrossRef]
65. Courteille, L.; Tardieu, L.; Boukhelifa, N.; Lutton, E.; Lagacherie, P. What is the best way to communicate the uncertainty of a digital soil mapping product? Some lessons from an end-users survey. *Geoderma* **2025**, *459*, 117302. [CrossRef]
66. Stumpf, F.; Schmidt, K.; Goebes, P.; Behrens, T.; Schönbrodt-Stitt, S.; Wadoux, A.; Xiang, W.; Scholten, T. Uncertainty-guided sampling to improve digital soil maps. *CATENA* **2017**, *153*, 30–38. [CrossRef]
67. Hengl, T.; Leenaars, J.G.; Shepherd, K.D.; Walsh, M.G.; Heuvelink, G.B.; Mamo, T.; Tilahun, H.; Berkhout, E.; Cooper, M.; Fegraus, E.; et al. Soil nutrient maps of Sub-Saharan Africa: Assessment of soil nutrient content at 250 m spatial resolution using machine learning. *Nutr. Cycl. Agroecosyst.* **2017**, *109*, 77–102. [CrossRef] [PubMed]

land

MDPI

Article

Assessment of Soil Organic Matter and Its Microbial Role in Selected Locations in the South Bohemia Region (Czech Republic)

David Kabelka, Petr Konvalina, Marek Kopecký *, Eva Klenotová and Jaroslav Šíma

Department of Agroecosystems, Faculty of Agriculture and Technology, University of South Bohemia in České Budějovice, 370 05 České Budějovice, Czech Republic; dkabelka@fzt.jcu.cz (D.K.); konvalina@fzt.jcu.cz (P.K.); klenoe00@fzt.jcu.cz (E.K.); simaja00@fzt.jcu.cz (J.Š.)
* Correspondence: mkopecky@fzt.jcu.cz

Abstract: Organic matter has a very important function in soil, without which, soil formation processes cannot take place properly. It can be divided and classified based on several aspects; the most general division is between the living and non-living parts of organic matter. The results presented in this paper specifically refer to the living microbial part of organic matter. This research was carried out in the years 2021–2024 in the South Bohemia region located in the Czech Republic. Two types of land use (four permanent grassland areas, two forest areas) were evaluated. Based on laboratory soil analyses, some significant dependencies were found. For example, in grasslands with statistically identical pH, there was a dependence (p-value 0.05) between soil organic carbon (SOC), carbon of microbial biomass (MBC) and microbial basal respiration (MBR). Additionally, coniferous forest experimental locations had a lower pH, which, in turn, slowed the activity of microorganisms and promoted the accumulation of SOC in the soil. The results from this experiment support the current knowledge of organic matter and are important for a better understanding of the soil organic matter cycle.

Keywords: soil science; microorganisms; organic carbon; microbial biomass; basal respiration

1. Introduction

Organic matter, only being a small part of soil, plays a key role because the decomposition, mineralisation and subsequent humification of organic material are among the most important soil processes affecting soil fertility and quality [1,2]. It is an important material derived from plant and animal sources [3] and has an effect on the following aspects of soil: (1) improving structure: it helps aggregate soil particles, which improves the stability of soil aggregates and increases soil porosity [4]; (2) nutrient source: organic matter contains nutrients that it releases when it decomposes [5]; (3) increasing soil retention: it gives soil a better ability to retain water [6]; (4) reduction of erosion and surface runoff [7,8]; and (5) providing a source of energy for microorganisms: it is the main source of feed and nutrients for soil microorganisms [9].

The total amount of organic matter in soil can be influenced in the long term by management practices [10]. The use of organic fertilisers, intercrops, deep-rooted crops and ploughing puts crop residues back into the soil and contributes to enriching soil with organic matter [7,10,11]. Conversely, a lack of organic matter input into soil and the loss of organic matter are behind some serious degradation processes [12].

The basic general division of organic matter is into its non-living and living parts. The non-living part of soil organic matter is divided according to the degree of decomposition [13] and the presence of non-humified soil organic matter comprising undecomposed organic matter. This includes the remains of plants (leaves, roots) and animals whose original structure is retained [9,14]. Another part consists of partially decomposed organic matter that has already been altered by soil organisms [15]. The next part is fully decomposed organic matter, which, in later stages, is referred to as humus [16]. This decomposed organic matter no longer shows signs of its original structure [17].

The living part of soil organic matter can be divided as follows: microflora [18], microfauna, mesofauna, macrofauna and megafauna [19]. This classification includes a range from smallest to largest, i.e., from micrometres to centimetres and, in some cases, up to metres [20]. Microfauna and microflora inhabit water-filled pores and water films around soil particles [21]. Mesofauna occur in existing air-filled pore spaces and are largely confined by these spaces. Macrofauna have the ability to create their own spaces through their activities and, like megafauna, can significantly influence the coarse structure of soils [22,23]. The biodiversity in soils is impressively large [24,25] and, therefore, species with similar biology and morphology are often grouped together for classification purposes [26]. The most abundant species are among the microbial component of soil [19]. When assessing the living microbial part of organic matter, the basic parameters evaluated include microbial basal respiration (MBR) and microbial biomass carbon (MBC). These parameters have been monitored by a number of authors [27–29] and are, therefore, also evaluated in this paper.

Of course, many variables enter into the whole process of microorganism activity [30,31]. For example, temperature [32,33], soil moisture [34,35], pH [36], nutrient availability [37,38], redox potential [39] and carbon/nitrogen ratio [40] are known to be the main factors that contribute to microbial activity.

Differences can also be expected for different types of land use [41,42]. Microbial activity is particularly important for land use types where soil is the basis for management, i.e., mainly agricultural land and forests, possibly also in other ecological systems [43]. The concept of agricultural land is general and includes different types of land cover. In soil environments where disturbance is frequent (typically conventional arable land management), it is more difficult to draw general conclusions, as a number of different agricultural operations can be used during cultivation that have a significant impact on soil parameters [44]. For this reason, grasslands and forests are the subject of the current assessment as they have a relatively stable soil environment and there is an assumption of less inter-annual variation.

The aim of this study was to verify the following hypotheses: (1) soil microbial activity will vary in permanent grassland and forest locations; (2) microbial activity assessed by MBR and MBC is dependent on the amount of soil organic carbon (SOC); and (3) the amount of SOC in soil is influenced by the pH level.

2. Materials and Methods

Six soil samples from South Bohemia (49°05′00.0″ N, 14°40′00.0″ E) in the Czech Republic were tested every year in 2021–2024 (Figure 1). Four soil samples were from agricultural land (grassland) and two were from coniferous forest soil. Soil texture at each location listed in Table 1 was determined using the hydrometric method. This method is based on the principle of measuring the sedimentation rate of soil particles in a water environment. A hydrometer (an instrument for measuring the density of the suspension) was used to record the concentration of particles in the water at various time intervals. This makes it possible to measure the proportion of each fraction and thus determine the soil texture. The subsequent classification was performed using the texture triangle [45], which is also currently used in the Czech Republic.

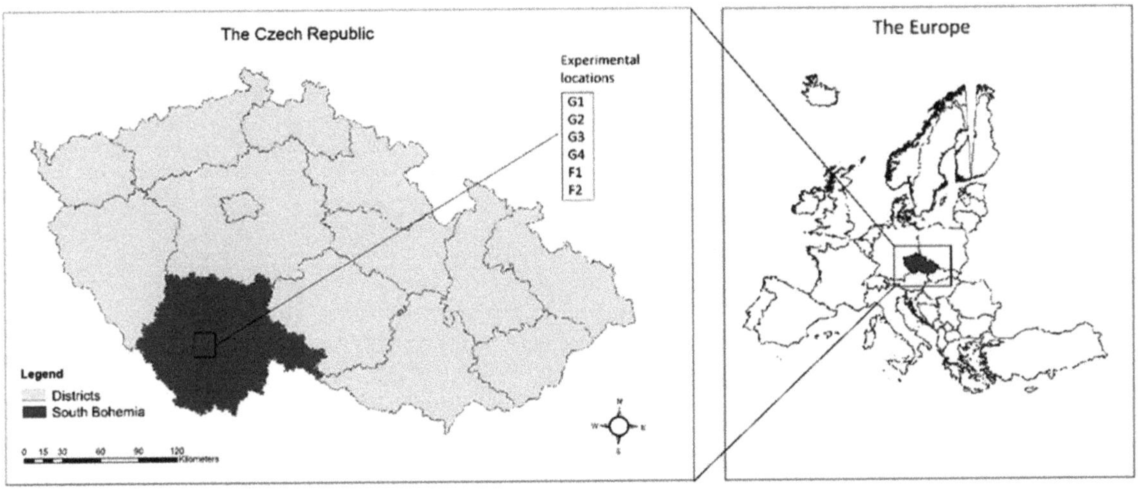

Figure 1. Map of the Czech Republic and the district of South Bohemia.

Table 1. Experimental locations.

Locations	Type of Location	Soil Texture (USDA *)
G1	Grassland	Sandy Loam
G2	Grassland	Sandy Loam
G3	Grassland	Sandy Loam
G4	Grassland	Loam
F2	Forest (coniferous)	Sandy Loam
F1	Forest (coniferous)	Sandy Loam

* USDA—U.S. Department of Agriculture: texture triangle.

One composited soil sample was taken from each locality with a spade and placed in paper bags. The composited sample consisted of a total of 5 samples from a 10 × 10 m area. The depth of sampling was 0–15 cm and the amount of soil sampled from each location was approximately 3 kg. It should be mentioned that forest locations usually have an organic horizon, which has a strong influence on the amount of SOC in the soil. When taking soil samples, the emphasis was always on doing everything the same way every year. First, the organic horizon was carefully removed and then the topsoil was taken. Nevertheless, it was difficult to ensure the uniformity of the sampling. This problem can be clearly seen in Table 2, where the standard deviation is higher for forest locations than for grassland. In the case of grassland, the top layer containing live plants was removed. Subsequently, soil sampling was carried out using a spade.

After sampling, material from the composite soil samples was taken to the laboratory and used for subsequent analyses. Samples were air-dried so that they could be sieved and then particles larger than 2 mm were removed. The soil intended for pH and SOC analysis was dried at 60 °C to a constant weight before analysis. Part of the remaining soil was homogenised and sieved through a 0.25 mm sieve. Soil sieving was conducted because some analyses were performed on soil textures less than <2 mm (pH, MBR, MBC), while SOC was analysed on a fraction < 0.25 mm.

Table 2. Results from the experimental plots (2021–2024).

Locations	pH		SOC		MBR		MBC		MMQ—qCO_3	
	(KCl)	SD	(%)	SD	($\mu g\ CO_2 \cdot g^{-1} \cdot h^{-1}$)	SD	($\mu g \cdot g^{-1}$)	SD	($\times 10^{-4}$ MBR:MBC)	SD
G1	4.90	0.19	3.93	0.18	0.3034	0.0203	756	65	4.1	0.6
G2	4.77	0.28	2.52	0.25	0.1621	0.0112	354	38	4.6	0.6
G3	4.66	0.12	2.36	0.23	0.1589	0.0160	301	23	5.3	0.5
G4	4.70	0.09	3.44	0.35	0.2271	0.0243	577	37	4.0	0.7
F1	3.22	0.05	7.37	1.10	0.1021	0.0284	141	18	7.4	2.2
F2	3.32	0.09	7.28	1.74	0.1333	0.0143	181	42	7.9	2.7

Standard deviation (SD), soil organic carbon (SOC), carbon of microbial biomass (MBC), microbial basal respiration (MBR), microbial metabolic quotient (MMQ).

Individual analyses of soil samples were carried out using the following procedures:

- Soil pH (ISO 10390:2005 [46])— was determined using a 1 M KCl solution. In 250 mL plastic bottles, 10 g of soil was weighed and 50 mL of KCl was added. Subsequently, the samples were shaken for 60 min on a shaker. The pH was then measured using the SI Analytics Lab 875P (Xylem Analytics, Weilhem, Germany). The measurements were carried out in two repetitions each time.
- SOC—the basis of the analysis was the method of Tyurin [47], where the process of determination was as follows: 0.1 g of soil (<0.25 mm) was weighed in triplicate for each location. Subsequently, 0.3 M $K_2Cr_2O_7$ and concentrated H_2SO_4 were mixed to form a chromosulfur mixture. The titration cups with the weighed soil were added to 12.5 mL of the chromo sulphur mixture and then the samples were burned in an oven at 135° for 30 min. Then, titration was performed using 0.2 M Mohr's salt ($(NH_4)_2Fe(SO_4)_2 \cdot 6H_2O + H_2SO_4$) on the Mettler Toledo DL55 titrator (Mettler Toledo, Schwerzenbach, Switzerland).
- MBR—the analysis was based on the ISO standard 16072:2002 [48], but some modifications were made in the determination (soil moistening). The procedure was as follows: 150 g of soil moistened to 50% retention capacity was pre-incubated for 3 days and then 25 g of soil was weighed into breathable bags (in three replications). The breathable bags were hung in a 750 mL sealed glass bottle with 20 mL of 0.05 M NaOH at the bottom for 3 days. Before titration, 2 mL of 0.5 M $BaCl_2$ and phenolphthalein were added. Subsequently, titration was performed on the Mettler Toledo DL55 titrator (Mettler Toledo, Schwerzenbach, Switzerland) using 0.1 M HCl.
- MBC—the analysis was performed according to the ISO standard 14240-2:1997 [49] with some modifications (the weighing and moistening of the soil): 250 g of soil moistened to 50% retention capacity was pre-incubated for 3 days and then 25 g of soil was weighed (in six replications). Three samples were fumigated in a desiccator using boiling chloroform and left in the dark for 24 h. To the remaining samples, 200 mL of 0.5 M K_2SO_4 was added, and the samples were shaken for 30 min. The same process was carried out for the fumigated samples after 24 h. MBC was determined by dichromate oxidation. All samples were filtered and 8 mL was taken. Subsequently, 2 mL 0.07 M $K_2Cr_2O_7$ and 15 mL acid mixture (H_2SO_4, H_3PO_4) were added. After refluxing for 30 min, titration was carried out using 0.04 M Mohr's salt ($(NH_4)_2Fe(SO_4)_2 \cdot 6H_2O + H_2SO_4$) on the Mettler Toledo DL55 titrator (Mettler Toledo, Schwerzenbach, Switzerland).
- Microbial metabolic quotient (MMQ)—this was the ratio between the MBR and the MBC, where the MBR was divided by the MBC.

A basic statistical evaluation was performed on the results to obtain significant correlations using one-way ANOVA. The Pearson correlation coefficient was also determined for the processed data. This correlation coefficient was evaluated if there were at least two distinct locations in the statistical evaluation.

3. Results

3.1. Soil Reaction (pH) in Experimental Locations

The basic analysis performed was the determination of pH. The results of the measurements are shown in Figure 2 and Table 2. Statistically significantly higher values were measured at the grassland locations compared to the forest locations (Table 3). The highest pH across the grasslands was measured at G1 (4.90). The lowest value was at G3 (4.64). Locations G2 and G2 fell within this range in their values. Statistical evaluation of all grasslands showed no significant difference. They were therefore identical in terms of pH. Statistical agreement was also valid for the forest locations, where the mean pH for F1 was 3.22 and that for F2 was 3.32 (Table 3). The values and standard deviations (Table 2) across research years could be considered relatively stable in the locations. The highest standard deviation was measured at G2 (0.28), but this was not a significant fluctuation. Lower values were measured at the other locations.

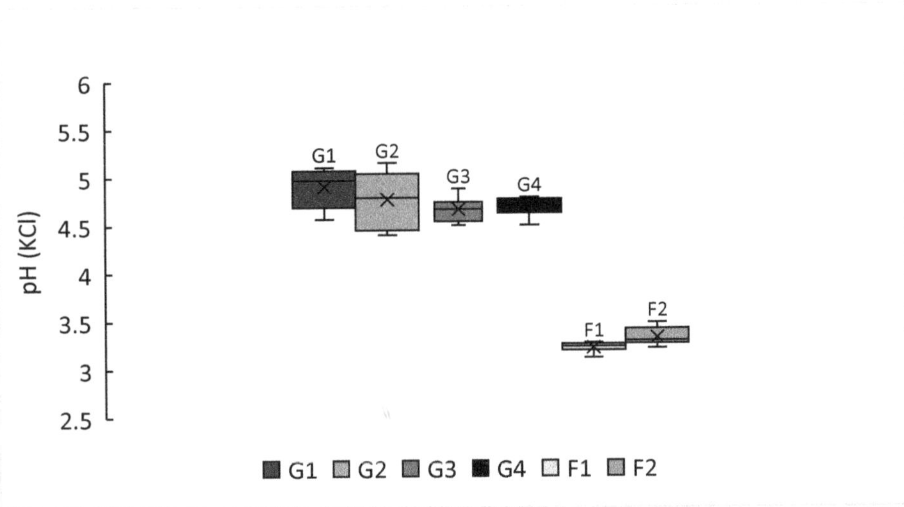

Figure 2. Soil reaction in experimental locations (box plots).

Table 3. Statistical evaluation of selected parameters (one-way ANOVA, *p*-value <0.05).

Parameters	Locations	G1	G2	G3	G4	F1	F2
pH	G1	x	6.08×10^{-1}	7.14×10^{-2}	1.64×10^{-1}	1.27×10^{-11}	1.27×10^{-11}
	G2	x	x	8.14×10^{-1}	9.56×10^{-1}	1.27×10^{-11}	1.27×10^{-11}
	G3	x	x	x	9.99×10^{-1}	1.27×10^{-11}	1.27×10^{-11}
	G4	x	x	x	x	1.27×10^{-11}	1.27×10^{-11}
	F1	x	x	x	x	x	8.19×10^{-1}
	F2	x	x	x	x	x	x

Table 3. *Cont.*

Parameters	Locations	G1	G2	G3	G4	F1	F2
SOC	G1	x	3.75×10^{-3}	9.14×10^{-4}	7.58×10^{-1}	2.50×10^{-11}	2.81×10^{-11}
	G2	x	x	9.98×10^{-1}	1.44×10^{-1}	2.29×10^{-11}	2.29×10^{-11}
	G3	x	x	x	5.33×10^{-2}	2.29×10^{-11}	2.29×10^{-11}
	G4	x	x	x	x	2.30×10^{-11}	2.30×10^{-11}
	F1	x	x	x	x	x	1.00×10^{-0}
	F2	x	x	x	x	x	x
MBR	G1	x	3.22×10^{-11}	3.22×10^{-11}	5.85×10^{-11}	3.22×10^{-11}	3.22×10^{-11}
	G2	x	x	9.99×10^{-1}	2.56×10^{-9}	2.71×10^{-8}	1.55×10^{-2}
	G3	x	x	x	5.70×10^{-10}	1.25×10^{-7}	4.37×10^{-2}
	G4	x	x	x	x	3.22×10^{-11}	3.23×10^{-11}
	F1	x	x	x	x	x	7.31×10^{-3}
	F2	x	x	x	x	x	x
MBC	G1	x	2.29×10^{-11}	2.29×10^{-11}	2.30×10^{-11}	2.29×10^{-11}	2.29×10^{-11}
	G2	x	x	3.26×10^{-2}	2.29×10^{-11}	2.29×10^{-11}	2.31×10^{-11}
	G3	x	x	x	2.29×10^{-11}	2.45×10^{-11}	2.16×10^{-8}
	G4	x	x	x	x	2.29×10^{-11}	2.29×10^{-11}
	F1	x	x	x	x	x	1.96×10^{-1}
	F2	x	x	x	x	x	x
MMQ	G1	x	9.45×10^{-1}	3.90×10^{-1}	1.00×10^{-0}	2.83×10^{-5}	9.86×10^{-7}
	G2	x	x	9.05×10^{-1}	9.00×10^{-1}	7.48×10^{-4}	3.29×10^{-5}
	G3	x	x	x	3.11×10^{-1}	1.96×10^{-2}	1.34×10^{-3}
	G4	x	x	x	x	1.64×10^{-5}	5.55×10^{-7}
	F1	x	x	x	x	x	9.54×10^{-1}
	F2	x	x	x	x	x	x

Locations that were statistically equal are highlighted. Soil organic carbon (SOC), carbon of microbial biomass (MBC), microbial basal respiration (MBR), microbial metabolic quotient (MMQ). The light grey colour in the table indicates statistically identical locations.

The results from this analysis were very important for the subsequent evaluation because it was possible to divide all assessed locations into two groups with statistically identical pH. Forest locations represented one group and grasslands represented the second group. This made it possible to evaluate the effect of different pH on the other assessed parameters (SOC, MBC, MBR, MMQ). As shown in the following Sections 3.2 and 3.3, this fact proved to be very important for some parameters.

3.2. Relationship Between pH and SOC

SOC, like pH, is an essential parameter in the assessment of soil parameters. The highest amount of SOC was found at the forest locations, where the average value was 7.37% at F1 and 7.28% at F2. In this context, it should be added that the method of soil sampling mattered because as the amount of organic horizon (for forest locations) in the soil sample increased, so did the amount of SOC. It was problematic to ensure the same amount of organic horizon in the soil samples. This can be clearly seen in the standard deviations (Table 2), which were higher for forest locations than for grasslands. These facts need to be taken into account when comparing the results with other studies. In the statistical evaluation, the values from forest locations (F1 and F2) were identical (Table 3). When comparing them with grasslands, they were significantly different (p-value < 0.05). The individual grassland locations, however, did not have statistically the same SOC content, confirming that the whole issue was much more complex. The highest SOC value was measured at G1 (3.93%). This location was statistically identical to G4 (3.44%) but significantly different from G2 (2.52%) and G3 (2.36%). Locations G2, G3 and G4 could be considered identical (Table 3).

Correlations were determined at all locations but also separately for grasslands and forests (Figure 3). In the case of the forest locations (F1, F2), which were statistically identical, correlations were not assessed further. When all locations were assessed, the strongest correlation was obtained between pH and SOC. Specifically, a strong inverse correlation (−0.913) was found, meaning that the lower the pH, the higher the SOC content of the soil. For this statement, the depth of sampling needs to be considered (see Discussion chapter).

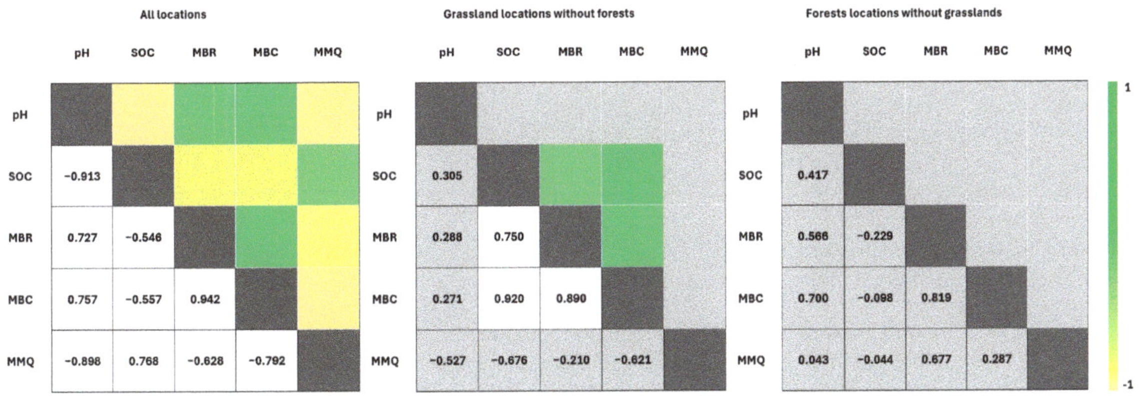

Figure 3. Correlation heat map for the evaluated parameters. Fields that are light grey were not evaluated due to statistical equality at all locations (p-value 0.05). Soil organic carbon (SOC), carbon of microbial biomass (MBC), microbial basal respiration (MBR), microbial metabolic quotient (MMQ).

3.3. Microbial Activity (MBR, MBC) as a Function of SOC

Soil microbial activity was measured using two basic indicators (MBR, MBC). The lowest MBR values were determined at both forest locations, which were also identical according to the statistical evaluation. The MBR values were higher for the grasslands. For MBC, the situation was similar to that of MBR. Lower MBC values were measured for forest locations compared to grasslands.

Based on the results (MBR, MBC), it was confirmed that microbial activity in soil differed between grasslands and forests (hypothesis 1), but statistical differences were also found for individual grassland locations (see, Table 3). In the case of MBR, only G2 and G3 were statistically identical. For MBC, no statistical concordance was confirmed and, therefore, each grassland location could be considered different.

The dependencies between MBR, MBC and SOC were assessed using the correlation coefficient and trend lines. When MBR and MBC were evaluated, there was a strong relationship (R^2 = 0.9461) between these two parameters. The dependence was confirmed when evaluating all locations, but only at grasslands (Figure 3). This points to the fact that different soil pH did not significantly affect this dependence. Thus, at the locations evaluated, it was valid that MBC increased with increasing MBR.

The above statement did not apply to the relationship between SOC and MBC. In assessing the correlation across all locations (Figure 3), an inverse correlation (−0.557) was found, but, when looking for a definite trend between SOC and MBC (Figure 4), it is clear that no trend (R^2 = 0.2914) was observed. The situation was completely different if the forest locations were omitted and only grassland locations with statistically equally high pH were evaluated. Then, a strong direct correlation (0.920) between SOC and MBC was evident and a significant trend (R^2 = 0.907) could be established (Figure 4).

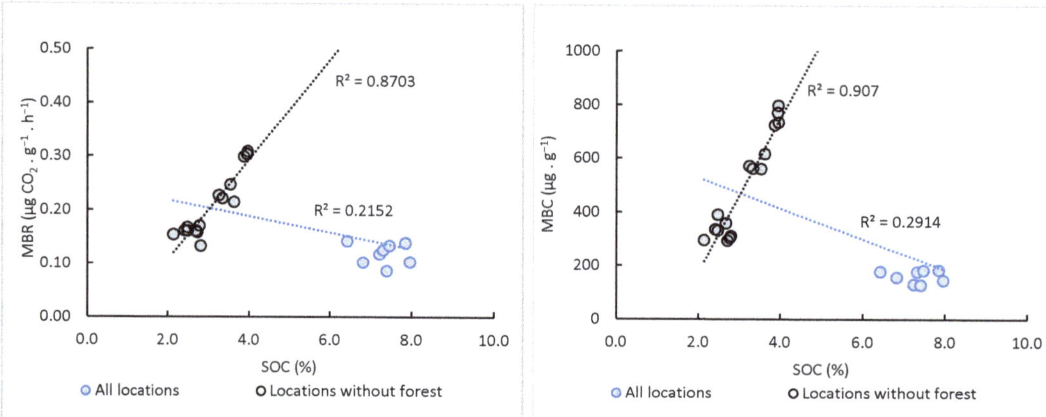

Figure 4. Relationship between SOC and microbial parameters. Soil organic carbon (SOC), carbon of microbial biomass (MBC), microbial basal respiration (MBR).

A similar situation, slightly less marked, existed for the relationship between SOC and MBR. When forest locations were included, an inverse correlation (−0.546) was found and the resulting trend (R^2 = 0.2152) was significantly negatively affected. When assessing only grassland locations, a strong direct correlation (0.750) was found and a significant trend (R^2 = 0.8703) was also established.

From the results, the conclusion was that hypothesis 2 (microbial activity assessed by MBR and MBC is SOC-dependent) was only valid for locations with the same pH. When different locations (grasslands and forests) were evaluated, the hypothesis was not confirmed.

In this paper, the efficiency of the microbial environment expressed in terms of MMQ was also calculated. The highest values were determined for forest locations, which were statistically significantly different from grasslands. It was also confirmed that all grasslands were statistically identical, as were forest locations (Table 3). Figure 3 shows that the highest correlation value (−0.898) was achieved between MMQ and pH when all locations were evaluated. Thus, the results indicate that the MMQ value increased with decreasing pH. An assessment of only grasslands or forests was not carried out for statistical concordance.

4. Discussion

Soil pH is one of the basic indicators that is monitored in a comprehensive soil assessment [50–52]. Its value influences a number of soil properties: physical, chemical and biological [53,54]. The level of pH is determined by the composition of the soil but is also influenced by the plants themselves, as the organic matter from each species can have significantly different compositions [55–57]. This can be clearly seen in Figure 2, where the pH at the two forest locations differed significantly from the pH at the grassland locations. The results confirm the generally accepted conclusion that coniferous forest locations need and have a lower pH than agricultural land in the upper soil layer (0–15 cm). This conclusion is in agreement with [58–60], but is only valid for topsoil, because at deeper soil levels, the differences in pH for individual land uses tend to decrease, as previously reported [61,62].

When assessing the pH of agricultural land, it must be taken into account that changes may occur as a result of agronomic operations. These may affect the pH. Some agrotechnical operations are directly carried out to adjust the pH to the optimum value for the crop [63]. The classic case is the liming of soils, which increases the pH [64]. However, the pH can also be lowered by, for example, applying certain types of mineral or organic fertilisers [65]. The way the soil is cultivated also has an influence. All these agrotechnical operations that are carried out affect not only pH but also other soil parameters including SOC, MBR and MBC [66,67]. The greatest changes in soil parameters can be expected on arable land, where the soil is generally subjected to more frequent cultivation than grasslands and forests. In this context, it is important to note that no agrotechnical operations (liming, organic fertilisation) were carried out on the experimental locations during the research that would have significantly altered the pH. This is apparent from Table 2 (standard deviation) and Figure 2 (box plots).

Similar to pH, the amount of SOC is influenced by a number of parameters. Some of the basic ones include the type of farming [68], soil texture [69] or species composition [70]. In evaluating the relationship of SOC with other parameters, certain dependencies were found. A very strong correlation was determined between pH and SOC, but it must be added that the results were valid for the upper layer in which the sampling took place (0–15 cm). For deeper soil levels, this statement may not be valid. This was also confirmed by some studies in which sampling was carried out in lower layers [43]. The same relationship (ph–SOC) in upper layer was found by [71] in a tropical forest soil environment, [72] in grasslands and forests, and [59] in steppe and cropland locations. Based on the results (Table 2), it can be confirmed with some simplification that, in this case, higher SOC was measured in forest locations due to a more acidic soil environment. This conclusion is also clearly visible in Figure 3, which shows the correlation coefficient between the evaluated parameters. For higher pH (above 8), this may no longer be applicable, as [73] reported a hump-back model between soil organic carbon and pH in grasslands. On arable land, there may even be reverse dependence due to soil cultivation, as noted by [74], i.e., a decrease in organic carbon as pH decreases. This means that hypothesis 3 involving the dependence of SOC on pH was only partially confirmed because in significantly different soil environments than those in this study, the relationship may not always be valid. This observed pH–SOC relationship can be significantly disturbed by agrotechnical operations. It is therefore essential to know the management practices that have been applied to the land.

There were also some findings for SOC and microbial activity (MBR, MBC). In the case of MBC, the lowest values were found at forest locations (F1, F2). The values in this study are in agreement with many authors [75–77] who determined the MBC for forest soils. However, values at forest locations can vary considerably. The main reason is due to the complicated soil sampling. It is necessary to separate soil from forest floor during sampling because [78,79] reported that MBC can vary significantly between forest floor and forest soil. For grassland, some authors [80,81] gave similar values to those in this paper, but, in some studies [82,83], there were higher values.

The situation was similar for MBR, but not the same. Almost all locations were statistically different, but several individual MBR values from forest locations were comparable to those of some grassland locations (Figure 4). This finding is in agreement with [84], who also compared grasslands and forests. When evaluating the relationship between MBR and MBC, it can be concluded that a higher MBR in soil indicates a higher MBC (Figure 5), as confirmed by other authors [85,86].

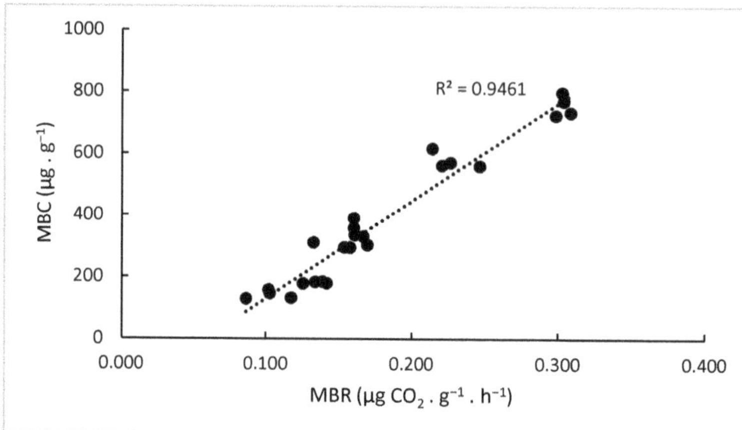

Figure 5. Dependence between basic parameters related to microbial activity (MBR, MBC). Carbon of microbial biomass (MBC), microbial basal respiration (MBR).

From the results of the SOC–MBC and SOC–MBR relationships, it can be further assumed that the more SOC in the soil, the more microorganisms. This is in line with other authors [87–89], but only applicable when the pH is the same. When forest locations are included, this statement cannot be confirmed. The same conclusion may apply to the MBR. A strong correlation between SOC and MBR at grassland locations was reported by [90,91] and also by [92], who assessed only forest locations.

MMQ is considered an important indicator of soil health [93] and the highest correlation was found for pH. Similar findings were reported in a study by [94], who analysed 24 studies and concluded that MMQ often declines with increasing pH in a stable environment. The results indicate that organic matter is more efficiently used by microorganisms in grassland locations, where MMQ is lower compared to forest locations. Higher ratios may reflect that the microbial community is exposed to stressful conditions such as inappropriate pH (Table 2). This means that microorganisms must expend more energy to maintain basic life functions and use less for growth [95]. Soil pH is not the only parameter that affects MMQ, but there are other soil parameters such as clay content, amount of MBC [94] or soil moisture [96].

The conclusion regarding the relationship between SOC and MBC, SOC and MBR or pH and MMQ highlights the issue of overgeneralisation. In some cases, it is not possible to make general conclusions that apply across locations. This is also the case for microbial activity, which is influenced by a number of factors. These factors may vary due to different land cover types and it is therefore always important to take into account basic parameters (e.g., pH) that may differ between land use types.

5. Conclusions

This paper summarised the knowledge concerning organic matter and its microbial role. The results show that some significant trends and dependencies can be found for the parameters evaluated (SOC, MBR, MBC). When assessing microbial activity, both parameters MBR and MBC appear to be suitable indicators in relation to SOC. However, it is also important to assess other parameters such as the pH in the soil environment. Different pH levels can significantly affect the resulting trends, which was evident when forest locations were included. The findings support the current knowledge of soil organic matter and are useful for other authors in evaluating and interpreting results related to microbial activity. Information on microbial activity is also important from the point of view of soil quality. This can be influenced (both negatively and positively) in the long term by the way it is managed. Therefore, management practices that contribute to the enrichment of the diversity of microflora and microfauna in the ecosystem should be promoted in order to increase landscape stability and improve soil health, crop health and agricultural production. From this perspective, it would be interesting to better clarify the dynamics of pH, SOC, MBR and MBC in other ecological systems and land use types. Microbial activity through the decomposition of organic matter influences nutrient cycles, and a deeper understanding of this could provide more support for sustainable management practices and development.

Author Contributions: Conceptualisation, D.K.; methodology, D.K. and M.K.; validation, P.K. and M.K.; formal analysis, D.K. and M.K.; investigation, D.K. and P.K.; resources, D.K, J.Š. and E.K.; data curation D.K. and E.K.; writing—original draft preparation, D.K. and E.K.; writing—review and editing, D.K., M.K. and P.K.; visualisation, J.Š.; supervision, D.K. All authors have read and agreed to the published version of the manuscript.

Funding: This research was financially supported by the Grant Agency of the University of South Bohemia in České Budějovice, Czech Republic (no. GA JU 122/2025/Z).

Data Availability Statement: The original contributions presented in this study are included in the article. Further inquiries can be directed to the corresponding author.

Conflicts of Interest: The authors declare no conflicts of interest.

References

1. Bünemann, E.K.; Bongiorno, G.; Bai, Z.; Creamer, R.E.; De Deyn, G.; De Goede, R.; Fleskens, L.; Geissen, V.; Kuyper, T.W.; Mäder, P.; et al. Soil quality–A critical review. *Soil Biol. Biochem.* **2018**, *120*, 105–125. [CrossRef]
2. Karlen, D.L.; Cambardella, C.A. Conservation strategies for improving soil quality and organic matter storage. In *Structure and Organic Matter Storage in Agricultural Soils*; Carter, M.R., Stewart, B.A., Eds.; CRC Press: Boca Raton, FL, USA, 2020; pp. 395–420.
3. Kozak, C.; Leithold, J.; do Prado, L.L.; Knapik, H.G.; de Rodrigues Azevedo, J.C.; Braga, S.M.; Fernandes, C.V.S. Adaptive monitoring approach to assess dissolved organic matter dynamics during rainfall events. *Environ. Monit. Assess.* **2021**, *193*, 423. [CrossRef] [PubMed]
4. Mustafa, A.; Minggang, X.; Shah, S.A.A.; Abrar, M.M.; Nan, S.; Baoren, W.; Zejiang, C.; Saeed, Q.; Naveed, M.; Mehmood, K.; et al. Soil aggregation and soil aggregate stability regulate organic carbon and nitrogen storage in a red soil of southern China. *J. Environ. Manag.* **2020**, *270*, 110894. [CrossRef] [PubMed]
5. Gerke, J. The central role of soil organic matter in soil fertility and carbon storage. *Soil Syst.* **2022**, *6*, 33. [CrossRef]
6. Lal, R. Soil organic matter and water retention. *Agron. J.* **2020**, *112*, 3265–3277. [CrossRef]
7. Chowaniak, M.; Głąb, T.; Klima, K.; Niemiec, M.; Zaleski, T.; Zuzek, D. Effect of tillage and crop management on runoff, soil erosion and organic carbon loss. *Soil Use Manag.* **2020**, *36*, 581–593. [CrossRef]
8. Du, X.; Jian, J.; Du, C.; Stewart, R.D. Conservation management decreases surface runoff and soil erosion. *Int. Soil Water Conserv. Res.* **2022**, *10*, 188–196. [CrossRef]
9. Paul, E.A. The nature and dynamics of soil organic matter: Plant inputs, microbial transformations, and organic matter stabilization. *Soil Biol. Biochem.* **2016**, *98*, 109–126. [CrossRef]
10. Bai, Z.; Caspari, T.; Gonzalez, M.R.; Batjes, N.H.; Mäder, P.; Bünemann, E.K.; de Goede, R.; Brussaard, L.; Xu, M.; Ferreira, C.S.S.; et al. Effects of agricultural management practices on soil quality: A review of long-term experiments for Europe and China. *Agric. Ecosyst. Environ.* **2018**, *265*, 1–7.
11. Wang, H.; Xu, J.; Liu, X.; Zhang, D.; Li, L.; Li, W.; Sheng, L. Effects of long-term application of organic fertilizer on improving organic matter content and retarding acidity in red soil from China. *Soil Tillage Res.* **2019**, *195*, 104382. [CrossRef]
12. Obalum, S.E.; Chibuike, G.U.; Peth, S.; Ouyang, Y. Soil organic matter as sole indicator of soil degradation. *Environ. Monit. Assess.* **2017**, *189*, 176. [CrossRef] [PubMed]
13. Kolář, L.; Kužel, S.; Horáček, J.; Čechová, V.; Borová-Batt, J.; Peterka, J. Labile fractions of soil organic matter, their quantity and 540 quality. *Plant Soil Environ.* **2009**, *55*, 245–251. [CrossRef]
14. Ochs, M. Influence of humified and non-humified natural organic compounds on mineral dissolution. *Chem. Geol.* **1996**, *132*, 119–124. [CrossRef]

15. Bani, A.; Pioli, S.; Ventura, M.; Panzacchi, P.; Borruso, L.; Tognetti, R.; Tonon, G.; Brusetti, L. The role of microbial community in the decomposition of leaf litter and deadwood. *Appl. Soil Ecol.* **2019**, *126*, 75–84. [CrossRef]
16. Osman, K.T.; Osman, K.T. Soil organic matter. In *Soils: Principles, Properties and Management*; Springer: Dordrecht, The Netherlands, 2013; pp. 89–96.
17. Zavarzina, A.G.; Danchenko, N.N.; Demin, V.V.; Artemyeva, Z.S.; Kogut, B.M. Humic substances: Hypotheses and reality (a review). *Eurasian Soil Sci.* **2021**, *54*, 1826–1854. [CrossRef]
18. Prashar, P.; Shah, S. Impact of fertilizers and pesticides on soil microflora in agriculture. *Sustain. Agric. Rev.* **2016**, *19*, 331–361.
19. Coleman, D.C.; Geisen, S.; Wall, D.H. Soil fauna: Occurrence, biodiversity, and roles in ecosystem function. In *Soil Microbiology, Ecology and Biochemistry*, 5th ed.; Paul, E.A., Frey, S.D., Eds.; Elsevier: Amsterdam, The Netherlands, 2024; pp. 131–159.
20. Moreira, F.M.; Huising, E.J.; Bignell, D.E. *A Handbook of Tropical Soil Biology: Sampling and Characterization of Below-Ground Biodiversity*; Taylor & Francis: London, UK, 2012; pp. 1–256.
21. van Vliet, P.C.J.; Hendrix, P.F. Role of fauna in soil physical processes. In *Soil Biological Fertility: A Key to Sustainable Land Use in Agriculture*; Abbott, L.K., Murphy, D.V., Eds.; Kluwer Academic Publishers: Dordrecht, The Netherlands, 2003; pp. 61–80.
22. Bottinelli, N.; Jouquet, P.; Capowiez, Y.; Podwojewski, P.; Grimaldi, M.; Peng, X. Why is the influence of soil macrofauna on soil structure only considered by soil ecologists? *Soil Tillage Res.* **2015**, *146*, 118–124. [CrossRef]
23. Jeffery, S.; Gardi, C.; Jones, A.; Montanarella, L.; Marmo, L.; Miko, L.; Ritz, K.; Peres, G.; Römbke, J.; van der Putten, H.W. *European Atlas of Soil Biodiversity*; European Commission, Publications Office of the European Union: Luxembourg, 2010; pp. 1–122.
24. Fierer, N.; Lennon, J.T. The generation and maintenance of diversity in microbial communities. *Am. J. Bot.* **2011**, *98*, 439–448. [CrossRef]
25. Potapov, A.M.; Beaulieu, F.; Birkhofer, K.; Bluhm, S.L.; Degtyarev, M.I.; Devetter, M.; Goncharov, A.A.; Gongalsky, K.B.; Klarner, B.; Korobushkin, D.I.; et al. Feeding habits and multifunctional classification of soil-associated consumers from protists to vertebrates. *Biol. Rev.* **2022**, *97*, 1057–1117. [CrossRef]
26. Geisen, S.; Mitchell, E.A.; Adl, S.; Bonkowski, M.; Dunthorn, M.; Ekelund, F.; Fernández, L.D.; Jousset, A.; Krashevska, V.; Singer, D.; et al. Soil protists: A fertile frontier in soil biology research. *FEMS Microbiol. Rev.* **2018**, *42*, 293–323. [CrossRef]
27. Bouguerra, S.; Gavina, A.; Natal-da-Luz, T.; Sousa, J.P.; Ksibi, M.; Pereira, R. The use of soil enzymes activity, microbial biomass, and basal respiration to assess the effects of cobalt oxide nanomaterial in soil microbiota. *Appl. Soil Ecol.* **2022**, *169*, 104246. [CrossRef]
28. Yang, Y.; Cheng, S.; Fang, H.; Guo, Y.; Li, Y.; Zhou, Y. Interactions between soil organic matter chemical structure and microbial communities determine the spatial variation of soil basal respiration in boreal forests. *Appl. Soil Ecol.* **2023**, *183*, 104743. [CrossRef]
29. Das, S.; Deb, S.; Sahoo, S.S.; Sahoo, U.K. Soil microbial biomass carbon stock and its relation with climatic and other environmental factors in forest ecosystems: A review. *Acta Ecol. Sin.* **2023**, *43*, 933–945. [CrossRef]
30. Cao, H.; Chen, R.; Wang, L.; Jiang, L.; Yang, F.; Zheng, S.; Wang, G.; Lin, X. Soil pH, total phosphorus, climate and distance are the major factors influencing microbial activity at a regional spatial scale. *Sci. Rep.* **2016**, *6*, 25815. [CrossRef] [PubMed]
31. Maurya, S.; Abraham, J.S.; Somasundaram, S.; Toteja, R.; Gupta, R.; Makhija, S. Indicators for assessment of soil quality: A mini-review. *Environ. Monit. Assess.* **2020**, *192*, 1–22. [CrossRef]
32. Curiel Yuste, J.; Baldocchi, D.D.; Gershenson, A.; Goldstein, A.; Misson, L.; Wong, S. Microbial soil respiration and its dependency on carbon inputs, soil temperature and moisture. *Glob. Chang. Biol.* **2007**, *13*, 2018–2035. [CrossRef]
33. García-Palacios, P.; Crowther, T.W.; Dacal, M.; Hartley, I.P.; Reinsch, S.; Rinnan, R.; Rousk, J.; van den Hoogen, J.; Ye, J.S.; Bradford, M.A. Evidence for large microbial-mediated losses of soil carbon under anthropogenic warming. *Nat. Rev. Earth Environ.* **2021**, *2*, 507–517. [CrossRef]
34. Borowik, A.; Wyszkowska, J. Soil moisture as a factor affecting the microbiological and biochemical activity of soil. *Plant Soil Environ.* **2016**, *62*, 250–255. [CrossRef]
35. Singh, S.; Mayes, M.A.; Shekoofa, A.; Kivlin, S.N.; Bansal, S.; Jagadamma, S. Soil organic carbon cycling in response to simulated soil moisture variation under field conditions. *Sci. Rep.* **2021**, *11*, 10841. [CrossRef]
36. Naz, M.; Dai, Z.; Hussain, S.; Tariq, M.; Danish, S.; Khan, I.U.; Qi, S.; Du, D. The soil pH and heavy metals revealed their impact on soil microbial community. *J. Environ. Manag.* **2022**, *321*, 115770. [CrossRef]
37. Stark, S.; Männistö, M.K.; Eskelinen, A. Nutrient availability and pH jointly constrain microbial extracellular enzyme activities in nutrient-poor tundra soils. *Plant Soil* **2014**, *383*, 373–385. [CrossRef]
38. Penn, C.J.; Camberato, J.J. A critical review on soil chemical processes that control how soil pH affects phosphorus availability to plants. *Agriculture* **2019**, *9*, 120. [CrossRef]
39. Mattila, T.J. Redox potential as a soil health indicator–how does it compare to microbial activity and soil structure? *Plant Soil* **2024**, *494*, 617–625. [CrossRef]
40. Angst, G.; Mueller, K.E.; Nierop, K.G.; Simpson, M.J. Plant-or microbial-derived? A review on the molecular composition of stabilized soil organic matter. *Soil Biol. Biochem.* **2021**, *156*, 108189. [CrossRef]

41. Moghimian, N.; Hosseini, S.M.; Kooch, Y.; Darki, B.Z. Impacts of changes in land use/cover on soil microbial and enzyme activities. *Catena* **2017**, *157*, 407–414. [CrossRef]
42. Lepcha, N.T.; Devi, N.B. Effect of land use, season, and soil depth on soil microbial biomass carbon of Eastern Himalayas. *Ecol. Process.* **2020**, *9*, 65. [CrossRef]
43. Van Leeuwen, J.P.; Djukic, I.; Bloem, J.; Lehtinen, T.; Hemerik, L.; De Ruiter, P.C.; Lair, G.J. Effects of land use on soil microbial biomass, activity and community structure at different soil depths in the Danube floodplain. *Eur. J. Soil Biol.* **2017**, *79*, 14–20. [CrossRef]
44. Valboa, G.; Lagomarsino, A.; Brandi, G.; Agnelli, A.E.; Simoncini, S.; Papini, R.; Vignozzi, N.; Pellegrini, S. Long-term variations in soil organic matter under different tillage intensities. *Soil Tillage Res.* **2015**, *154*, 126–135. [CrossRef]
45. Soil Science Division Staff. *Soil Survey Manual*; Agriculture Handbook No. 18; United States Department of Agriculture: Washington, DC, USA, 2017; pp. 1–603.
46. *ISO 10390-2005*; Soil Quality—Determination of pH. International Standard: Geneva, Switzerland, 2005.
47. Tyurin, I.V. A new modification of the volumetric method of determining soil organic matter by means of chromic acid. *Pochvovedenie* **1931**, *26*, 36–47.
48. *ISO 16072-2002*; Soil Quality—Laboratory Methods for Determination of Microbial Soil Respiration. International Standard: Geneva, Switzerland, 2002.
49. *ISO 14240-2-1997*; Soil Quality—Determination of Soil Microbial Biomass. Part 2: Fumigation-Extraction Method. International Standard: Geneva, Switzerland, 1997.
50. Tkaczyk, P.; Mocek-Płóciniak, A.; Skowrońska, M.; Bednarek, W.; Kuśmierz, S.; Zawierucha, E. The mineral fertilizer-dependent chemical parameters of soil acidification under field conditions. *Sustainability* **2020**, *12*, 7165. [CrossRef]
51. Shetty, R.; Prakash, N.B. Effect of different biochars on acid soil and growth parameters of rice plants under aluminium toxicity. *Sci. Rep.* **2020**, *10*, 12249. [CrossRef] [PubMed]
52. Hartemink, A.E.; Barrow, N.J. Soil pH-nutrient relationships: The diagram. *Plant Soil* **2023**, *486*, 209–215. [CrossRef]
53. Neina, D. The role of soil pH in plant nutrition and soil remediation. *Appl. Environ. Soil Sci.* **2019**, *1*, 5794869. [CrossRef]
54. Junior, E.C.; Gonçalves, A.C., Jr.; Seidel, E.P.; Ziemer, G.L.; Zimmermann, J.; Oliveira, V.H.D.; Schwantes, D.; Zeni, C.D. Effects of liming on soil physical attributes: A review. *J. Agric. Sci.* **2020**, *12*, 278. [CrossRef]
55. Bending, G.D.; Turner, M.K.; Jones, J.E. Interactions between crop residue and soil organic matter quality and the functional diversity of soil microbial communities. *Soil Biol. Biochem.* **2002**, *34*, 1073–1082. [CrossRef]
56. Tang, C.; Yu, Q. Impact of chemical composition of legume residues and initial soil pH on pH change of a soil after residue incorporation. *Plant Soil* **1999**, *215*, 29–38. [CrossRef]
57. McCauley, A.; Jones, C.; Jacobsen, J. Soil pH and organic matter. *Nutr. Manag. Modul.* **2009**, *8*, 1–12.
58. South, D.B. Optimum pH for growing pine seedlings. *Tree Plant. Notes* **2017**, *60*, 49–62.
59. Yang, L.; Jia, W.; Shi, Y.; Zhang, Z.; Xiong, H.; Zhu, G. Spatiotemporal differentiation of soil organic carbon of grassland and its relationship with soil physicochemical properties on the northern slope of Qilian mountains, China. *Sustainability* **2020**, *12*, 9396. [CrossRef]
60. Ohno, T.; Fernandez, I.J.; Hiradate, S.; Sherman, J.F. Effects of soil acidification and forest type on water soluble soil organic matter properties. *Geoderma* **2007**, *140*, 176–187. [CrossRef]
61. Jobbágy, E.G.; Jackson, R.B. Patterns and mechanisms of soil acidification in the conversion of grasslands to forests. *Biogeochemistry* **2003**, *64*, 205–229. [CrossRef]
62. Shiwakoti, S.; Zheljazkov, V.D.; Gollany, H.T.; Kleber, M.; Xing, B.; Astatkie, T. Macronutrient in soils and wheat from long-term agroexperiments reflects variations in residue and fertilizer inputs. *Sci. Rep.* **2020**, *10*, 3263. [CrossRef] [PubMed]
63. Li, Y.; Cui, S.; Chang, S.X.; Zhang, Q. Liming effects on soil pH and crop yield depend on lime material type, application method and rate, and crop species: A global meta-analysis. *J. Soils Sediments* **2019**, *19*, 1393–1406. [CrossRef]
64. Holland, J.E.; Bennett, A.E.; Newton, A.C.; White, P.J.; McKenzie, B.M.; George, T.S.; Pakeman, R.J.; Bailey, J.S.; Fornara, D.A.; Hayes, R.C. Liming impacts on soils, crops and biodiversity in the UK: A review. *Sci. Total Environ.* **2018**, *610*, 316–332. [CrossRef]
65. Fangueiro, D.; Hjorth, M.; Gioelli, F. Acidification of animal slurry–a review. *J. Environ. Manag.* **2015**, *149*, 46–56. [CrossRef]
66. Busari, M.A.; Salako, F.K. Effect of tillage, poultry manure and NPK fertilizer on soil chemical properties and maize yield on an Alfisol at Abeokuta, south-western Nigeria. *Niger. J. Soil Sci.* **2013**, *23*, 206–218.
67. Gajda, A.M.; Czyż, E.A.; Klimkowicz-Pawlas, A. Effects of different tillage intensities on physicochemical and microbial properties of a Eutric Fluvisol Soil. *Agronomy* **2021**, *11*, 1497. [CrossRef]
68. Šimon, T.; Javůrek, M.; Mikanova, O.; Vach, M. The influence of tillage systems on soil organic matter and soil hydrophobicity. *Soil Tillage Res.* **2009**, *105*, 44–48. [CrossRef]
69. Plante, A.F.; Conant, R.T.; Stewart, C.E.; Paustian, K.; Six, J. Impact of soil texture on the distribution of soil organic matter in physical and chemical fractions. *Soil Sci. Soc. Am. J.* **2006**, *70*, 287–296. [CrossRef]

70. Semenov, V.M.; Kravchenko, I.K.; Ivannikova, L.A.; Kuznetsova, T.V.; Semenova, N.A.; Gispert, M.; Pardini, J. Experimental determination of the active organic matter content in some soils of natural and agricultural ecosystems. *Eurasian Soil Sci.* **2006**, *39*, 251–260. [CrossRef]
71. Lu, X.; Vitousek, P.M.; Mao, Q.; Gilliam, F.S.; Luo, Y.; Turner, B.L.; Zhou, G.; Mo, J. Nitrogen deposition accelerates soil carbon sequestration in tropical forests. *Proc. Natl. Acad. Sci. USA* **2021**, *118*, e2020790118. [CrossRef] [PubMed]
72. Yu, P.; Li, Q.; Jia, H.; Li, G.; Zheng, W.; Shen, X.; Diabate, B.; Zhou, D. Effect of cultivation on dynamics of organic and inorganic carbon stocks in Songnen plain. *Agron. J.* **2014**, *106*, 1574–1582. [CrossRef]
73. Zhang, J.; Wu, X.; Shi, Y.; Jin, C.; Yang, Y.; Wei, X.; Mu, C.; Wang, J. A slight increase in soil pH benefits soil organic carbon and nitrogen storage in a semi-arid grassland. *Ecol. Indic.* **2021**, *130*, 108037. [CrossRef]
74. Pietri, J.A.; Brookes, P.C. Relationships between soil pH and microbial properties in a UK arable soil. *Soil Biol. Biochem.* **2008**, *40*, 1856–1861. [CrossRef]
75. Vance, E.D.; Brookes, P.C.; Jenkinson, D.S. Microbial biomass measurements in forest soils: The use of chloroform fumigation–incubations methods for strongly acid soils. *Soil Biol. Biochem.* **1987**, *19*, 697–702. [CrossRef]
76. Mahia, J.; Perez Ventura, L.; Cabaneiro, A.; DiazRavina, M. Soil microbial biomass under pine forest in the northwestern Spain: Influence of stand age, site index and parent material. *Investig. Agrar. Sist. Recur. For.* **2006**, *15*, 152–159.
77. Tian, Y.; Haibara, K.; Toda, H.; Ding, F.; Liu, Y.; Choi, D. Microbial biomass and activity along a natural pH gradient in forest soils in a karst region of the upper Yangtze River, China. *J. For. Res.* **2008**, *13*, 205–214. [CrossRef]
78. Bolat, İ.; Ömer, K.A.R.A.; Tunay, M. Effects of seasonal changes on microbial biomass and respiration of forest floor and topsoil under Bornmullerian fir stand. *Eurasian J. For. Sci.* **2015**, *3*, 1–13. [CrossRef]
79. Bauhus, J.; Khanna, P.K. The significance of microbial biomass in forest soils. In *Going Underground—Ecological Studies in Forest Soils*; Rastin, N., Bauhus, J., Eds.; Research Signpost: Trivandrum, India, 1999; pp. 77–110.
80. Zhang, R.; Bai, Y.; Zhang, T.; Henkin, Z.; Degen, A.A.; Jia, T.; Guo, C.; Long, R.; Shang, Z. Driving factors that reduce soil carbon, sugar, and microbial biomass in degraded alpine grasslands. *Rangel. Ecol. Manag.* **2019**, *72*, 396–404. [CrossRef]
81. Melendez Gonzalez, M.; Crofts, A.L.; McLaren, J.R. Plant biomass, rather than species composition, determines ecosystem properties: Results from a long-term graminoid removal experiment in a northern *Canadian grassland. J. Ecol.* **2019**, *107*, 2211–2225. [CrossRef]
82. Cui, J.; Holden, N.M. The relationship between soil microbial activity and microbial biomass, soil structure and grassland management. *Soil Tillage Res.* **2015**, *146*, 32–38. [CrossRef]
83. Lovell, R.D.; Jarvis, S.C.; Bardgett, R.D. Soil microbial biomass and activity in long-term grassland: Effects of management changes. *Soil Biol. Biochem.* **1995**, *27*, 969–975. [CrossRef]
84. Dang, N.; Wu, H.; Liu, H.; Ma, R.; Wang, C.; Xu, L.; Wang, Z.; Jiang, Y.; Li, H. Ecological strategies of soil microbes along climatic gradients: Contrasting patterns in grassland and forest ecosystems. *Plant Soil* **2024**, *505*, 645–665. [CrossRef]
85. Wang, W.J.; Dalal, R.C.; Moody, P.W.; Smith, C.J. Relationships of soil respiration to microbial biomass, substrate availability and clay content. *Soil Biol. Biochem.* **2003**, *35*, 273–284. [CrossRef]
86. Dube, F.; Zagal, E.; Stolpe, N.; Espinosa, M. The influence of land-use change on the organic carbon distribution and microbial respiration in a volcanic soil of the Chilean Patagonia. *For. Ecol. Manag.* **2009**, *257*, 1695–1704. [CrossRef]
87. McGonigle, T.P.; Turner, W.G. Grasslands and croplands have different microbial biomass carbon levels per unit of soil organic carbon. *Agriculture* **2017**, *7*, 57. [CrossRef]
88. Richter, A.; Huallacháin, D.Ó.; Doyle, E.; Clipson, N.; Van Leeuwen, J.P.; Heuvelink, G.B.; Creamer, R.E. Linking diagnostic features to soil microbial biomass and respiration in agricultural grassland soil: A large-scale study in Ireland. *Eur. J. Soil Sci.* **2018**, *69*, 414–428. [CrossRef]
89. Bastida, F.; Eldridge, D.J.; García, C.; Kenny Png, G.; Bardgett, R.D.; Delgado-Baquerizo, M. Soil microbial diversity–biomass relationships are driven by soil carbon content across global biomes. *ISME J.* **2021**, *15*, 2081–2091. [CrossRef]
90. Makova, J.; Javorekova, S.; Medo, J.; Majerčíková, K. Characteristics of microbial biomass carbon and respiration activities in arable soil and pasture grassland soil. *J. Cent. Eur. Agric.* **2011**, *12*, 745–758. [CrossRef]
91. Rovira, P.; Sauras-Yera, T.; Poch, R.M. Constraints on Organic Matter Stability in Pyrenean Subalpine Grassland Soils: Physical Protection, Biochemical Quality, and the Role of Free Iron Forms. *Environments* **2024**, *11*, 126. [CrossRef]
92. Cheng, F.; Peng, X.; Zhao, P.; Yuan, J.; Zhong, C.; Cheng, Y.; Cui, C.; Zhang, S. Soil microbial biomass, basal respiration and enzyme activity of main forest types in the Qinling Mountains. *PLoS ONE* **2013**, *8*, e67353. [CrossRef]
93. Ashraf, M.N.; Waqas, M.A.; Rahman, S. Microbial metabolic quotient is a dynamic indicator of soil health: Trends, implications and perspectives. *Eurasian Soil Sci.* **2022**, *55*, 1794–1803. [CrossRef]
94. Wardle, D.A.; Ghani, A.A. A critique of the microbial metabolic quotient (qCO_2) as a bioindicator of disturbance and ecosystem development. *Soil Biol. Biochem.* **1995**, *27*, 1601–1610. [CrossRef]

95. Dilly, O. Microbial energetics in soils. In *Microorganisms in Soils: Roles in Genesis and Functions*; Varma, A., Buscot, F., Eds.; Springer: Berlin/Heidelberg, Germany, 2005; pp. 123–138.
96. Braman, S.; Tenuta, M.; Entz, M.H. Selected soil biological parameters measured in the 19th year of a long term organic-conventional comparison study in Canada. *Agric. Ecosyst. Environ.* **2016**, *233*, 343–351. [CrossRef]

land

MDPI

Article

Impact of Management Practices on Soil Organic Carbon Content and Microbial Diversity Under Semi-Arid Conditions

Nadia Bekhit [1], Fatiha Faraoun [2,*], Faiza Bennabi [1], Abbassia Ayache [2], Fawzia Toumi [1], Rawan Mlih [3,4,*], Viktoriia Lovynska [5] and Roland Bol [3]

1 Laboratory of Spaces Eco-Development, Department of Environmental Sciences, Faculty of Nature and Life Sciences, Djilalli Liabes University, Ex-ITMA, PB 89, Sidi Bel Abbes 22000, Algeria; bekhit.na@gmail.com (N.B.); faiza.bennabi@univ-sba.dz (F.B.); fawzia.toumi@univ-sba.dz (F.T.)
2 Laboratory of Plant Biodiversity, Conservation and Valorization, Department of Environmental Sciences, Faculty of Nature and Life Sciences, Djilalli Liabes University, Ex-ITMA, PB 89, Sidi Bel Abbes 22000, Algeria; abbassia.ayache@univ-sba.dz
3 Institute of Bio- and Geosciences, Agrosphere (IBG-3), Forschungszentrum Jülich GmbH, 52425 Jülich, Germany; r.bol@fz-juelich.de
4 Institute of Water and Environment (IWE), Al Azhar University-Gaza, Gaza P860, Palestine
5 Laboratory of Forestry and Forest Management, Dnipro State Agrarian and Economic University, 49600 Dnipro, Ukraine; glub@ukr.net
* Correspondence: fatiha.faraoun@univ-sba.dz (F.F.); r.mlih@fz-juelich.de (R.M.)

Abstract: Globally, arid and semi-arid agricultural land is characterized by low soil organic carbon (SOC) content. This impacts on the abundance and diversity of soil microorganisms in such environments. We therefore examined SOC and bacterial community structure dynamics in the single plots of the conventional (PC), improved fertilization (PA) and unimproved control (PT) at El Hmadna experimental station (Northwest Algeria) during five-time intervals T(0), T(15), T(70), T(104) and T(147 days). The SOC content was determined using the modified Walkley and Black method. The 16S rRNA genes were isolated from soils and sequenced using the Illumina sequencing platform. Over time, OC levels increased by more than 15%, especially in the improved plot. The highest OC stock was observed for the unmanaged control plot (47 Mg ha^{-1}), also associated with higher bacterial biomass. However, taxonomic analysis revealed that bacterial diversity was higher in PA and PC, with Actinobacteria (42%) and Firmicutes (15%) dominating. Soil salinity did negatively influence SOC but the imposed management practices such as organic amendments did improve both carbon retention and bacterial diversity. The results underline the importance of imposing sustainable agricultural practices to improve carbon sequestration and soil health in semi-arid regions.

Keywords: semi-arid area; organic carbon dynamics; bacterial microflora; physicochemical properties; soil management practices; clay soil

1. Introduction

Arid and semi-arid lands constitute approximately one-third of global terrestrial surface according to the United Nations Committee to Combat Desertification (UNCCD). These lands suffer from degraded and vulnerable soil, limited vegetation cover and low soil organic matter (SOM) with less than 4% [1]. Future projections for climate change in semi-arid areas indicate that global temperature will rise and the temporal and spatial patterns of precipitation will change [2]. Consequently, crop production and food security for one-third of the global population living in these areas will deteriorate [3]. Nevertheless, arid and semi-arid soils' total potential of soil organic carbon (SOC) sequestration can reach up to 20 Pg C over 50 years [4]. Numerous studies indicate that this sequestration can be achieved through adopting land use and management practices emphasizing enhancing land vegetation cover and increasing SOM contents [5–7]. These practices not only aim to mitigate the effects of climate change but also to restore degraded soils and strengthen the resilience of agroecosystems in these vulnerable regions [4,7].

In addition to their role in ameliorating the physicochemical properties of the soil [8,9], SOM is a key factor limiting microbial growth and activity in the soil [10]. The interaction between SOM and soil microbiology enormously impacts the ecosystem's functionality [11]. A decline in SOM content results in a decrease in soil microbial biomass, a crucial indicator of soil health and thus a loss in the essential ecosystem functions and disturbance of the vital processes provided by soil [12].

In semi-arid environments, microbial biomass plays a critical role in nitrogen fixation [13,14] and soil carbon cycling [15]. The main function of microbial biomass is to decompose SOM to release the essential nutrients from their organic sources [16,17]. On the other hand, the microbial-derived residues produced upon their decomposition such as amino sugars, proteins, lipids, biopolymers and other molecules are considered a vital source of OC and N in the soil [18,19].

Typically, semi-arid soils display reduced microbial biomass and diversity. Among the dominant microbial communities detected in arid and semi-arid lands are Proteobacteria, Actinobacteria, Planctomycetota and Acidobacteriota [20,21]. The accumulation and diversity of microbial community structure in these environments are affected by lack of moisture, extreme variations in precipitation, high temperature and soil salinity [22,23]. However, these stresses can be influenced by environmental parameters such as climate, land use type and soil management practices [24–26]. Moreover, shifts in microbial communities under different management practices can provide important insights into the capacity of these soils to recover their biological functions and enhance carbon sequestration [11,25].

Adopting management practices such as fertilization or changing land use have significant and long-lasting effects on microbial activity in semi-arid soils [5,27,28]. Mineral and organic fertilization modifies C and N sources for microbial communities through a substantially higher input of organic matter below and above ground [6,29]. In a meta-analysis study, Allison and Martiny (2008) [30] found that 84% of 38 studies reported that microbial community composition is sensitive to N, P and K fertilization. Wardle (1992) [31] indicated that C and N are limiting elements for soil microorganism accumulation in the soil. In another meta-analysis based on data from long-term fertilization trials in cropping systems, Geisseler and Sow (2014) [32] found that mineral fertilizer application results in a 15.1% increase in microbial biomass above levels in unfertilized control treatments. Sui (2019) [33] found that the beta diversity of the microbial community in semi-arid lands can be significantly affected by soil pH, available P, N and OC due to changes in land use. In their study, they indicated that arable land consistently showed higher alpha diversity for bacteria, Acidobacteria and fungi compared to other land use types. Plant species and biomass directly impact the availability of carbon and other nutrients in the soils, thus affecting microbial biomass's diversity, distribution and community composition [34].

Additionally, organic agricultural practices can significantly influence soil bacterial and fungal biomass; it was found that their diversity increases under no-tillage practices [35]. The activity of the microbial biomass is high in undisturbed soils compared to tilled soils as repeated tillage results in loss of SOM through rapid decomposition from improved aeration and exposure of protected OM to microbial mineralization [36,37]. Furthermore, undisturbed soil or fallow land management favors soil carbon sequestration and the microbial community structure [38]. Despite this knowledge, integrating microbial community assessments with SOC dynamics under various land use intensities and soil amendments in semi-arid agriculture remains underexplored and represents a research gap that this study aims to address.

Studies concerning microbial communities and diversity responses to the OC dynamics in semi-arid lands are scarce. In Algeria, few studies addressed the relationship between bacterial microflora and OC accumulation and dynamics, particularly on clayey and saline agricultural soils. Previous studies focused on characterizing soil microbial diversity in hypersaline natural ecosystems, other studies explored the relationships between microbial biomass and OM accumulation [39–42]. Consequently, there is a lack of precise data on microbial density and diversity of agricultural soils, their relationship with OC dynamics and the impact of management practices on bacterial structure and diversity.

We therefore used genomic DNA analysis to examine changes in soil bacteria communities and their relationships with soil physiochemical properties in three plots receiving different management practices as follows: 1. a fallow or undisturbed land, used in this study as a control plot; 2. a conventional plot that receives regular mineral fertilization NPK; and 3. an amended plot that received NPK, sand, cattle manure, gypsum and ferrous sulfate amendments over a 3-year project.

Thus, our specific objectives of this study are: (i) to assess the effect of conventional and improved management practices on OC accumulation and dynamics in semi-arid soil; (ii) to elucidate the changes in soil bacterial structure and diversity affected by SOC and soil properties; and (iii) to investigate the role of bacterial communities in SOC storage.

The outcomes of this study will help quantify the influence of land management on soil bacterial community structure and diversity in northwestern Algeria. The research is of significant value for improving soil management practices in Algeria. Additionally, the findings are expected to provide actionable insights for sustainable soil restoration strategies and contribute to global knowledge on the interplay between microbial ecology and carbon dynamics in dryland agriculture.

2. Materials and Methods

2.1. Description of the Study Area

The experimental site is in the National Institute of Agronomic Research of Algeria (INRAA) experimental station in El Hmadna, region of Relizane (Northwest Algeria). The site coordinates are $35°55'26''$ N and $0°44'57''$ E at an altitude of 48 m (Figure 1). The station, which is surrounded by two mountain ranges, the Dahra to the north and the Ouarsenis to the south, covers an area of 77 ha and is part of Bas-Cheliff zone, which covers 2750 km^2 [43]. The climate is Mediterranean, characterized by hot summers, cold winters and poorly distributed annual rainfall. This area is also known as the "oven of the Tell" due to the extreme heat in the summer season [44]. The soil of the study area is a clay texture with high salinity [45].

2.2. Description of the Plots

The study was conducted on three plots, as illustrated in Figure 1. Table 1 summarizes the experimental design, highlighting the management history, applied treatments and the main baseline soil physical and chemical characteristics of each plot. These parameters represent the stable properties that did not vary during the observation period.

2.3. Sampling Strategy

Five soil samples (0–30 cm depth) were collected from each plot to form a composite sample. Physicochemical analyses included OC content, texture, bulk density, salinity, pH, cation exchange capacity (CEC), total and active limestone, total nitrogen, available phosphorus and potassium. Soil monitoring covered five time intervals: T(0), T(15), T(70), T(104) and T(147). Bacterial microflora was analyzed at two key stages: T(0), representing the initial state, and T(104), corresponding to the peak of microbial activity [46]. For each plot, a composite soil sample was prepared by homogenizing five subsamples collected randomly to ensure representativeness. The composite samples were then stored at 4 °C in sterile containers prior to analysis. The study covered both wet (November–April) and dry (May–October) seasons. Table 2 provides an overview of the sampling timeline, detailing the chronological progression and the specific objectives assigned to each sampling point.

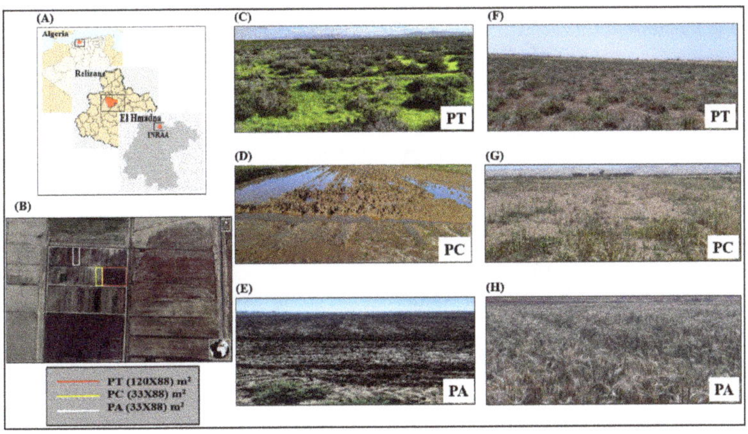

Figure 1. (**A**,**B**) Geographical location of the experimental plots. (**C–E**) Plots in the wet period (December 2021). (**F–H**) Plots in the dry period before harvest (May 2022).

Table 1. Main characteristics of the experimental plots.

Plot	Treatments Applied During the Study	Prior Management (Before 2021)	Crop 2021–2022	Vegetationcover	Particle Size				Bulk Density (g/cm^3)	$CaCO_3$ (%)
					Clay (%)	Fine Silt (%)	Coarse Silt (%)	Sand (%)		
PA	Plowing + NPK (15-15-15) 100 kg/plot	Algerian-Chinese project (3 years): NPK + cattle manure + sand + gypsum + ferrous sulfate	Rainfed soft wheat (*Triticum aestivum* L.)	–	41	28	4	27	1.3	22
					Clayey soil					
PC	Plowing + NPK (15-15-15) 100 kg/plot	NPK (15-15-15) 100 kg/plot	Rainfed soft wheat (*Triticum aestivum* L.)	–	63	28	5	4	1.4	20
					Clayey soil					
PT	None (uncultivated fallow since 1942)	No agronomic intervention	None	*Suaeda fruticosa*	61	30	2	7	1.4	21
					Clayey soil					

PT: control plot, PC: conventional plot, PA: amended plot.

Table 2. Soil sampling schedule, growth stages, seasons and sampling purposes during the experimental period.

Stage (Days)	Date	Growth Stage	Season	Sampling Purpose
T(0)	14 September 2021	Before sowing	Dry	Physicochemical + bacterial analysis
T(15)	29 December 2021	Germination/emergence	Wet	Physicochemical analysis
T(70)	23 February 2022	Tillering	Wet	Physicochemical analysis
T(104)	29 March 2022	Bolting/heading	Wet	Physicochemical + bacterial analysis
T(147)	11 May 2022	Maturation	Dry	Physicochemical analysis

2.4. Physicochemical Analysis

Mineral fractions were measured using the "Robinson" method [47]. The principle of this method is to remove all cement-like carbonates, oxides and organic substances by oxidation with hydrogen peroxide. After dispersion in sodium hexametaphosphate, the samples were pipetted at different times and depths, following different sedimentation intervals. Sampling time and depth were calculated using Stokes' law. All particle size data were expressed as a percentage of fine earth (<2 mm). pH was measured using a pH meter (Hanna Instruments, Netherlands) for suspensions with a 1:2.5 (m/v) ratio of fine earth and water. Salinity was measured by electrical conductivity (EC) in millisiemens per centimeter (mS cm^{-1}) using a conductivity meter (Hanna Instruments, Netherlands). A suspension of fine earth and water with a ratio of 1:5 (m/v) was set for this analysis [48]. The bulk density (BD) in g cm^{-3} was calculated by direct sampling using a metal cylinder with a height of 5 cm and a diameter of 5.5 cm [49]. Samples taken by the cylinder were dried and passed through a 2 mm diameter sieve to measure coarse particle content (expressed in %). OC content was determined by the modified Walkley and Black method, based on the oxidation of OC by potassium dichromate ($K_2Cr_2O_7$) in sulfuric acid [50]. Total limestone ($CaCO_3$) content was determined by a Bernard calcimeter. The CO_2 released by the reaction was measured by a gas burette [51]. Active limestone was measured according to the method described by Drouineau (1942) [52]. The excess ammonium oxalate was determined by titration with a solution of potassium permanganate in a sulfuric solution. Total nitrogen was determined by the Kjeldahl method described by Bremner (1965) [53]. Phosphorus content was measured according to Olsen (1954) [54] using a UV/visible spectrometer. The potassium content was measured based on dilute ammonium acetate with a flame spectrophotometer using a method described by Nash (1971) [55]. CEC was determined using the sodium acetate method reported by Rhoades (1982) [56].

2.5. Organic Carbon Stock Calculation (OCS)

The soil OCS was calculated for 0 to 30 cm depths using the equation recommended by the FAO [57,58] as follows:

$$\text{OCS}\left(\text{Mg C ha}^{-1}\right)\ 0.1 \times \text{C} \times \text{BD} \times \text{T} \times (1 - \text{CP})$$

where OCS is organic carbon stock (Mg C ha^{-1}), C is carbon content (g C kg^{-1} soil), BD is bulk density (g cm^{-3}), T is the thickness of the soil horizon (cm) and CP is the gravel content or coarse particles (g g^{-1} soil).

2.6. Biological Soil Analysis

2.6.1. DNA Extraction and Polymerase Chain Reaction (PCR) Amplification

The genomic DNA was extracted from 250 g of soil using a DNeasy® PowerSoil® Pro Soil Kit (sourced from QIAGEN GmbH, Hilden, Germany) according to the manufacturer's instructions. This kit is effective in removing PCR inhibitors as it employs the second generation of QIAGEN Inhibitor Removal Technology® (IRT). Cell lysis occurs through both mechanical and chemical methods. Total genomic DNA is captured onto a silica membrane in a centrifuge column format, then washed, eluted and is ready for NGS, PCR and other molecular analyses, following procedures as previously described [59,60].

PCR amplification of the targeted 16S rRNA gene regions was performed using specific primers combined with barcodes. PCR products of appropriate size were selected by 2% agarose gel electrophoresis. Equal quantities of PCR products from each sample were pooled, end-repaired, A-tailed and ligated with Illumina adapters, according to standard protocols [61].

To ensure data quality and reliability, quality control (QC) was implemented at every step of the process, as shown in Figure S1. After sequencing, raw data containing low-quality sequences ("dirty data") were processed to generate clean data through merging and filtering [60,62]. OTUs were clustered using UPARSE [62]. The representative sequence of each OTU was assigned a taxonomic classification using the SILVA 138 database [63] and the abundance distribution was calculated accordingly. The detailed workflow is shown in Figure S2.

2.6.2. Sequencing of the 16S rRNA Gene

Libraries were sequenced on an Illumina paired-end platform to generate 250 bp paired-end raw reads. The process of DNA library preparation, including PCR product selection, end-repair, A-tailing, adapter ligation and purification, was carried out following standard Illumina protocols, as summarized in Figure S3, which illustrates the key steps of the library construction workflow. The amplicon was then sequenced on the Illumina paired-end platform to produce 250 bp raw paired-end reads (Raw PE), which were subsequently merged and pre-processed to obtain clean tags [59]. Chimeric sequences were detected and removed using UCHIME [64], resulting in efficient tags used for downstream analysis.

Chimera checking and removal steps were performed to minimize false diversity estimates [61]. Final taxonomic annotation was performed using the SILVA 138 database, ensuring reliable taxonomic resolution [63].

2.6.3. Statistical Analyses

Matched reads were assigned to samples based on their unique barcodes and truncated by cutting the barcode and primer sequences. Matched reads were merged using FLASH (Version 1.2.7), a high-speed and accurate analysis tool designed to merge matched reads [59]. End reads when at least some of the reads overlap with the read generated from the opposite end of the same DNA fragment, and splice sequences were called raw tags. Quality filtering on the raw tags was performed under specific filtering conditions to obtain high-quality clean tags according to the Qiime (Version 1.7.0) quality-controlled process [60]. The tags were compared with the reference SILVA138 database using the UCHIME algorithm to detect chimeric sequences [61]. The chimeric sequences were then removed [62]. After obtaining the Effective Tags, sequence analysis was performed by Uparse software (Version 7.0.1090) [63]. Sequences with a similarity $\geq 97\%$ were assigned to the same OTUs. The representative sequence of each OTU was examined for further annotation. For each representative sequence (Version 1.7.0) Qiime [64] in the Mothur method was performed against the SSUrRNA database of the SILVA138 database [65] for species annotation at each taxonomic rank (threshold: 0.8~1) (kingdom, phylum, class, order, family, genus, species) [66]. MUSCLE (Version 3.8.31) was used to understand the phylogenetic relationship between all representative OTU sequences [67]. The software can compare multiple sequences quickly. OTU abundance information was normalized using a sequence number standard corresponding to the sample with the fewest sequences. Subsequent analyses of alpha diversity were performed based on this normalized output data. Alpha diversity is applied in biodiversity complexity analysis for a sample across six indices, including Observed-species, Chao1, Shannon, Simpson, ACE and Good-coverage. All these indices were calculated with QIIME (Version 1.7.0) and displayed with R software (Version 2.15.3). Linear regression was carried out to estimate the temporal OC response of each plot over the experimental period. Statistical analyses were performed using Python program (version 3.12.0), using the Pandas, Numpy, Mtplotlib.pyplot, Sklearn.linear_model and Sklearn.metrics libraries. Statistical significance of coefficients was assessed using the *p*-value and the student *t*-test.

3. Results

3.1. Temporal Changes in Soil Physicochemical Properties

The temporal variation of key soil physicochemical properties across the different plots and sampling stages is presented in Table S1. The pH is neutral to slightly alkaline (7.0–8.1) at all sampling stages. The EC varied considerably throughout the different sampling periods. Plots PT and PC vary from excessively salty to slightly salty, while PA ranges from very salty to slightly salty. CEC is high to very high in all plots, which is typical for clay-dominated soils where cation exchange between soil and plant is relatively limited. The C/N ratio varies remarkably throughout the different samples. At T(0), T(15) and T(147), all plots recorded low to very low C/N ratios, indicating rapid OM decomposition. On the other hand, at T(70) and T(104), C/N is very high, revealing reduced biological activity and slow OM decomposition. The phosphorus levels varied from high to very high for all samples; similarly, the potassium levels were high in the three plots.

3.2. Organic Carbon Variations

At T(0), OC content remained constant across all three plots; however, at T(15), it increased significantly in PC and PA, coinciding with plowing operations that incorporated fresh plant residues into the soil. This period also experienced 86.5 mm of cumulative rainfall and an average temperature of 19 °C; these conditions are favorable for OM mineralization. At T(70), OC content was higher in PA than in PC (Figure 2A), likely due to greater plant biomass, increased photosynthesis and enhanced rhizodeposition. Despite low rainfall (~5 mm), PA benefited from prior soil improvement practices that enhanced soil aeration and water retention, helping crops withstand the precipitation deficit. At T(104), a significant increase in OC levels was observed in all three plots (PA > PT > PC), while at T(147), OC levels rose further, particularly in PA. The last samples from PA and PC were taken before the harvest when the wheat crop was in its final vegetative phase, which may have increased carbon intake. Rainfall (~92 mm) during this stage may have also enhanced primary production, resulting in increased carbon inputs into the soil.

The dispersion of OC values is greater in the PA plot. The PT plot has the lowest OC concentration on average and the least dispersion (Figure 2B). Figure 2C,D indicate that PT displays the highest OCS values, with a median of 50 Mg ha^{-1} and low data dispersion. PA shows intermediate values, with a median of 43 Mg ha^{-1} and greater variability. PC has the lowest OCS, with a median of 40 Mg ha^{-1} and a reduced total extent. PT seems to promote more OC storage than the other plots. PA also tends to favor greater OC accumulation than the PC plot.

A difference in plot means is noted, with PA > PC > PT. The variance reveals greater PC data homogeneity, while PA and PT show greater dispersion around the mean. The standard deviation indicates more stable and less dispersed data for PC, while PA shows the greatest variability. PT has a slightly higher standard deviation than PC and PT suggests moderate variability (Table S4). The *p*-values were well above the 0.05 threshold in all comparisons (PT vs. PC, PT vs. PA, PC vs. PA) (Table S5). This means that there is no statistically significant difference between the means of the PT, PC and PA plots.

Linear regression analysis reveals a strong positive relationship between sampling periods and OC values for PA (Figure 2E), with a very well-fitted model (R^2 = 0.96) and a significant increase in OC of 0.0063 units per period (p = 0.003). In contrast, for PT, although 69% of variability was explained (R^2 = 0.69), the relationship was not significant (p = 0.08), while for PC, the model was unreliable (R^2 = 0.30, p = 0.3), suggesting a limited influence of time. Other factors could explain these variations, such as temperature, rainfall, soil texture, salinity, tillage intensity, fertilization and microbial activity.

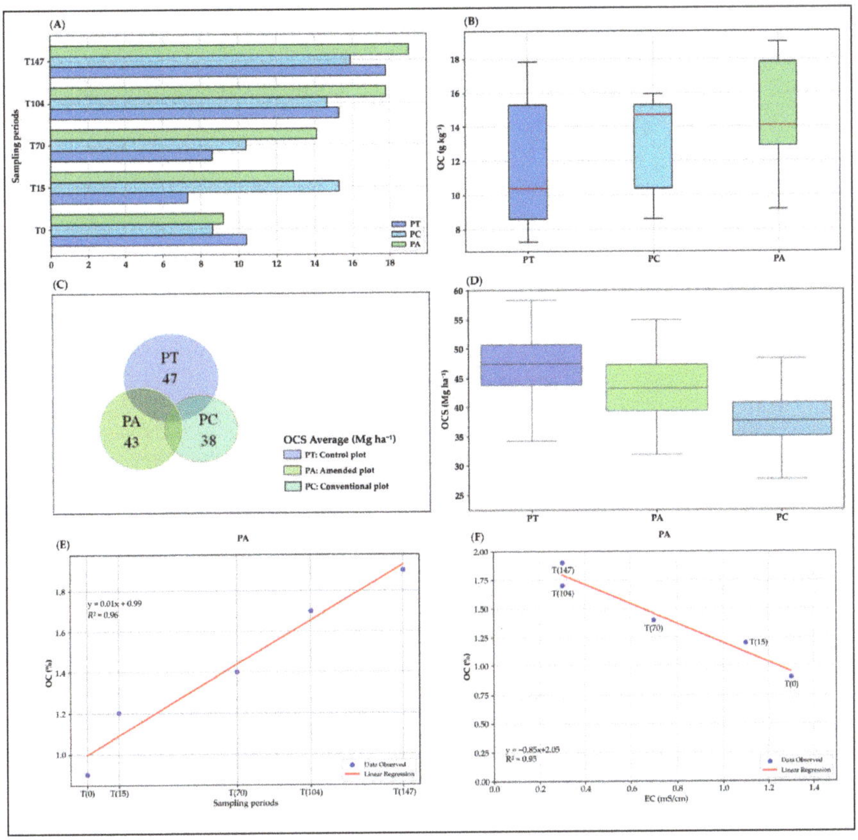

Figure 2. (**A**) Variation of soil OC at each sampling stage. (**B**) Difference of soil OC variation among plots. (**C**) Variation in average of soil OCS. (**D**) Boxplots for difference of OCS variation. (**E**) Linear regression between sampling periods and OC values for PA. (**F**) Linear regression between EC and OC for PA. PT: control plot, PC: conventional plot, PA: amended plot.

Linear regression revealed a strong and statistically significant negative relationship between EC and OC for PA (Figure 2F), with 95% of OC variability explained by EC ($R^2 = 0.95$, $p = 0.005$). In contrast, PT showed a moderate but non-significant negative correlation ($R^2 = 0.51$, $p = 0.17$), while PC displayed a very weak and non-significant association ($R^2 = 0.07$, $p = 0.65$). These results highlight PA as the only site where EC strongly and reliably predicts OC changes, whereas the relationship is uncertain in PT and negligible in PC.

3.3. *Composition and Structure of Bacterial Microflora*

Identification of Operational Taxonomic Units (OTUs)

During the OTU construction process, basic information for each sample, including efficient tag data, low-frequency tag data and tag annotation data, was collected. The detailed summary of these steps is presented in Table S2, which shows the sequencing depth and data quality for all samples.

An average of 87,061 raw reads per sample was obtained. Following quality control procedures, an average of 77,981 valid reads per sample was retained for further analysis. The sequences were clustered into OTUs, resulting in a total of 2267 OTUs, which were taxonomically assigned using the SILVA138 database, as illustrated in Figure 3.

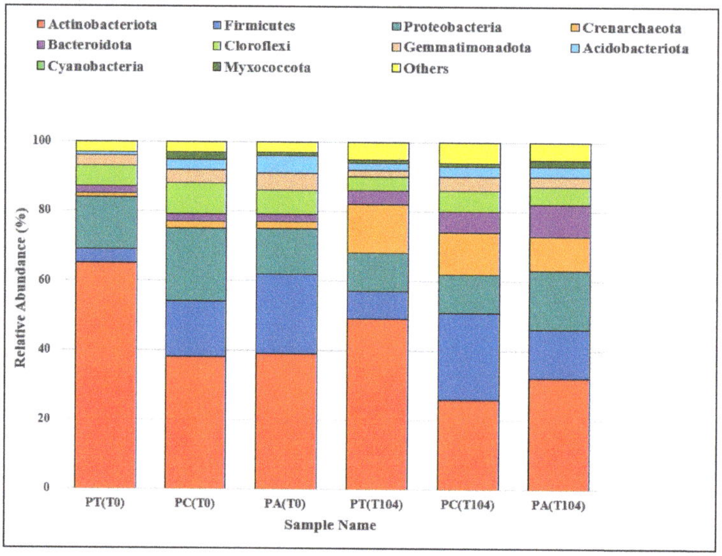

Figure 3. Taxa relative abundance in phylum. PT: control plot, PC: conventional plot, PA: amended plot.

Regarding the domain distribution, more than 98% of OTUs at T(0) were classified within the Bacteria domain, while only 2% were affiliated with the Archaea domain. At T(104), the proportion of Archaea increased to 12%, while the Bacteria domain represented 88% of the OTUs. The detailed summary of the tag counts and OTU numbers for each sample at both T(0) and T(104) is provided in Figure S4, which also highlights the distribution of sequencing effort and OTU richness across the PT (control), PC and PA plots.

Detailed analysis of the alpha diversity metrics presented in Table S3, including observed species, Shannon, Simpson, Chao1, ACE, Goods coverage and phylogenetic diversity (PD whole tree) indices reveals a close link between cropping practices and the structuring of microbial communities. As seen in Figure 3, the analysis showed the relative abundance of the different phyla. Complementing these results, the Krona display shown in Figure S5 provides a detailed visualization of the percentage contribution of each microbial group. It can be seen that at T(0), the dominant communities were Actinobacteriotas (47%), followed by Firmicutes (14%) and Proteobacteria (16%), while the least represented included Cyanobacteria (<0.1%) and Myxococcota (1%). At T(104), the composition remained similar, with Actinobacteriotas dominating (36%), followed by Firmicutes (16%) and Proteobacteria (13%), while Cyanobacteria (0.1%) and Myxococcota (1%) remained the least abundant. Overall, the relative abundances of the different microbial groups ranged from <0.1% to 47%, depending on the taxonomic group and sampling condition.

Figure 4 provides more details on the microbial structure of the four most abundant phyla. The OTUs with the highest indicator values for soils from the three plots at T(0) and T(104) were dominated by the phylum Actinobacteriota with the genus *Rubrobacter* and the classes Actinobacteria, Acidimicrobiia and Thermoleophilia. Then came the Firmicutes with the class Bacilli, followed by the Proteobacteria, with the classes Alphaproteobacteria and Gammaproteobacteria. Archaea are dominated by the Crenarchaeota, notably the *Nitrososphaeraceae* family.

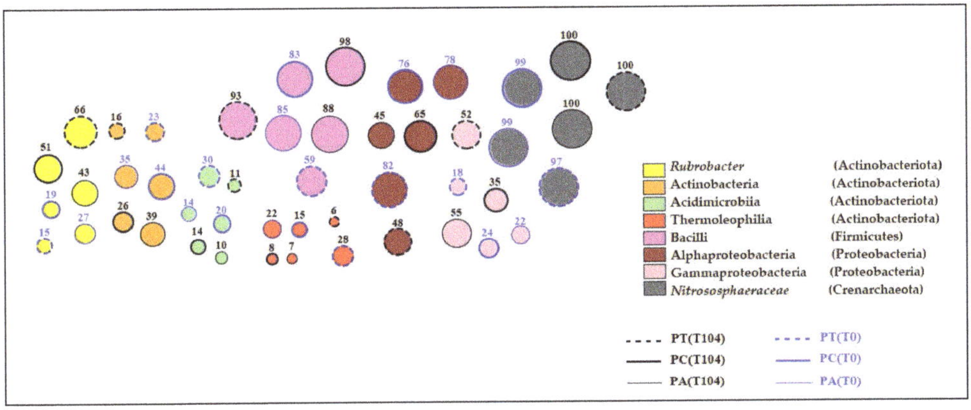

Figure 4. Abundance of taxonomic ranks. PT: control plot, PC: conventional plot, PA: amended plot.

Analyses of beta diversity, as illustrated in Figure S6, reveal significant shifts in microbial community composition between T(0) and T(104). The UPGMA cluster tree based on Weighted UniFrac distance (Figure S6A) shows that the different treatments influenced not only the presence or absence of certain phyla but also their relative abundance, reflecting modifications in the fine structure and composition of bacterial communities across the plots.

In parallel, the UPGMA cluster tree based on Unweighted UniFrac distance (Figure S6B) emphasizes that the treatments affected the overall diversity of bacterial communities by altering both species richness and their phylogenetic structure. Together, these analyses underline the combined effects of management practices and seasonal variation on the microbial community assemblages at both taxonomic and phylogenetic levels.

4. Discussion

4.1. Organic Carbon Dynamics and Soil Physicochemical Properties

As detailed in Table S1, the OC levels increased notably at T(15) across the plots. This increase can be attributed to plowing, which enhances OM decomposition in the presence of optimal temperature and humidity, favoring carbon loss [68–70]. Additionally, the rise in OC may be linked to mineral fertilization amendments containing nitrogen, e.g., NPK or from atmospheric nitrogen fixed by N-fixing plants [71,72]. In addition, Table S1 shows a significant increase in total nitrogen content, especially in the PC plot, which likely supported enhanced OC storage through improved nitrogen availability [73]. Regarding salinity, the EC values reported in Table S1 highlight a marked reduction at T(104), particularly in the PA plot, likely due to heavy rainfall events during this period. This decrease in salinity was further supported by improved soil management practices in PA that facilitated better drainage and salt leaching, thereby contributing to the slowdown of OC mineralization [74]. At T(147), EC values from Table S1 confirm non-saline conditions in both PA and PC plots, which would further limit carbon mineralization.

Moreover, our linear regression analysis, supported by the EC and OC data from Table S1, revealed a clear negative correlation between salinity and OC levels, indicating that reduced salinity conditions are conducive to enhanced OC accumulation in these semi-arid soils.

4.2. Organic Carbon Stock Variations

The fluctuations in OC averages have a direct impact on OCS. Higher concentrations and dispersion were observed in the PT plot (Figure 2C,D). The PT plot represents a fallow undisturbed natural ecosystem since 1942, therefore, C mineralization is negligible [75]. Previous studies indicated that soils with permanent vegetation cover and zero-till are characterized by a significant concentration of OM in the topsoil, higher microbial biomass (i.e., total tags for PT > 91400) and slower carbon mineralization [69,76]. Moreover, the high clay and carbonate contents might have influenced SOC resistance to mineralization through soil aggregate formation and stabilization [77–79]. The precipitation rate (≈53 mm) at T(104) may have also contributed to the decrease in salinity and led to an increase in microbial activity and stimulating OM mineralization. Noteworthy is that the C/N ratio indicates low microbial activity at T(104) for all plots, which may have contributed to SOC storage.

4.3. Variation in Soil Bacterial Biomass and Diversity

The results showed that bacterial biomass was lower during the wet period at T(104), with a total number of markers of 84,319 compared with 89,803 at T(0), despite a higher level of OCS at T(104) (53 Mg ha^{-1}) compared with T(0) (31 Mg ha^{-1}). On the other hand, bacterial diversity was higher during the wet period at T(104), with OTU giving indicator values of 2443 compared with 2091 for the dry period at T(0). The variation in microbial biomass and diversity is the result of several factors. In the wet period, elevated moisture and nutrient levels increase environmental and habitat diversity, creating more favorable conditions for microbial diversity. However, increased competition and fragmentation of by-product resources limit the dominance of certain species, thus reducing total biomass [80]. In dry periods, the more stressful conditions select a smaller number of resistant species, resulting in higher biomass but reduced diversity [81]. OC, although more abundant in wet periods, may be less accessible, while in dry periods, it is potentially more concentrated and labile, favoring higher bacterial biomass [82,83]. In the dry period at T(0), soil salinity in all three plots is high (1.5 mS cm^{-1}), while it decreases significantly in the wet period (0.3 mS cm^{-1}). The combined stress of salinity and drought reduces microbial diversity but favors the dominance of tolerant species, therefore maintaining high biomass [84,85].

The high density of the Actinobacteriota phylum in the PT plot may be attributed to the high vegetation cover [86]. Our results agree with previous studies showing that Actinobacteriota are of the dominant microbial communities in arid environments [87–89]. Unpredictably, the decline in OCS at T(0) despite the high density of Actinobacteriota may be explained by the high salinity levels at that time, which may have restricted the activity of the bacterial community [90]. Firmicutes dominated in the anthropized plots, i.e., PC and PA, and were mainly represented by the Bacilli class (Figures S5 and 4), which is probably due to soil management practices such as plowing and fertilization [91]. In contrast, the undisturbed PT plot showed a very low density of Firmicutes. Firmicutes can also resist unfavorable environmental conditions, such as high temperatures, drought and nutrient-poor soils [92]. In addition, the morphology and physiology of Firmicutes allow them to live in hypersaline and especially humus-bearing environments [42]. Among Firmicutes, the Bacilli class includes a significant concentration of halophilic species requiring very high OM rates. All three plots had high EC at T(0); however, at T(104) salinity decreased due to the high rainfall during this period, while OC content increased significantly, which may have amplified the density of the Bacilli class (Figures 4 and S5). Furthermore, the dominance of Firmicutes in the PC and PA plots at both T(0) and T(104) could be attributed to tillage practices affecting bacterial community composition. Firmicutes play a key role in soil fertility by contributing to the mineralization of OM and thus affecting the residence time of the OC in the soil [93]. This was evident in the PC plot, which has the highest density of Firmicutes and demonstrates low OCS levels.

Proteobacteria are among the most abundant phyla after Actinobacteriota and Firmicutes. They are mainly represented by the classes Alphaproteobacteria and Gammaproteobacteria. Proteobacteria are nitrogen-fixing bacteria [94] and play a role in the carbon and sulfur cycles [94]. The Alphaproteobacteria class can adapt to dry, nutrient-poor soils, thrive in saline soils and is found in abundance in undisturbed and stable ecosystems [84]. Alphaproteobacteria were abundant in all three plots, particularly during the dry period at T(0) (Figure 4). In contrast, Gammaproteobacteria were dominant in the three plots during the wet period at T(104), when OC levels were higher compared with T(0). This phylum prefers moist soils rich in OM, tolerates saline soils and is more abundant in disturbed ecosystems [80,82].

Ammonia-oxidizing archaea are represented in our study by the Crenarchaeota class, dominated by the *Nitrososphaeraceae* family. The latter are abundant in warm, humid soils [95]. *Nitrososphaeraceae* are involved in the nitrification process [96]. They are most active in soils containing medium to high levels of OC (1–3%) [96]. Their activity is optimal in moderately humid environments with low to moderate salinity [97]. This explains why *Nitrososphaeraceae* are more abundant at T(104), which also explains the higher nitrogen levels at T(0). Biological nitrogen removal through nitrification could lead to OC loss [98], but this is not true for plots at T(104), where OCS is higher than at T(0). The hypothesis may still relate to soil salinity, which is very high at T(0) and may have significantly contributed to OC loss. High OCS values were observed at PT, particularly at T(104), where the reduction in salinity, probably due to rainfall (~53 mm), may have mitigated its limiting effect on microbial activity and OM mineralization. Despite unfavorable conditions at T(0), such as low OC content, high salinity and temperature, bacterial biomass was surprisingly higher; the results are in line with a previous study showing bacterial resilience in carbon-deficient soils [98]. The taxa were dominated by Actinobacteriota and Firmicutes, probably influenced by the clay texture and sand amendments. The observed higher bacterial biomass in PT at both T(0) and T(104) may likely be attributed to the dense root systems of permanent vegetation, which could enhance rhizosphere resource availability for microbial communities [99,100].

4.4. The Role of Actinobacteria and Firmicutes in Soil Organic Carbon Dynamics

Our results reveal contrasting dynamics between Actinobacteria and Firmicutes in regulating SOC under semi-arid conditions, shaped by both environmental factors and management practices. The dominance of Actinobacteria (42%, mainly Rubrobacter) in the undisturbed control plot (PT) was associated with higher SOC stocks (47 Mg ha^{-1}), suggesting a key role for this group in the long-term stabilization of soil OM. This observation aligns with the well-documented capacity of Actinobacteria to degrade complex polymers such as cellulose and lignin through the production of extracellular enzymes like cellulases and peroxidases [18]. Moreover, their resilience under dry conditions (T0), despite high salinity levels (EC = 1.8 mS/cm), can be attributed to their robust cell walls and production of hydrophobic metabolites [101]. Conversely, Firmicutes (15%, predominantly Bacilli) were more abundant in the managed plots (PA, PC), where their presence was negatively correlated with SOC stocks (40–43 Mg ha^{-1}). This supports the hypothesis that Firmicutes may contribute to the accelerated turnover of labile carbon via fermentative pathways [81]. Their increased abundance under the influence of organic amendments and tillage reflects their sensitivity to anthropogenic disturbances [27].

Although the statistical analysis did not reveal significant differences ($p > 0.05$) in SOC between plots, a 15% increase was observed in the PA plot. This apparent discrepancy can be attributed to (1) the relatively short duration of the experiment, which may be insufficient for SOC changes to reach statistically detectable levels and (2) the inherently high spatial variability in semi-arid soils, which often limits the statistical power of short-term studies. From a biological perspective, even a non-significant 15% increase in SOC could have meaningful functional implications in these systems. Previous studies have demonstrated that even modest increases in SOC (5–15%) can enhance aggregate stability [102], stimulate microbial activity [103] and improve water retention [104]. These effects are particularly relevant in soils with low organic matter content, where minor improvements can positively impact crop productivity [103,104]. Therefore, while these findings remain preliminary, they highlight the potential of organic amendments to progressively improve soil quality in semi-arid areas. Long-term monitoring is, however, needed to confirm the sustainability of these trends.

Seasonal dynamics further revealed that during the wet period (T104), microbial diversity increased alongside a marked emergence of nitrifying Archaea (Nitrososphaeraceae), while overall microbial biomass appeared to decrease, likely due to intensified competition [105]. In contrast, during the dry period (T0), the community structure was dominated by stress-tolerant taxa such as Actinobacteria, leading to reduced SOC mineralization. Importantly, the strong negative correlation observed between EC and SOC in the PA plot ($R^2 = 0.95$, $p = 0.005$) highlights the inhibitory effect of salinity on microbial-driven carbon sequestration processes, consistent with the findings of Rath and Rousk (2015) [84] on the suppressive effects of salts on enzymatic activities.

Overall, these findings suggest that undisturbed systems (PT) maintain microbial communities that favor SOC retention over the long term; organic amendment practices (PA), although subject to variability, may contribute to mitigating SOC losses by supporting higher microbial diversity; salinity emerges as a critical limiting factor, acting independently of soil moisture regimes and potentially constraining microbial activity and SOC dynamics.

5. Conclusions

This study evaluated how soil management practices influence SOC dynamics and microbial diversity under semi-arid conditions in Northwest Algeria. Land management practices significantly influenced bacterial diversity, biomass and SOC dynamics in Algeria's semi-arid soils. During dry periods in the experiment, the PC plot showed high bacterial diversity (due to clay content) but lower biomass from salinity stress, while the undisturbed PT plot maintained the highest biomass. During wet periods, amended PA plots surpassed others in diversity due to improved soil conditions, whereas PC plots showed stress-adapted biomass increases. Undisturbed PT plots demonstrated resilient microbial communities and superior long-term SOC storage, while amended PA soils enhanced microbial diversity and soil properties. Conventional PC practices reduced SOC and favored stress-resistant bacteria. These findings underscore that sustainable practices (organic amendments, reduced tillage) optimize carbon sequestration, microbial health and ecosystem resilience in semi-arid regions like Algeria and beyond. Additionally, this study highlights the need to integrate drought and salinity into SOC models, assess long-term impacts of organic amendments and explore Actinobacteria and Firmicutes interactions across climates variability.

Supplementary Materials: The following supporting information can be downloaded at: https://www.mdpi.com/article/10.3390/land14051126/s1, Figure S1. The procedures from DNA extraction to sequencing, Figure S2. The procedures from DNA samples to final data, Figure S3. Workflow of library construction, Figure S4. Summary of the Tags and number of OTUs in each sample for T(0) and T(104) samples. PT: control plot, PC: conventional plot, PA: amended plot, Figure S5. KRONA display, Figure S6. UPGMA cluster tree: (A) based on Weighted Unifrac distance; (B) based on Unweighted Unifrac distance, Table S1. Soil physicochemical characteristics of different plots at each sampling stage, Table S2. Quality control statistics (QC), Table S3. Alpha diversity indice, Table S4. Descriptive statistics of OC values for studied plots, Table S5. Student's t-Test: Comparing OC differences between plots.

Author Contributions: Conceptualization, F.F., F.B., A.A. and F.T.; methodology, F.F., F.B. and N.B.; software, F.T. and A.A.; validation, R.B.; formal analysis, F.F. and N.B.; investigation, F.F., N.B. and F.B.; resources, N.B. and F.F.; data curation, N.B.; writing—original draft preparation, N.B., F.F. and R.M.; writing—review and editing, R.M., R.B. and V.L.; visualization, F.F. and N.B.; supervision, R.B. and R.M.; project administration, F.F. and F.B.; funding acquisition, F.F. and F.B. All authors have read and agreed to the published version of the manuscript.

Funding: This research was supported through personal funding by the co-authors (Fatiha Faraoun and Faiza Bennabi). The authors gratefully acknowledge the Department of Environmental Sciences, Faculty of Natural and Life Sciences, for providing access to laboratory facilities during this study.

Data Availability Statement: The author confirms that the data supporting the findings of this study are available within the article and its Supplementary Materials. Additional data can be supplied by the corresponding author upon reasonable request.

Acknowledgments: All our gratitude and thanks are addressed to Gacemi Abdelhamid, Director of the INRAA (El Hmadna), and Bellague Djamel, Khaldi Abdelkrim, Lariche Abdelaziz and Chedjerat Abed for facilitating these research activities.

Conflicts of Interest: The authors declare no conflicts of interest.

Abbreviations

Abbreviation	Definition
16S rRNA	16S ribosomal RNA (gene used for bacterial studies)
BD	Bulk density
C/N	Carbon-to nitrogen ratio
$CaCO_3$	Calcium carbonate
CEC	Cation exchange capacity
DNA	Deoxyribonucleic acid
EC	Electrical conductivity (measure of soil salinity)
GC%	Guanine-cytosine content (DNA base composition)
INRAA	National Institute of Agronomic Research of Algeria
IRT	Inhibitor Removal Technology (DNeasy® PowerSoil® kit)
NGS	Next-generation sequencing
NPK	Nitrogen-phosphorus-potassium fertilizer
OC	Organic carbon
OCS	Organic carbon stock
OM	Organic matter
OTU	Operational taxonomic unit
PA	Amended plot (improved management practices)
PC	Conventional plot (mineral fertilization)
PCR	Polymerase chain reaction
PE	Paired-end sequencing
PT	Control plot (uncultivated since 1942)
QC	Quality control
Q20/Q30	Sequencing quality scores (1/100 or 1/1000 error rate)
SOC	Soil organic carbon
SSUrRNA	Small subunit ribosomal RNA (taxonomic annotation database)
UPGMA	Unweighted pair group method with arithmetic mean (clustering method)

References

1. Garcia, C.; Moreno, J.L.; Hernandez, T.; Bastida, F. *The Biology of Arid Soils*; Steven, B., Ed.; De Gruyter: Berlin, Germany, 2017; pp. 15–30, ISBN 9783110419047.
2. Mirzabaev, A.; Stringer, L.C.; Benjaminsen, T.A.; Gonzalez, P.; Harris, R.; Jafari, M.; Stevens, N.; Tirado, C.M.; Zakieldeen, S. Cross-Chapter Paper 3: Deserts, Semiarid Areas and Desertification. *Clim. Chang.* **2022**, 2195–2231.
3. Jia, Q.; Li, M.; Dou, X. Climate Change Affects Crop Production Potential in Semi-Arid Regions: A Case Study in Dingxi, Northwest China, in Recent 30 Years. *Sustainability* **2022**, *14*, 3578. [CrossRef]
4. Lal, R. Carbon Sequestration in Dryland Ecosystems. *Environ. Manag.* **2004**, *33*, 528–544. [CrossRef] [PubMed]
5. Ramesh, T.; Bolan, N.S.; Kirkham, M.B.; Wijesekara, H.; Kanchikerimath, M.; Rao, C.S.; Sandeep, S.; Rinklebe, J.; Ok, Y.S.; Choudhury, B.U.; et al. Chapter One—Soil Organic Carbon Dynamics: Impact of Land Use Changes and Management Practices: A Review. In *Advances in Agronomy*; Sparks, D.L., Ed.; Academic Press: Cambridge, MA, USA, 2019; Volume 156, pp. 1–107, ISBN 0065-2113.
6. Brahim, N.; Ibrahim, H.; Mlih, R.; Bouajila, A.; Karbout, N.; Bol, R. Soil OC and N Stocks in the Saline Soil of Tunisian Gataaya Oasis Eight Years after Application of Manure and Compost. *Land* **2022**, *11*, 442. [CrossRef]
7. Amelung, W.; Bossio, D.; de Vries, W.; Kögel-Knabner, I.; Lehmann, J.; Amundson, R.; Bol, R.; Collins, C.; Lal, R.; Leifeld, J.; et al. Towards a global-scale soil climate mitigation strategy. *Nat. Commun.* **2020**, *11*, 5427. [CrossRef] [PubMed]
8. Tisdall, J.M. Formation of Soil Aggregates and Accumulation of Soil Organic Matter. In *Structure and Organic Matter Storage in Agricultural Soils*; CRC Press: Boca Raton, FL, USA, 2020; pp. 57–96.
9. Murphy, B.W. Impact of Soil Organic Matter on Soil Properties—A Review with Emphasis on Australian Soils. *Soil Res.* **2015**, *53*, 605–635. [CrossRef]
10. Foster, R.C. Microenvironments of Soil Microorganisms. *Biol. Fertil. Soils* **1988**, *6*, 189–203. [CrossRef]
11. Mohammadi, K.; Heidari, G.; Khalesro, S.; Sohrabi, Y. Soil Management, Microorganisms and Organic Matter Interactions: A Review. *Afr. J. Biotechnol.* **2011**, *10*, 19840.
12. Kaviya, N.; Upadhayay, V.K.; Singh, J.; Khan, A.; Panwar, M.; Singh, A.V. Role of Microorganisms in Soil Genesis and Functions. In *Mycorrhizosphere Pedogenesis*; Springer: Berlin/Heidelberg, Germany, 2019; pp. 25–52.
13. Ramond, J.-B.; Jordaan, K.; Díez, B.; Heinzelmann, S.M.; Cowan, D.A. Microbial Biogeochemical Cycling of Nitrogen in Arid Ecosystems. *Microbiol. Mol. Biol. Rev.* **2022**, *86*, e00109–e00121. [CrossRef]
14. Fisk, L.M.; Barton, L.; Maccarone, L.D.; Jenkins, S.N.; Murphy, D.V. Seasonal Dynamics of Ammonia-Oxidizing Bacteria but Not Archaea Influence Soil Nitrogen Cycling in a Semi-Arid Agricultural Soil. *Sci. Rep.* **2022**, *12*, 7299. [CrossRef]
15. Smith, J.L.; Halvorson, J.J.; Bolton, H., Jr. Spatial Relationships of Soil Microbial Biomass and C and N Mineralization in a Semi-Arid Shrub-Steppe Ecosystem. *Soil Biol. Biochem.* **1994**, *26*, 1151–1159. [CrossRef]
16. Khatoon, H.; Solanki, P.; Narayan, M.; Tewari, L.; Rai, J.; Hina Khatoon, C. Role of Microbes in Organic Carbon Decomposition and Maintenance of Soil Ecosystem. *Int. J. Chem. Stud.* **2017**, *5*, 1648–1656.
17. Raza, T.; Qadir, M.F.; Khan, K.S.; Eash, N.S.; Yousuf, M.; Chatterjee, S.; Manzoor, R.; ur Rehman, S.; Oetting, J.N. Unrevealing the Potential of Microbes in Decomposition of Organic Matter and Release of Carbon in the Ecosystem. *J. Environ. Manag.* **2023**, *344*, 118529. [CrossRef]
18. Kögel-Knabner, I. The Macromolecular Organic Composition of Plant and Microbial Residues as Inputs to Soil Organic Matter. *Soil Biol. Biochem.* **2002**, *34*, 139–162. [CrossRef]
19. Mlih, R.K.; Gocke, M.I.; Bol, R.; Berns, A.E.; Fuhrmann, I.; Brahim, N. Soil Organic Matter Composition in Coastal and Continental Date Palm Systems: Insights from Tunisian Oases. *Pedosphere* **2019**, *29*, 444–456. [CrossRef]
20. Pan, Y.; Kang, P.; Qu, X.; Zhang, H.; Li, X. Response of the Soil Bacterial Community to Seasonal Variations and Land Reclamation in a Desert Grassland. *Ecol. Indic.* **2024**, *165*, 112227. [CrossRef]
21. Ayangbenro, A.S.; Babalola, O.O. Reclamation of Arid and Semi-Arid Soils: The Role of Plant Growth-Promoting Archaea and Bacteria. *Curr. Plant Biol.* **2021**, *25*, 100173. [CrossRef]
22. Moussa, A.S.; Van Rensburg, L.; Kellner, K.; Bationo, A. Soil Microbial Biomass in Semi-Arid Communal Sandy Rangelands in the Western Bophirima District, South Africa. *Appl. Ecol. Environ. Res.* **2007**, *5*, 43–56. [CrossRef]
23. Pasternak, Z.; Al-Ashhab, A.; Gatica, J.; Gafny, R.; Avraham, S.; Minz, D.; Gillor, O.; Jurkevitch, E. Spatial and Temporal Biogeography of Soil Microbial Communities in Arid and Semiarid Regions. *PLoS ONE* **2013**, *8*, e69705. [CrossRef]
24. Abdul Rahman, N.S.N.; Abdul Hamid, N.W.; Nadarajah, K. Effects of Abiotic Stress on Soil Microbiome. *Int. J. Mol. Sci.* **2021**, *22*, 9036. [CrossRef]
25. Lüneberg, K.; Schneider, D.; Siebe, C.; Daniel, R. Drylands Soil Bacterial Community Is Affected by Land Use Change and Different Irrigation Practices in the Mezquital Valley, Mexico. *Sci. Rep.* **2018**, *8*, 1413. [CrossRef] [PubMed]
26. Ding, G.-C.; Piceno, Y.M.; Heuer, H.; Weinert, N.; Dohrmann, A.B.; Carrillo, A.; Andersen, G.L.; Castellanos, T.; Tebbe, C.C.; Smalla, K. Changes of Soil Bacterial Diversity as a Consequence of Agricultural Land Use in a Semi-Arid Ecosystem. *PLoS ONE* **2013**, *8*, e59497. [CrossRef]

27. García-Orenes, F.; Morugán-Coronado, A.; Zornoza, R.; Scow, K. Changes in Soil Microbial Community Structure Influenced by Agricultural Management Practices in a Mediterranean Agro-Ecosystem. *PLoS ONE* **2013**, *8*, e80522. [CrossRef] [PubMed]
28. Szoboszlay, M.; Dohrmann, A.B.; Poeplau, C.; Don, A.; Tebbe, C.C. Impact of Land-Use Change and Soil Organic Carbon Quality on Microbial Diversity in Soils across Europe. *FEMS Microbiol. Ecol.* **2017**, *93*, fix146. [CrossRef]
29. van der Bom, F.; Nunes, I.; Raymond, N.S.; Hansen, V.; Bonnichsen, L.; Magid, J.; Nybroe, O.; Jensen, L.S. Long-Term Fertilisation Form, Level and Duration Affect the Diversity, Structure and Functioning of Soil Microbial Communities in the Field. *Soil Biol. Biochem.* **2018**, *122*, 91–103. [CrossRef]
30. Allison, S.D.; Martiny, J.B.H. Resistance, Resilience, and Redundancy in Microbial Communities. *Proc. Natl. Acad. Sci. USA* **2008**, *105*, 11512–11519. [CrossRef] [PubMed]
31. Wardle, D. A Comparative Assessment or Factors Which Influence Microbial Biomass Carbon and Nitrogen Levels in Soil. *Biol. Rev. Camb. Philos. Soc.* **1992**, *67*, 321–358. [CrossRef]
32. Geisseler, D.; Scow, K.M. Long-Term Effects of Mineral Fertilizers on Soil Microorganisms—A Review. *Soil Biol. Biochem.* **2014**, *75*, 54–63. [CrossRef]
33. Shi, X.; Fan, J.; Zhang, J.; Shen, Y. Enhanced Phosphorus Removal in Intermittently Aerated Constructed Wetlands Filled with Various Construction Wastes. *Environ. Sci. Pollut. Res.* **2017**, *24*, 22524–22534. [CrossRef]
34. Domeignoz-Horta, L.A.; Cappelli, S.L.; Shrestha, R.; Gerin, S.; Lohila, A.K.; Heinonsalo, J.; Nelson, D.B.; Kahmen, A.; Duan, P.; Sebag, D.; et al. Plant Diversity Drives Positive Microbial Associations in the Rhizosphere Enhancing Carbon Use Efficiency in Agricultural Soils. *Nat. Commun.* **2024**, *15*, 8065. [CrossRef]
35. Wu, T.; Milner, H.; Díaz-Pérez, J.C.; Ji, P. Effects of Soil Management Practices on Soil Microbial Communities and Development of Southern Blight in Vegetable Production. *Appl. Soil Ecol.* **2015**, *91*, 58–67. [CrossRef]
36. Krauss, M.; Berner, A.; Perrochet, F.; Frei, R.; Niggli, U.; Mäder, P. Enhanced Soil Quality with Reduced Tillage and Solid Manures in Organic Farming—A Synthesis of 15 Years. *Sci. Rep.* **2020**, *10*, 4403. [CrossRef] [PubMed]
37. Ahmad, S.; Hussain, I.; Ghaffar, A.; Rahman, M.H.U.; Saleem, M.Z.; Yonas, M.W.; Hussnain, H.; Ikram, R.M.; Arslan, M. Organic Amendments and Conservation Tillage Improve Cotton Productivity and Soil Health Indices under Arid Climate. *Sci. Rep.* **2022**, *12*, 14072. [CrossRef] [PubMed]
38. Li, G.; Wu, C.; Gao, W. Effects of Short-Term Fallow Managements on Soil Microbial Properties: A Case Study in China. *Appl. Soil Ecol.* **2018**, *125*, 128–137. [CrossRef]
39. Afaf, N.; Zohra, I.; Faiza, B.Z.; Abdelkader, B. Diversity of Arbuscular Mycorrhizal Fungi in Two Perturbed Ecosystems (Dune and Saline Soil) in West Algeria. *Int. J. Agric. Crop Sci.* **2015**, *8*, 380.
40. Karabi, M.; Hamdi, A.B.; Zenkhri, S. Microbial Diversity and Organic Matter Fractions under Two Arid Soils in Algerian Sahara. In *Proceedings of the AIP Conference Proceedings*; AIP Publishing: Melville, NY, USA, 2016; Volume 1758.
41. Bencherif, K.; Boutekrabt, A.; Fontaine, J.; Laruelle, F.; Dalpè, Y.; Sahraoui, A.L.-H. Impact of Soil Salinity on Arbuscular Mycorrhizal Fungi Biodiversity and Microflora Biomass Associated with Tamarix Articulata Vahll Rhizosphere in Arid and Semi-Arid Algerian Areas. *Sci. Total Environ.* **2015**, *533*, 488–494. [CrossRef]
42. Menasria, T. Biodiversité Microbienne Dans les Milieux Extrêmes Salés du Nord-Est Algérien. Ph.D. Thesis, Université de Batna, Batna, Algeria, 2020.
43. Bellague, D.; M'Hammedi-Bouzina, M.; Abdelguerfi, A. Measuring the Performance of Perennial Alfalfa with Drought Tolerance Indices. *Chil. J. Agric. Res.* **2016**, *76*, 273–284. [CrossRef]
44. Chedjerat, A. Comportement de Seize Cultivars de Luzerne Pérenne (*Médicago sativa* L.) Conduits en Pluvial et en Irrigué Dans les Conditions du Bas Chélif. Ph.D. Thesis, Ecole Nationale Supérieure Agronomique, Algiers, Algeria, 2017.
45. Douaoui, A.; Hartani, T.; Lakehal, M. La Salinisation Dans La Plaine Du Bas-Cheliff: Acquis et Perspectives. In Proceedings of the Economies d'eau en Systèmes IRrigués au Maghreb, Marrakech, Maroc, 29–31 May 2006; Deuxième atelier Régional du Projet SIRMA.
46. Pietikäinen, J.; Pettersson, M.; Bååth, E. Comparison of Temperature Effects on Soil Respiration and Bacterial and Fungal Growth Rates. *FEMS Microbiol. Ecol.* **2005**, *52*, 49–58. [CrossRef]
47. Gee, G.W.; Bauder, J.W. Particle-Size Analysis. In *Methods of Soil Analysis. Part 4: Physical Methods*; Dane, J.H., Topp, G.C., Eds.; Soil Science Society of America: Madison, WI, USA, 2002; pp. 255–293. [CrossRef]
48. Rhoades, J.D. Salinity: Electrical Conductivity and Total Dissolved Solids. In *Methods of Soil Analysis: Part 3—Chemical Methods*; Sparks, D.L., Ed.; Soil Science Society of America: Madison, WI, USA, 1996; pp. 417–435. [CrossRef]
49. Blake, G.R. Bulk Density. In *Methods of Soil Analysis: Part 1—Physical and Mineralogical Methods*, 2nd ed.; Klute, A., Ed.; Soil Science Society of America: Madison, WI, USA, 1986; pp. 363–375. [CrossRef]
50. Gillman, G.P.; Sinclair, D.F.; Beech, T.A. Recovery of Organic Carbon by the Walkley and Black Procedure in Highly Weathered Soils. *Commun. Soil Sci. Plant Anal.* **1986**, *17*, 885–892. [CrossRef]
51. Sparks, D.L.; Page, A.L.; Helmke, P.A.; Loeppert, R.H. *Methods of Soil Analysis, Part 3: Chemical Methods*; John Wiley & Sons: Hoboken, NJ, USA, 2020; Volume 14, ISBN 0891188258.

52. Drouineau, G. Dosage Rapide Du Calcaire Actif Du Sol: Nouvelles Données Sur La Separation et La Nature Des Fractions Calcaires. *Ann. Agron* **1942**, *12*, 441–450.
53. Bremner, J.M. Total Nitrogen. In *Methods of Soil Analysis; Agronomy Monographs*; Wiley: Hoboken, NJ, USA, 1965; pp. 1149–1178, ISBN 9780891182047.
54. Olsen, S.R. *Estimation of Available Phosphorus in Soils by Extraction with Sodium Bicarbonate*; US Department of Agriculture: Washington, DC, USA, 1954.
55. Nash, V.E. Potassium Release Characteristics of Some Soils of the Mississippi Coastal Plain as Revealed by Various Extracting Agents. *Soil Sci.* **1971**, *111*, 313–317. [CrossRef]
56. Rhoades, J.D. Cation Exchange Capacity. In *Methods of Soil Analysis: Part 2—Chemical and Microbiological Properties*, 2nd ed.; Page, A.L., Miller, R.H., Keeney, D.R., Eds.; American Society of Agronomy: Madison, WI, USA, 1982; pp. 149–157. [CrossRef]
57. Lefèvre, C.; Rekik, F.; Alcantara, V.; Wiese, L. *Soil Organic Carbon: The Hidden Potential*; CABI: Wallingford, UK, 2017; ISBN 9251096813.
58. Poeplau, C. *Measuring and Modelling Soil Carbon Stocks and Stock Changes in Livestock Production Systems: Guidelines for Assessment*; Version 1-Advanced Copy; FAO: Rome, Italy, 2019.
59. Magoč, T.; Salzberg, S.L. FLASH: Fast Length Adjustment of Short Reads to Improve Genome Assemblies. *Bioinformatics* **2011**, *27*, 2957–2963. [CrossRef] [PubMed]
60. Bokulich, N.A.; Subramanian, S.; Faith, J.J.; Gevers, D.; Gordon, J.I.; Knight, R.; Mills, D.A.; Caporaso, J.G. Quality-Filtering Vastly Improves Diversity Estimates from Illumina Amplicon Sequencing. *Nat. Methods* **2013**, *10*, 57–59. [CrossRef] [PubMed]
61. Haas, B.J.; Gevers, D.; Earl, A.M.; Feldgarden, M.; Ward, D.V.; Giannoukos, G.; Ciulla, D.; Tabbaa, D.; Highlander, S.K.; Sodergren, E. Chimeric 16S RRNA Sequence Formation and Detection in Sanger and 454-Pyrosequenced PCR Amplicons. *Genome Res.* **2011**, *21*, 494–504. [CrossRef]
62. Edgar, R.C. UPARSE: Highly Accurate OTU Sequences from Microbial Amplicon Reads. *Nat. Methods* **2013**, *10*, 996–998. [CrossRef]
63. Quast, C.; Pruesse, E.; Yilmaz, P.; Gerken, J.; Schweer, T.; Yarza, P.; Peplies, J.; Glöckner, F.O. The SILVA Ribosomal RNA Gene Database Project: Improved Data Processing and Web-Based Tools. *Nucleic Acids Res.* **2012**, *41*, D590–D596. [CrossRef] [PubMed]
64. Edgar, R.C.; Haas, B.J.; Clemente, J.C.; Quince, C.; Knight, R. UCHIME Improves Sensitivity and Speed of Chimera Detection. *Bioinformatics* **2011**, *27*, 2194–2200. [CrossRef]
65. Altschul, S.F.; Gish, W.; Miller, W.; Myers, E.W.; Lipman, D.J. Basic Local Alignment Search Tool. *J. Mol. Biol.* **1990**, *215*, 403–410. [CrossRef]
66. Wang, Q.; Garrity, G.M.; Tiedje, J.M.; Cole, J.R. Naive Bayesian Classifier for Rapid Assignment of RRNA Sequences into the New Bacterial Taxonomy. *Appl. Environ. Microbiol.* **2007**, *73*, 5261–5267. [CrossRef]
67. Edgar, R.C. MUSCLE: Multiple Sequence Alignment with High Accuracy and High Throughput. *Nucleic Acids Res.* **2004**, *32*, 1792–1797. [CrossRef]
68. Six, J.; Feller, C.; Denef, K.; Ogle, S.; de Moraes Sa, J.C.; Albrecht, A. Soil Organic Matter, Biota and Aggregation in Temperate and Tropical Soils-Effects of No-Tillage. *Agronomie* **2002**, *22*, 755–775. [CrossRef]
69. Juarez, S. Régulations Biotiques et Abiotiques de La Décomposition des Matières Organiques des Sols. Ph.D. Thesis, AgroParis-Tech, Paris, France, 2013.
70. Dignac, M.-F.; Derrien, D.; Barré, P.; Barot, S.; Cécillon, L.; Chenu, C.; Chevallier, T.; Freschet, G.T.; Garnier, P.; Guenet, B. Increasing Soil Carbon Storage: Mechanisms, Effects of Agricultural Practices and Proxies. A Review. *Agron. Sustain. Dev.* **2017**, *37*, 14. [CrossRef]
71. Larney, F.J.; Angers, D.A. The Role of Organic Amendments in Soil Reclamation: A Review. *Can. J. Soil Sci.* **2012**, *92*, 19–38. [CrossRef]
72. Macedo, M.O.; Resende, A.S.; Garcia, P.C.; Boddey, R.M.; Jantalia, C.P.; Urquiaga, S.; Campello, E.F.C.; Franco, A.A. Changes in Soil C and N Stocks and Nutrient Dynamics 13 Years after Recovery of Degraded Land Using Leguminous Nitrogen-Fixing Trees. *For. Ecol. Manag.* **2008**, *255*, 1516–1524. [CrossRef]
73. Chivenge, P.; Vanlauwe, B.; Gentile, R.; Six, J. Organic Resource Quality Influences Short- Term Aggregate Dynamics and Soil Organic Carbon and Nitrogen Accumulation. *Soil Biol. Biochem.* **2011**, *43*, 657–666. [CrossRef]
74. Mancer, H.; Bettiche, F.; Chaib, W.; Dekki, N.; Benaoun, S.; Rechachi, M.Z. Influence de La Salinité Des Eaux d'irrigation Sur La Minéralisation Du Carbone Organique Dans Le Sol. *J. Algérien Des Régions Arid.* **2020**, *14*, 48–55.
75. Blevins, R.L.; Smith, M.S.; Thomas, G.W. Changes in Soil Properties under No-Tillage. In *No-Tillage Agriculture: Principles and Practices*; Springer: Boston, MA, USA, 1984; pp. 190–230.
76. Hill, R.L. Long-term Conventional and No-tillage Effects on Selected Soil Physical Properties. *Soil Sci. Soc. Am. J.* **1990**, *54*, 161–166. [CrossRef]
77. Feller, C.; Fritsch, E.; Poss, R.; Valentin, C. Effet de La Texture Sur Le Stockage et La Dynamique Des Matières Organiques Dans Quelques Sols Ferrugineux et Ferrallitiques (Afrique de l'Ouest, En Particulier). *Cah. ORSTOM* **1991**, *26*, 25–36.

78. Jobbágy, E.G.; Jackson, R.B. The Vertical Distribution of Soil Organic Carbon and Its Relation to Climate and Vegetation. *Ecol. Appl.* **2000**, *10*, 423–436. [CrossRef]
79. Rowley, M.C.; Grand, S.; Verrecchia, É.P. Calcium-Mediated Stabilisation of Soil Organic Carbon. *Biogeochemistry* **2018**, *137*, 27–49. [CrossRef]
80. Fierer, N.; Jackson, R.B. The diversity and biogeography of soil bacterial communities. *Proc. Natl. Acad. Sci. USA* **2006**, *103*, 626–631. [CrossRef] [PubMed]
81. Schimel, J.; Balser, T.C.; Wallenstein, M. Microbial stress-response physiology and its implications for ecosystem function. *Ecology* **2007**, *88*, 1386–1394. [CrossRef] [PubMed]
82. Six, J.; Frey, S.D.; Thiet, R.K.; Batten, K.M. Bacterial and fungal contributions to carbon sequestration in agroecosystems. *Soil Sci. Soc. Am. J.* **2006**, *70*, 555–569. [CrossRef]
83. Faust, K.; Raes, J. Microbial interactions: From networks to models. *Nat. Rev. Microbiol.* **2012**, *10*, 538–550. [CrossRef]
84. Rath, K.M.; Rousk, J. Salt effects on the soil microbial decomposer community and their role in organic carbon cycling: A review. *Soil Biol. Biochem.* **2015**, *81*, 108–123. [CrossRef]
85. Lennon, J.T.; Jones, S.E. Microbial seed banks: The ecological and evolutionary implications of dormancy. *Nat. Rev. Microbiol.* **2011**, *9*, 119–130. [CrossRef]
86. Lopez, S. Déterminisme de la Diversité Bactérienne Rhizosphérique des Hyperaccumulateurs de Nickel. Ph.D. Thesis, Université de Lorraine, Nancy, France, 2018.
87. Hatimi, A.; Tahrouch, S. Caractérisations Chimique, Botanique et Microbiologique Du Sol Des Dunes Littorales Du Souss-Massa. *Biomatec Echo* **2007**, *2*, 85–97.
88. Makhalanyane, T.P.; Valverde, A.; Gunnigle, E.; Frossard, A.; Ramond, J.-B.; Cowan, D.A. Microbial Ecology of Hot Desert Edaphic Systems. *FEMS Microbiol. Rev.* **2015**, *39*, 203–221. [CrossRef]
89. Wang, Y. The Bacterial Communities of Sand-like Surface Soils of the San Rafael Swell (Utah, USA) and the Desert of Maine (USA). Ph.D. Thesis, Université Paris-Saclay, Paris, France, 2015.
90. Purbalisa, W.; Hendrayanti, D.; Yusuf, W.A. Biodiversity, Roles, and Potency of Bacteria in Agricultural Land. *J. Presipitasi: Media Komun. Dan Pengemb. Tek. Lingkungan* **2022**, *19*, 520–531.
91. Yang, T.; Lupwayi, N.; Marc, S.-A.; Siddique, K.H.M.; Bainard, L.D. Anthropogenic Drivers of Soil Microbial Communities and Impacts on Soil Biological Functions in Agroecosystems. *Glob. Ecol. Conserv.* **2021**, *27*, e01521. [CrossRef]
92. Lienhard, P.; Terrat, S.; Prévost-Bouré, N.C.; Nowak, V.; Régnier, T.; Sayphoummie, S.; Panyasiri, K.; Tivet, F.; Mathieu, O.; Levêque, J. Pyrosequencing Evidences the Impact of Cropping on Soil Bacterial and Fungal Diversity in Laos Tropical Grassland. *Agron. Sustain. Dev.* **2014**, *34*, 525–533. [CrossRef]
93. Ling, L.; Fu, Y.; Jeewani, P.H.; Tang, C.; Pan, S.; Reid, B.J.; Gunina, A.; Li, Y.; Li, Y.; Cai, Y. Organic Matter Chemistry and Bacterial Community Structure Regulate Decomposition Processes in Post-Fire Forest Soils. *Soil Biol. Biochem.* **2021**, *160*, 108311. [CrossRef]
94. Mhete, M.; Eze, P.N.; Rahube, T.O.; Akinyemi, F.O. Soil Properties Influence Bacterial Abundance and Diversity under Different Land-Use Regimes in Semi-Arid Environments. *Sci. Afr.* **2020**, *7*, e00246. [CrossRef]
95. Jung, M.-Y.; Well, R.; Min, D.; Giesemann, A.; Park, S.-J.; Kim, J.-G.; Kim, S.-J.; Rhee, S.-K. Isotopic Signatures of N2O Produced by Ammonia-Oxidizing Archaea from Soils. *ISME J.* **2014**, *8*, 1115–1125. [CrossRef]
96. Tourna, M.; Stieglmeier, M.; Spang, A.; Könneke, M.; Schintlmeister, A.; Urich, T.; Engel, M.; Schloter, M.; Wagner, M.; Richter, A. Nitrososphaera Viennensis, an Ammonia Oxidizing Archaeon from Soil. *Proc. Natl. Acad. Sci. USA* **2011**, *108*, 8420–8425. [CrossRef]
97. Pester, M.; Schleper, C.; Wagner, M. The Thaumarchaeota: An emerging view of their phylogeny and ecophysiology. *Curr. Opin. Microbiol.* **2011**, *14*, 300–306. [CrossRef]
98. Six, J.; Conant, R.T.; Paul, E.A.; Paustian, K. Stabilization of Organic Matter by Soil Minerals: Implications for C-Saturation of Soils. *Plant Soil.* **2002**, *241*, 155–176. [CrossRef]
99. Ventura, M.; Canchaya, C.; Tauch, A.; Chandra, G.; Fitzgerald, G.F.; Chater, K.F.; van Sinderen, D. Genomics of Actinobacteria: Tracing the evolutionary history of an ancient phylum. *Microbiol. Mol. Biol. Rev.* **2007**, *71*, 495–548. [CrossRef]
100. Lauber, C.L.; Strickland, M.S.; Bradford, M.A.; Fierer, N. The influence of soil properties on the structure of bacterial and fungal communities across land-use types. *Soil Biol. Biochem.* **2008**, *40*, 2407–2415. [CrossRef]
101. Evans, S.E.; Wallenstein, M.D. Soil microbial community response to drying and rewetting stress: Does historical precipitation regime matter? *Biogeochemistry* **2012**, *109*, 101–116. [CrossRef]
102. Six, J.; Bossuyt, H.; Degryze, S.; Denef, K. A History of Research on the Link between (Micro)Aggregates, Soil Biota, and Soil Organic Matter Dynamics. *Soil Tillage Res.* **2004**, *79*, 7–31. [CrossRef]
103. Lehmann, J.; Kleber, M. The Contentious Nature of Soil Organic Matter. *Nature* **2015**, *528*, 60–68. [CrossRef] [PubMed]

104. Bot, A.; Benites, J. *The Importance of Soil Organic Matter: Key to Drought-Resistant Soil and Sustained Food Production*; FAO Soils Bulletin 80; FAO: Rome, Italy, 2005. Available online: https://www.fao.org/3/a0100e/a0100e.pdf (accessed on 12 May 2025).
105. Fierer, N.; Bradford, M.A.; Jackson, R.B. Toward an ecological classification of soil bacteria. *Ecology* **2007**, *88*, 1354–1364. [CrossRef]

land

Article

Effects of Management Practices on Soil Microbial Diversity and Structure on Eucalyptus Plantations

Yuanyuan Xue [1,2], Wei Liu [1,*], Qi Feng [1], Jutao Zhang [1], Lingge Wang [1,2], Zexia Chen [1,2], Xuejiao Li [1,2] and Meng Zhu [1,*]

[1] Key Laboratory of Ecological Safety and Sustainable Development in Arid Lands, Northwest Institute of Eco-Environment and Resources, Chinese Academy of Sciences, Lanzhou 730000, China; xueyuanyuan@nieer.ac.cn (Y.X.); qifeng@lzb.ac.cn (Q.F.); jutzhang@lzb.ac.cn (J.Z.); wanglingge@nieer.ac.cn (L.W.); chenzexia@nieer.ac.cn (Z.C.); lixuejiao@nieer.ac.cn (X.L.)

[2] University of Chinese Academy of Sciences, Beijing 100049, China

* Correspondence: weiliu@lzb.ac.cn (W.L.); zhumeng@lzb.ac.cn (M.Z.)

Abstract: Soil microbes are critical in regulating the growth and function of eucalyptus plantations. The mechanisms underlying soil microbial communities' response to different eucalyptus plantation management practices remain elusive. In this study, we compiled datasets containing 2744 observations across global eucalyptus-planted regions and analyzed the effects of five management practices (i.e., burning, residual removal, fertilization, mixed planting, and controlling planting years) on soil microbial biomass, diversity, and structures. Our results showed that fungal community alpha diversity responds more sensitively to management practices than bacterial community alpha diversity on eucalyptus plantations. Although the implementation of management practices significantly increased the content of most soil nutrients and microbial biomass elements (excluding burning), these practices did not necessarily improve soil microbial biomass and diversity, particularly among fungal communities. Burning, fertilization, and mixing eucalyptus with nitrogen-fixing species significantly decreased the diversity of fungal communities, which were mainly impacted by soil organic carbon and total potassium content. Compared to the four other management practices, mixing eucalyptus with nitrogen-fixing species favored the growth of bacterial communities and the storage of microbial biomass nitrogen, making it the most effective management practice. However, attention should also be paid to the protection of fungal communities. In addition, these management practices significantly changed microbial community structures, which were positively correlated with the microbial biomass elements carbon and nitrogen and, to a lesser extent, soil microbial alpha diversity. Our results highlight the importance of prioritizing mixing eucalyptus with nitrogen-fixing species as a management practice and safeguarding fungal community diversity during its implementation and suggest that microbial diversity development associated with soil organic carbon and potassium contents should be given priority in eucalyptus plantation management.

Keywords: soil microbial community; eucalyptus; management practices; meta-analysis

1. Introduction

Eucalyptus trees exhibit rapid growth, even in nutrient-poor soils, outperforming species such as Acacia, Pinus, and other common plantation species [1–4]. These trees are capable of efficiently absorbing nutrients from highly weathered soils in tropical and subtropical regions, enabling them to produce substantial biomass in a relatively short period [5]. Consequently, eucalyptus plantations have been successfully established across various soil types, providing both commercial and subsistence opportunities for rural smallholders [6]. Furthermore, owing to its significant carbon sequestration potential, eucalyptus plays a vital role in maintaining carbon balance under global climate change conditions [7,8]. These advantages have thus enabled eucalyptus to become one of the most widely utilized species for afforestation efforts globally, with a broad distribution across Australia, Asia, Africa, the Americas, and other regions [8]. However, the results of a number of studies have indicated that during the growth of eucalyptus, it can consume large amounts of water and nutrients, substantially depleting soil resources and impacting the global flux of nitrogen dioxide [9,10]. The allelopathic effects of eucalyptus can reduce understory plant biomass and diversity, which not only diminishes soil resource availability but also induces deterioration in local microclimates, leading to a loss of biodiversity [11–14]. At present, a series of management practices have been implemented around the globe to mitigate the adverse impacts of eucalyptus and optimize its benefits. These practices include burning, fertilization, mixed planting, residue removal, and controlling planting age [15–19]. The results of previous studies suggest that artificial interventions on eucalyptus plantations are beneficial for eucalyptus growth and soil fertility [17,19–21]. However, the results of other studies indicate that such interventions may negatively impact understory and soil biodiversity, thereby hindering the sustainable development of eucalyptus plantations [15,16,18,22]. At present, the effects of different management practices on eucalyptus plantations remain unclear.

Microorganisms play a vital role in ecosystems, with them being widely involved in the processes of carbon and nitrogen cycling, in addition to soil structure formation [15,17,19]. Compared to soil properties, they exhibit higher sensitivity to environmental changes, and variations in their biomass, diversity, and structure are considered early indicators of soil ecosystem changes [23,24]. A series of management practices implemented by personnel on eucalyptus plantations can directly or indirectly (by affecting soil properties and aboveground vegetation) impact the biomass, structure, and diversity of soil microorganism communities [18–21,25]. For example, burning and residual removal can effectively mitigate the accumulation of surface fuels while concurrently diminishing the influx of carbon and nitrogen nutrients into the soil via leaching from litter, resulting in a substantial reduction in microbial biomass and alterations in microbial community structures [16,22,24–26]. However, the results of some studies suggest that the thermal effects of fire can stimulate the germination of numerous plant species within ecosystems [27] and stimulate spore germination of heat-tolerant species, which induces soil microbial activity [28,29]. In contrast, removing accumulated litter from a forest area can enhance soil aeration and water infiltration, in addition to promoting the growth of a wider variety of plants [30]. These factors, in turn, provide a wider range of ecological niches and nutritional resources for soil microorganisms, ultimately enhancing soil microbial diversity and activity [25,30]. Fertilization and mixing nitrogen-fixing species with eucalyptus mainly impact the microbial community through the addition of soil nitrogen supply and the enhancement of soil nitrogen utilization efficiency [17–19,31,32]. The introduction of nitrogen-fixing species can enhance aboveground vegetation diversity by promoting photosynthesis, further altering microbial community diversity and structure [18,20,33]. However, although fertilization has the potential to enhance soil nutrient contents [34,35], its influence on microorganisms is not consistently beneficial [36–38]. In their study, Hasselquist et al. [38] found that the enrichment of the nitrogen fertilizer inhibits saprophytic fungi while activating certain mycorrhizal fungi. Regarding timber, the determination of eucalyptus rotation periods often hinges on considerations of economic profitability and carbon sequestration potential, with limited regard for their effects on soil quality [21,39]. Shorter rotation periods typically yield higher productivity but lead to soil nutrient imbalances [20,40,41]. Conversely, longer rotation periods allow for the gradual opening of eucalyptus canopies, which promotes understory biodiversity and accelerates litter decomposition rates [42]. These processes benefit soil nutrient reserves [39,43] and support the resilience and development of microbial communities [21]. Nonetheless, the results of some studies suggest that with the aging of eucalyptus stands, the allelopathic effects lead to notable declines in the diversity of undisturbed vegetation and the content of soil nutrients, thereby impacting soil enzyme activity, microbial diversity, and community structure [44–47].

The inconsistent results presented in the above studies may stem from variations in management practices, plantation age, and climatic conditions at the location under study. Individual studies, unfortunately, often lack the capacity to comprehensively account for all of these influencing factors. To address the above issues, some researchers have employed meta-analysis to evaluate the effects of artificial management practices on forest soil properties and microbial communities. For instance, Guo et al. [48] and Curtright et al. [49] assessed the impact of mixed-species planting on soil properties, enzyme activity, and microbial biomass in plantation forests. The results of studies by Wang et al. [26] and Zhou et al. [50] demonstrated that nitrogen fertilization led to a significant decrease in microbial diversity and shift in microbial community composition in plantation forests. Furthermore, the results of several studies have highlighted the effects of forest residue harvesting on the soil nutrient content and microbial biomass [51,52]. However, the authors of these studies predominantly focused on individual management practices within global forests, primarily investigating their influence on soil properties, enzyme activity, microbial diversity, and community structure. At present, there are no studies specifically focusing on eucalyptus plantations that examine in detail the effects of artificial management practices on microbial biomass, diversity, and community structure across different microbial taxa. In light of this gap, it is necessary to conduct a meta-analysis of multiple studies to determine the effect of management practices on soil microorganisms on eucalyptus plantations, specifically in the different microbial taxa. The results of this study can provide scientific guidance for subsequent eucalyptus plantation management strategies and contribute to sustainable development.

To accomplish the above research aims, we conducted a meta-analysis of 2744 observations, including soil properties, enzyme activity, microbial biomass, microbial community diversity, and structure. The management practices implemented on eucalyptus plantations included burning, fertilization, mixed planting, residue removal, and control of planting age, which are five common practices. The primary aims of this study were as follows: (1) to analyze the effects of different management practices on the biomass of different soil microbial taxa, diversity, and structure on eucalyptus plantations on a global scale and (2) to explore the key factors driving different soil microbial taxa biomass, diversity, and structure on eucalyptus plantations globally. Through the following study, we aim to provide scientific recommendations for the subsequent management of eucalyptus plantations.

2. Materials and Methods

2.1. Data Collection

We collected all articles published prior to May 2023 from Web of Science (https://apps.webofknowledge.com (accessed on 30 April 2023)) and the China Knowledge Resource Integrated Database (https://www.cnki.net (accessed on 30 April 2023)) addressing the following topics: eucalyptus plantations and soil microorganisms, soil fungi, and soil bacteria. In total, 2744 observations were collected from 103 articles (Figure 1 and Supplementary S1). To mitigate potential biases arising from variations across studies, we implemented stringent inclusion criteria during data collection. These criteria were as follows: (1) The article reports at least one microbial indicator, including microbial biomass elements (microbial biomass carbon (MBC), microbial biomass nitrogen (MBN), or microbial biomass phosphorus (MBP)), microbial composition biomass (PLFA, bacterial biomass, fungal biomass, Gram-positive bacterial biomass, Gram-negative bacterial biomass, or Actinomycetes biomass), alpha diversity index, beta diversity, and community structure; (2) the availability of sample size, mean, and standard deviation, either reported directly or calculated from tables, digitized charts, or textual descriptions; (3) the inclusion of both field experiments and laboratory cultures; and (4) for factorial experiments, the selection of only one management practice as the control and treatment factor, excluding any interaction factors. Each study underwent independent screening by two reviewers to determine compliance with these criteria. In cases of disagreement, additional reviewers were engaged to reach consensus. Data extraction from the figures presented in the papers was conducted using GetData Graph Digitizer software (version 2.20, http://www.getdata-graph-digitizer.com/index.php (accessed on 30 April 2023)) to extract the mean values and standard deviation.

The information collected by our group also includes site locations (latitude, longitude, and altitude), climate conditions (mean annual temperature (MAT) and mean annual precipitation (MAP)), soil physical properties (bulk density (BD) and soil moisture content (SMC)), soil chemical properties (total carbon (TC), soil organic carbon (SOC), total nitrogen (TN), total phosphorus (TP), total potassium (TK), and pH), soil enzymes (catalase, urease, and invertase), and experimental forcing factors (burning, fertilization, mixed planting, residue removal, and controlling planting age).

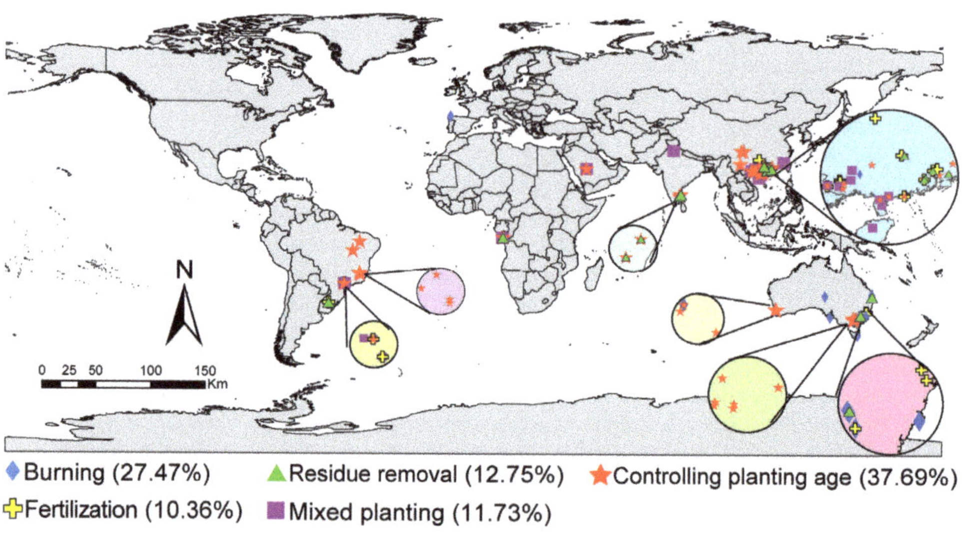

Figure 1. Soil sampling sites.

2.2. Data Calculations

For each case study, the effect of the practice on each variable was assessed and computed after transformation into natural logarithms (Table S1):

$$RR = \ln\left(\frac{\overline{x_t}}{\overline{x_c}}\right) = \ln(\overline{x_t}) - \ln(\overline{x_c}) \tag{1}$$

where $\overline{x_c}$ and $\overline{x_t}$ represent the mean values of the respective variable in the control and treatment groups, respectively. The change in soil pH demonstrated the impact of the practice on soil pH; the difference in pH between the treatment and control was calculated based on the method outlined in a previous meta-analysis [53]. The corresponding sample variance (v) was calculated as follows:

$$v = \frac{s_t^2}{n_t \overline{x_t}^2} + \frac{s_c^2}{n_c \overline{x_c}^2} \quad (2)$$

where n_c and n_t represent the replicate numbers of the treatment and control groups, respectively, and s_c and s_t represent the standard deviation of the treatment and control groups, respectively.

Microbial alpha diversity is generally evaluated using the Shannon and Richness indices [26,54]. Notably, the Chao, ACE, and OTU indices are commonly reported metrics for microbial richness in the literature.

The methodologies employed for analyzing microbial community structure in the reviewed articles predominantly encompassed redundancy analysis, principal coordinate analysis, principal component analysis, and non-metric multidimensional scaling analysis. These methods revealed discernible differences in soil microbial community structure between the control and treatment conditions [55]. The approach of Zhou et al. [56] was used to assess microbial community structure. Specifically, sample positions along the first two ordination axes were extracted, the "vegan" package was utilized to compute the Euclidean distance among different samples [57], and the RR of the microbial community structure ($RR_{structure}$) and the RR of beta diversity (RR_{Beta}) were calculated as follows:

$$RR_{structure} \ln\left(\frac{\overline{D_b}}{\overline{D_c + D_t}}\right) \quad (3)$$

$$RR_{\mathrm{Beta}} \ln\left(\frac{\overline{D_t}}{\overline{D_c}}\right) \quad (4)$$

where D_c is the distance within the control, D_t is the distance within the treatment, D_b is the distance between the control and the treatment, and $\overline{D_t}$, $\overline{D_c}$, and $\overline{D_b}$ represent the means of D_c, D_t, and D_b.

$RR_{structure} < 0$ indicates that the practice has no effect on microbial community structure; $RR_{structure} > 0$ indicates that there is a notable alteration in the community structure; and $RR_{Beta} > 0$ indicates that beta diversity was increased by the management practice.

2.3. Statistical Analyses

Egger's regression test and sensitivity analyses were conducted to test for potential publication bias and to determine the robustness of the meta-analysis (Table S2) [58]. Any outliers were excluded before analyses [51]. In addition, we assessed the heterogeneity of effect sizes using formal Cochran's Q-test to determine whether the variability in the observed effect sizes was larger than that expected by chance (Table S2) [59]. Based on Cochran's Q-test results, we chose either a random effects or fixed effects model to assess the impact of management practices on soil properties and microbial indicators using the "Metafor" package [60]. The significance of the difference between management practice treatment and control was determined for each variable based on whether the CI of the RR overlapped with zero [56].

Model selection relied on AIC correction. The relative importance of a given predictor was determined by summing the Akaike weights across all models in which the predictor appeared. Consequently, predictors featured in models with higher Akaike weights received greater importance values, indicating stronger overall support across all models. For this process, the "gmulti" package was utilized [61]. Four types of candidate predictors were considered: (1) climate factors, (2) soil properties, (3) soil enzymes, and (4) management practices. Lastly, we employed linear regression models to investigate the associations between the RR of microbial indicators and predictor variables, in addition to the relationship between the RR of microbial indicators (microbial biomass and alpha diversity) and beta diversity and structure. All statistical analyses were conducted in R software (Version 4.0.2).

3. Results

3.1. Variation in Soil Microbial Community Biomass, Diversity, and Structure

The soil microbial biomass elements of eucalyptus plantations were significantly influenced by the management practices employed (Figure 2). Both burning and residue removal significantly decreased MBC and MBN; in comparison, burning notably increased MBP ($p < 0.05$). Mixed planting significantly increased MBN ($p < 0.001$). Additionally, with the increase in planting age, MBC, MBN, and MBP significantly increased ($p < 0.05$). These findings demonstrate that implementing management practices on eucalyptus plantations does not always significantly increase the abundance of soil microbial biomass elements, with MBN being more sensitive to environmental changes compared to MBC.

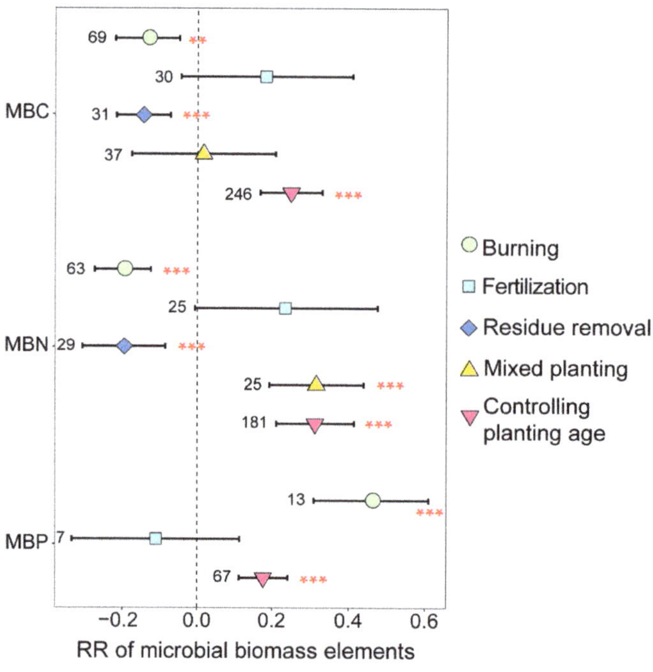

Figure 2. The response ratios of soil microbial biomass elements under different management practices. The values beside the bars are the corresponding number of observations. *** and ** indicates $p < 0.001$ and $p < 0.01$, respectively.

The microbial biomass of different species on eucalyptus plantations was also influenced by management practices employed, with the biomass of most species significantly increasing as a result of these practices, whereas fungal biomass significantly decreased (Figure 3). After burning, total microbial and bacterial biomass significantly increased; in comparison, fungal biomass significantly decreased ($p < 0.05$). Residue removal significantly decreased total microbial and fungal biomass ($p < 0.01$), and mixed planting significantly increased total microbial and bacterial biomass ($p < 0.05$). In addition, fertilization significantly increased Gram-negative bacterial biomass; in comparison, the increase in planting age significantly decreased Actinomycetes biomass ($p < 0.05$).

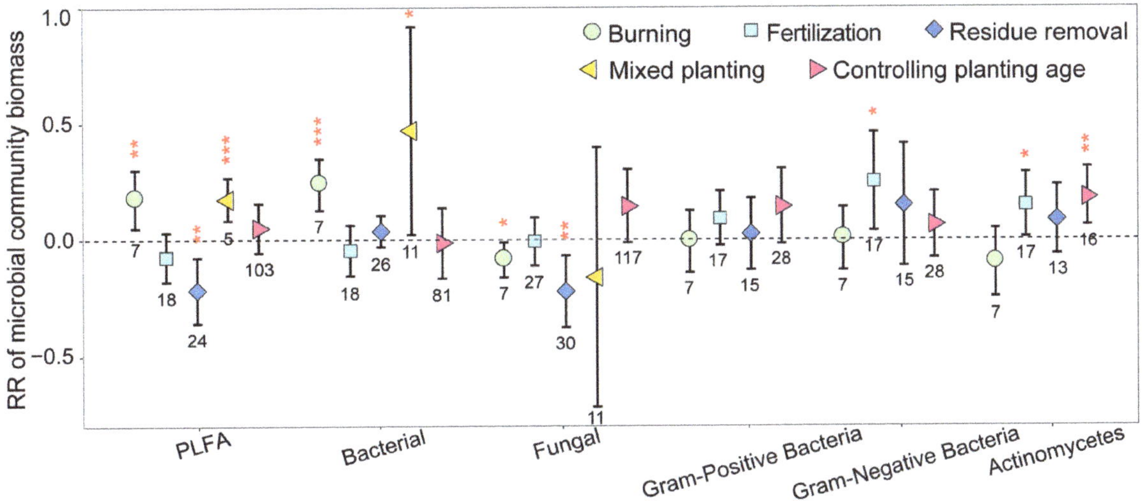

Figure 3. The response ratios of soil microbial biomass of different species under different management practices. The values shown at the bottom of the bars represent the corresponding number of observations. ***, ** and * indicates $p < 0.001$, $p < 0.01$ and $p < 0.05$, respectively.

The different management practices significantly decreased fungal community richness on eucalyptus plantations, with it being important to note that these practices may not consistently have a positive effect on the alpha diversity of bacterial communities (Figure 4). In addition, the management practices employed could possibly even reduce microbial beta diversity. Nonetheless, all management practices significantly impacted the soil microbial community structure (Figure 4).

As shown in Figure 4, burning significantly decreased the diversity and richness of soil fungal communities on eucalyptus plantations while having no effect on the diversity and richness of bacterial communities. These results suggest that the post-burning environment is unfavorable for the survival of fungal communities. The absolute values of the RR of bacterial and fungal Shannon indices were greater than those of the RR of bacterial and fungal richness indices, suggesting that rare species were more sensitive to burning. Similarly, the absolute values of the RR of bacterial and fungal richness indices on the mixed planting and older-age plantations were greater than those of the RR of bacterial and fungal Shannon indices, indicating that abundant species were more sensitive to the practice of mixed planting and increased planting age. In addition, with the increase in planting age, the richness of both soil fungal and bacterial communities significantly decreased ($p < 0.05$). In addition, we found that fungal community alpha diversity responds more strongly than bacterial community alpha diversity to different management practices.

Under all management practices, the RR of microbial community structures was greater than 0, particularly for fertilization, mixed planting, and increased planting age, which indicated that management practices significantly modified the microbial community structures, and microbial community structures were more sensitive than microbial alpha diversity to different management practices.

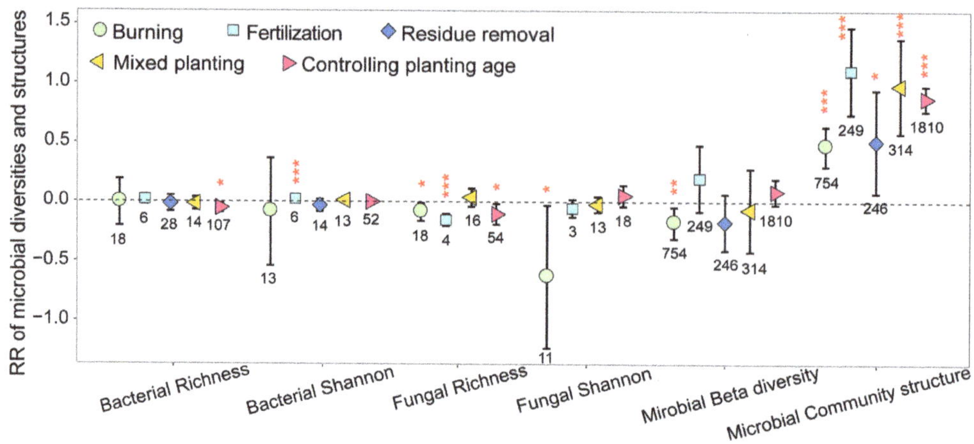

Figure 4. The response ratios of soil microbial alpha diversity and community structures under different management practices. The values shown at the bottom of the bars represent the corresponding number of observations. ***, ** and * indicates $p < 0.001$, $p < 0.01$ and $p < 0.05$, respectively.

3.2. Variation in Soil Properties and Enzymes

The implementation of different management practices on eucalyptus plantations significantly influenced the soil nutrient content while concurrently having no significant impact on soil aeration and the moisture content (Figure 5). Burning significantly increased soil TK, TP, and pH while significantly decreasing SMC ($p < 0.05$). Residue removal significantly decreased soil TK and TP; in comparison, mixed planting significantly increased soil TN ($p < 0.05$). Fertilization significantly increased soil TC, TN, and pH ($p < 0.05$). In addition, as planting age increased, soil BD, TC, TN, and SOC exhibited significant increases; in comparison, soil TK and pH showed significant decreases ($p < 0.05$). These results illustrated that most artificial management practices significantly increased the soil nutrient content on eucalyptus plantations.

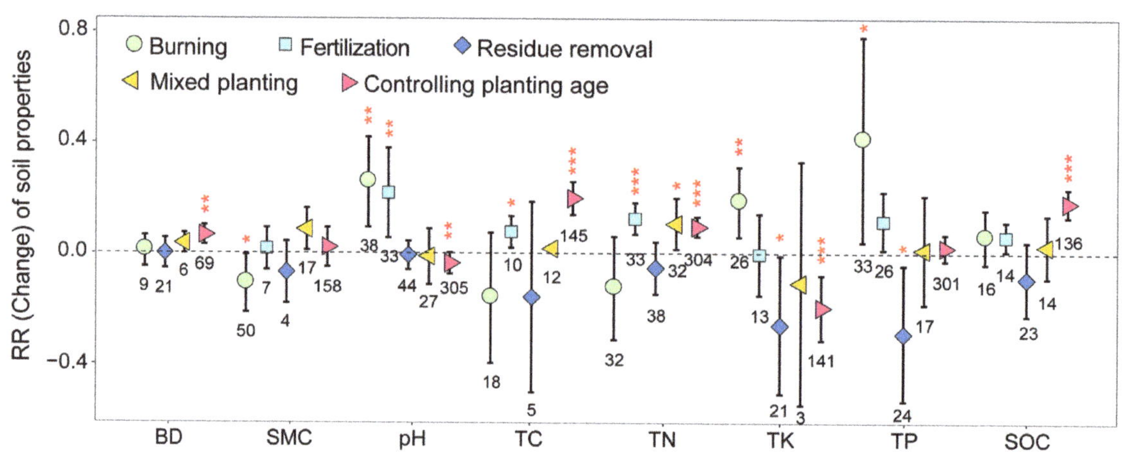

Figure 5. Response ratios of soil properties under different management practices. The values shown at the bottom of the bars represent the corresponding number of observations. BD: soil bulk density; SMC: soil moisture content; TC: total carbon; TN: total nitrogen; TP: total phosphorus; TK: total potassium; SOC: soil organic carbon. ***, ** and * indicates $p < 0.001$, $p < 0.01$ and $p < 0.05$, respectively.

The different management practices also exerted a significant influence on soil enzyme activity on eucalyptus plantations (Figure 6). Burning and fertilization significantly decreased soil catalase and urease activities while concurrently significantly increasing soil invertase activity ($p < 0.05$). Conversely, with the increase in planting age, catalase and urease contents significantly increased ($p < 0.05$).

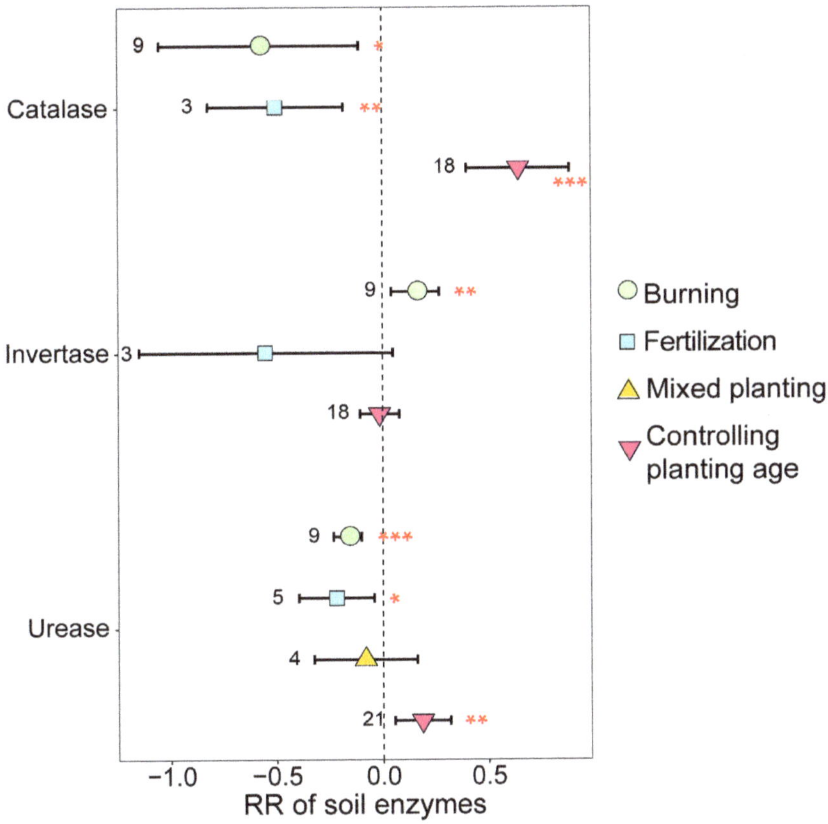

Figure 6. The response ratios of soil enzymes under different management practices. The values beside the bars are the corresponding number of observations. ***, ** and * indicates $p < 0.001$, $p < 0.01$ and $p < 0.05$, respectively.

3.3. Mechanisms of Soil Microbial Biomass, Diversity, and Structural Changes

Through model selection analysis, we identified that SOC and TK are crucial factors for soil microbes. Moreover, certain soil microbes were also affected by TP, MAT, and MAP (Figures 7 and 8).

We conducted further regression analyses and observed a significant negative correlation between the RR of SOC and the RR of MBC (Figure 9a, $p = 0.001$). With an increase in the RR of TK and SOC, there was a concurrent significant increase in the RR of fungal and Gram-negative bacteria biomass (Figure 9b,d, $p < 0.05$). In addition, with an increase in the RR of SOC, the RR of Actinomycetes and the total microbial biomass exhibited concurrent significant increases (Figure 9b, $p < 0.001$).

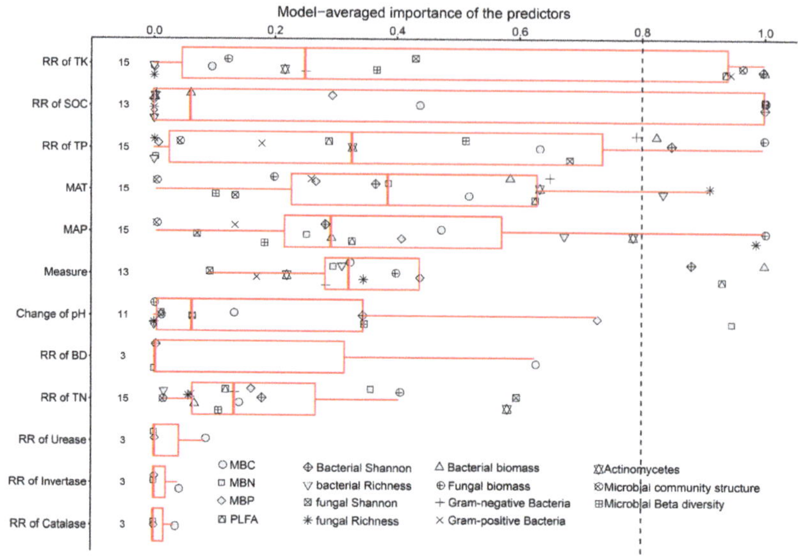

Figure 7. Factors influencing the response of microbial biomass, alpha diversity, and community structure. BD: soil bulk density; TN: total nitrogen; TP: total phosphorus; TK: total potassium; SOC: soil organic carbon; MAT: mean annual temperature; MAP: mean annual precipitation.

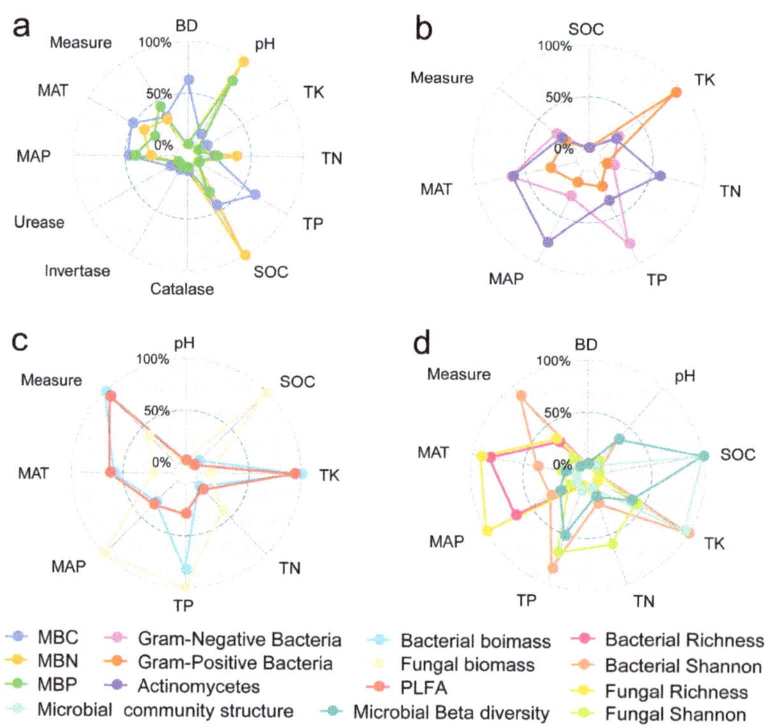

Figure 8. Factors influencing the response of microbial biomass elements (**a**), microbial biomass of different species (**b**,**c**), alpha diversity, and community structure (**d**). BD: soil bulk density; TN: total nitrogen; TP: total phosphorus; TK: total potassium; SOC: soil organic carbon; MAT: mean annual temperature; MAP: mean annual precipitation.

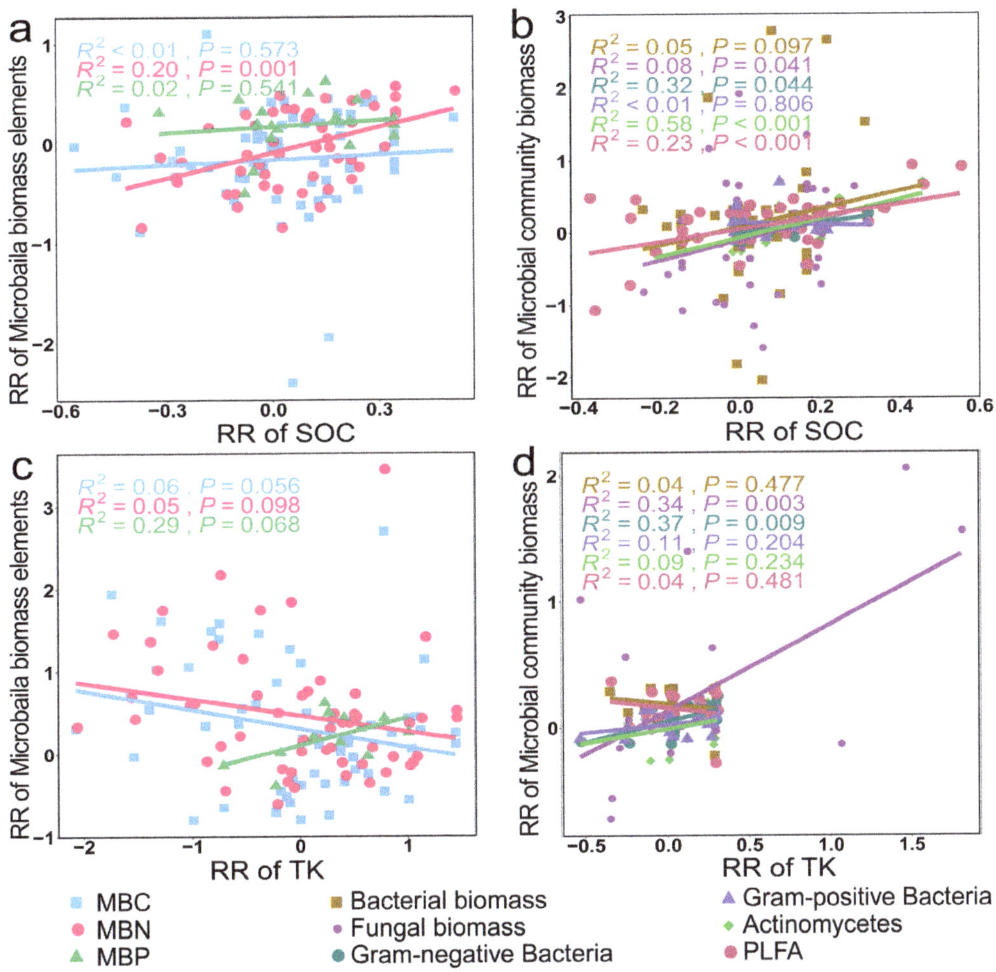

Figure 9. Relationships between the RR of microbial biomass elements and the microbial biomass of different species with the RR of SOC (**a**,**b**) and TK (**c**,**d**).

Soil microbial alpha diversity indices also exhibited significant correlations with SOC and TK (Figure 9a,c, $p < 0.01$). A discernible diminishing trend was observed in the RR of soil bacterial alpha diversity indices as the RR of SOC and TK increased (Figure 10a,c, $p < 0.05$). In particular, bacterial and fungal richness indices exhibited a negative correlation with SOC (Figure 10a, $p < 0.01$) while showing a concurrent positive correlation with TK (Figure 10c, $p < 0.001$).

As shown in Figure 11, we found a significant negative correlation between the RR of microbial community structures and the RR of MBC and MBN ($p < 0.01$). However, there was no significant correlation between the RR of microbial community structure and the RR of alpha diversity ($p > 0.05$).

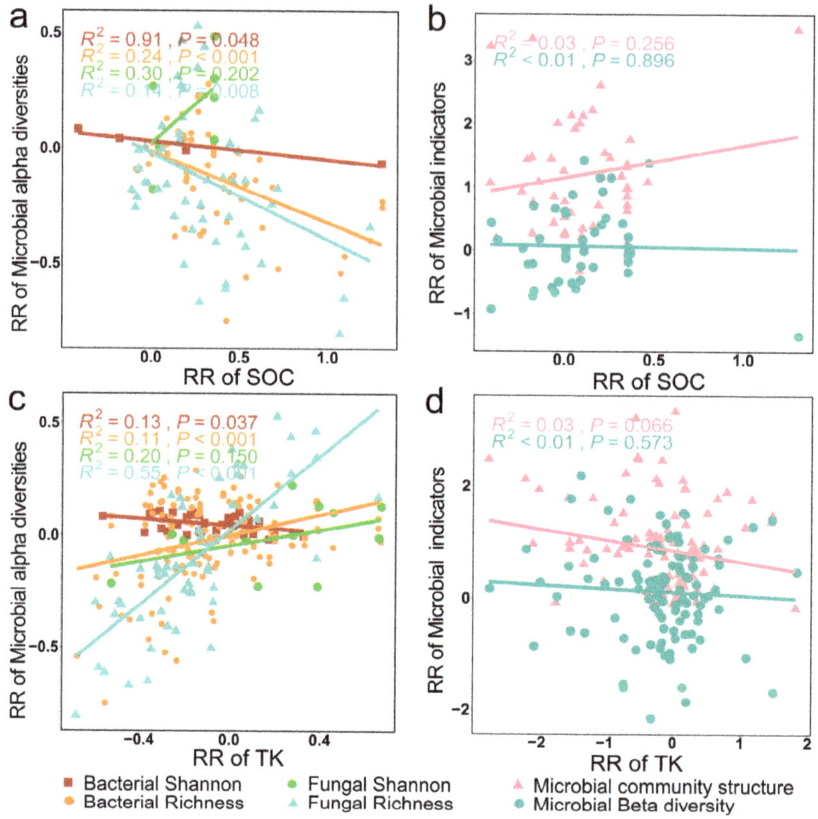

Figure 10. Relationships between the RR of microbial alpha diversity and microbial community structures with the RR of SOC (**a**,**b**) and TK (**c**,**d**).

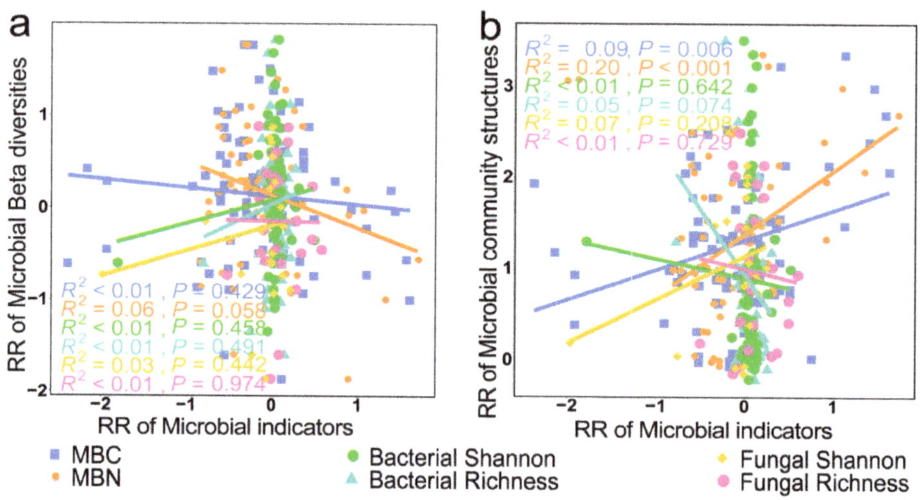

Figure 11. Relationships between the RR of microbial beta diversity (**a**) and microbial community structure (**b**) with the RR of microbial biomass elements and microbial alpha diversity.

4. Discussion

4.1. Factors Affecting Microbial Community Biomass, Diversity, and Structure

Both burning and residue removal had negative impacts on the soil MBC and MBN, particularly inhibiting the growth of fungal communities. The reduction in soil MBC and MBN was attributed to the decrease in SOC. The decrease in SOC after burning may be due to the reduction in peroxidase and urease enzyme contents as these enzymes play a vital role in soil carbon and nitrogen release, respectively [22]. The results of a previous study demonstrated that peroxidase and urease enzyme contents exhibited a significantly negative correlation with soil pH [62]. These findings indicate that, with the increase in soil pH, induced by the leaching of alkaline salts from the ash and charcoal in addition to the denaturation of organic acids, the peroxidase and urease enzyme contents will decrease, ultimately leading to a decline in SOC [22,63]. In contrast, the decrease in SOC following residue removal may be due to the direct loss of nutrients leached from the litter [15].

In addition, both burning and residue removal led to a reduction in fungal biomass; an important difference, however, was the fact that burning resulted in an increase in bacterial biomass. This discrepancy was attributed to bacteria showing a preference for utilizing readily available carbon substrates, whereas fungi displayed a tendency to degrade more complex carbon substrates [64]. Intense burning is often accompanied by deterioration in intricate carbon substrates such as cellulose and lignin [63], consequently augmenting the accessibility of simpler carbon substrates. From the above results, it can be concluded that the practice of burning promotes bacterial growth rather than fungal growth. Moreover, burning led to an increase in soil pH, and under elevated pH conditions, bacterial growth surpassed fungal growth [65,66]. Furthermore, burning significantly reduced the diversity and richness of fungal communities but had no significant effect on bacterial communities. This difference in response may be attributed to the greater fire tolerance of bacteria compared to fungi [22,67,68]. Soil temperatures above 70 °C are lethal to fungi [69]; in comparison, temperatures exceeding 200 °C may only eliminate certain bacterial species [70]. However, after residue removal, fungal biomass exhibited a significant decrease, while bacterial biomass remained relatively unchanged. This phenomenon may have arisen from the exposure of soil to air after the removal of surface residues, thereby transforming the anaerobic environment into one that is aerobic, which was unfavorable for fungal growth [18]. Consequently, this resulted in a substantial decrease in fungal biomass [71]. Overall, the complete removal of all residues and the practice of burning are detrimental to soil microbial growth, with burning negatively impacting the preservation of fungal community diversity.

Fertilization, mixing eucalyptus with nitrogen-fixing species, and controlling planting age primarily influenced soil health by impacting soil TC and TN contents. However, these management practices enhanced soil nutrient abundance through different pathways. Introducing nitrogen-fixing species alongside eucalyptus can stimulate the proliferation of nitrogen-fixing bacteria, which symbiotically fix atmospheric nitrogen into the soil through root symbiosis [72], thereby substantially elevating the TN content in eucalyptus plantation soil [17,19,32]. With the increasing age of plants, there was a gradual increase in the richness of understory plants and root yield, accompanied by accelerated litter decomposition rates. These processes contributed to enhanced soil aggregation and permeability [21,42,73], consequently leading to significant increases in TC, TN, and SOC on eucalyptus plantations [30,43]. Fertilization directly increases soil nutrient contents through the addition of nutrients to the soil.

The soil microbial biomass, diversity, and community structure exhibit contrasting responses to different management practices due to the varying methods of nutrient addition. For example, the biomass of Gram-negative bacteria significantly increased following fertilization, whereas the biomass of Gram-positive bacteria remained unchanged. This discrepancy may be attributed to the rapid response of Gram-negative bacteria to readily available carbon from fresh litter inputs; in comparison, Gram-positive bacteria must adapt to utilizing more complex carbon sources [74]. Mixing eucalyptus with nitrogen-fixing species led to a significant increase in MBN and total microbial biomass. This result was likely due to mixed planting increasing litter diversity, which provided sufficient carbon and nitrogen sources for microbial growth [75,76], in addition to creating a more diversified overall environment [77]. Moreover, we found that soil bacterial biomass significantly increased, whereas fungal biomass remained relatively stable. This discrepancy may have arisen from the greater changes in soil TN compared to TC, resulting in a decrease in soil C:N, which was more favorable for bacterial growth than fungal growth [20]. With the increase in planting age, the augmentation of the soil nutrient content and the secretion of allelochemicals notably enhanced soil microorganism activity [78,79], consequently resulting in a significant increase in soil MBC, MBN, and MBP. Actinomycetes, widely distributed in soil, play a crucial role in promoting the growth and development of eucalyptus [16,80,81]. In soils with older plantation ages, the biomass of Actinomycetes tended to be higher, potentially due to the increase in SOC with increasing plantation age, which in turn led to an increase in Actinomycetes biomass (Figure 9b). With the increase in planting age, there was a significant decrease in the richness of both fungal and bacterial communities. This finding could be attributed to the proliferation of environmental opportunists, such as organic decomposers, in response to abundant carbon resources. These opportunists may potentially suppress the reproduction of other species, leading to the disappearance of certain species and a reduction in richness [82]. Our findings also indicated a significant decrease in both bacterial and fungal community richness with increasing carbon resources (Figure 10a). Changes in bacterial and fungal community richness were also related to soil TK, exhibiting a significant decrease in richness as TK decreased (Figure 10c). In addition, ventilation also constrained bacterial community diversity [21]. As planting age increased, intensified canopy shading and litter accumulation created an anoxic environment, thereby diminishing bacterial community richness [18]. Overall, burning negatively affected fungal growth, diversity, MBC, and MBN while promoting bacterial community growth; residue removal hindered fungal growth and MBC and MBN storage; mixing eucalyptus with nitrogen-fixing species favored bacterial community growth and MBN storage; fertilization reduced fungal diversity but promoted bacterial diversity; and controlling stand age negatively affected both bacterial and fungal diversity while concurrently benefiting Actinomycetes growth and MBC, MBN, and MBP storage. From the above results, it can be concluded that mixing eucalyptus with nitrogen-fixing species is the most beneficial management practice; however, attention should be paid to properly protecting fungal communities, with residue removal being the least advantageous management practice.

The alpha diversity of fungal communities was more sensitive to the different management practices than that of bacterial communities, a finding that may be related to the different habits and physiological structures of fungi and bacteria. Fungi typically possess relatively complex cellular structures, including nuclei and cell walls, whereas bacteria typically possess less complex cellular structures and frequently establish biofilms in the soil environment. From the above findings, it is evident that compared to bacteria, fungi are less adaptable to environmental pressures [83]. Additionally, bacteria play a broader role in ecosystem biochemical cycles [84], whereas fungi are primarily involved in organic matter decomposition [85]; such factors result in bacteria being more adaptable to environmental changes compared to fungi [86].

Different management practices exert a significant impact on the structures of microbial communities, although it is important to note that there may not always be a direct correlation between microbial community structures and alpha diversity (Figure 11). Furthermore, microbial community structures were more sensitive to management practices than microbial alpha diversity (Figure 4). The reason for this finding is the fact that while certain species may have decreased in number or disappeared in response to the different management practices, another set of species may have increased in number or appeared, thereby altering the microbial community composition without necessarily affecting its diversity. However, significant changes in the internal composition of microbes may occur.

4.2. Limitations and Future Directions

While we were able to gather a substantial number of foundational data in the present study, few authors of existing studies have comprehensively addressed all of the indicators collected by our group. For the current dataset, constructing an accurate mechanistic framework to explore the changes in soil microbial communities on eucalyptus plantations following the implementation of various management practices remains a challenge. To address this limitation, we hope that the authors of future studies can expand our dataset. In addition, we hope that the authors of future studies can incorporate a broader range of environmental variables to more precisely evaluate the response of soil microbial communities to different management practices. Meanwhile, our study has limited focus on the influence of eucalyptus species on management practices. Although we selected common management practices applicable to most eucalyptus plantations and aimed to minimize the impact of eucalyptus species on these practices, we still hope to collect additional data in the future to address this drawback in our research. Moreover, we hope that future studies can include data on microbial functions and taxonomy, enabling a deeper investigation into how different management strategies influence the function of soil microbial communities. Such factors represent a critical component for accurately predicting and managing soil ecosystem functions and should not be overlooked.

5. Conclusions

On eucalyptus plantations, fungal community alpha diversity responds more sensitively to different management practices than bacterial community alpha diversity. Although the implementation of management practices significantly increased most soil nutrient contents and microbial biomass elements (except for burning), these practices did not necessarily improve soil microbial biomass and diversity, particularly among fungal communities. Specifically, burning negatively affected fungal growth, diversity, and microbial biomass carbon and nitrogen contents while simultaneously promoting bacterial community growth; residue removal hindered fungal growth and microbial biomass carbon and nitrogen storage; mixing eucalyptus with nitrogen-fixing species favored bacterial community growth and microbial biomass nitrogen storage; fertilization reduced fungal diversity but promoted bacterial diversity; and controlling stand age negatively affected both bacterial and fungal diversity while benefiting Actinomycetes growth and microbial biomass carbon, nitrogen, and phosphorus storage. Thus, mixing eucalyptus with nitrogen-fixing species is the most beneficial management practice; however, care should be taken to protect fungal communities, while residue removal is the least advantageous. Furthermore, our study indicated that variations of soil microbial diversity were primarily influenced by soil organic carbon and total potassium contents, suggesting that these should be prioritized when implementing management practices. Additionally, the implementation of management practices also significantly altered microbial community structures by positively affecting microbial biomass carbon and microbial biomass nitrogen instead of soil microbial alpha diversity. However, the impact of different practices on microbial community structure differed, burning mainly affected rare species structures, whereas mixed planting and increasing planting age mainly effected an abundant species structure. Therefore, different management practices require attention to different microbial community structures.

Supplementary Materials: The following supporting information can be downloaded at: https://www.mdpi.com/article/10.3390/land14040692/s1, Supplementary S1 is the source of the data, Table S1 is the results of the practice on each variable effect, and Table S2 is the results of potential publication bias analyses.

Author Contributions: Conceptualization: Y.X., W.L. and M.Z.; methodology: Y.X., W.L. and M.Z.; resources: J.Z., M.Z. and L.W.; software: L.W., Z.C. and X.L.; writing—original draft preparation, Y.X.; writing—review and editing, Y.X., W.L. and M.Z.; funding acquisition, W.L. and Q.F. All authors have read and agreed to the published version of the manuscript.

Funding: This research has been funded by the National Key R&D Program of China (Nos. 2022YFF1303301 and 2022YFF1302603), the National Natural Science Fund of China (Nos. U24A20178 and 42201133), Gansu Provincial Science and Technology Planning Project (Nos. 23ZDFA018, 23ZDKA017 and 22JR5RA072), Gansu Province Intellectual Property Plan Program (No. 23ZSCQD001), Excellent Doctoral Student Program of Gansu Province (No. 24JRRA110), and the National Grass Industry Technology Innovation Center (preparation) Key Innovation Platform Construction Project (CCPTZX2024GJ04).

Data Availability Statement: Data will be made available on request.

Conflicts of Interest: The authors declare that they have no known competing financial interests or personal relationships that could have appeared to influence the work reported in this paper.

References

1. Laclau, J.P.; Ranger, J.; Boullet, J.P.; Nzila, J.D.; Deleporte, P. Nutrient cycling in a clonal stand of Eucalyptus and an adjacent savanna ecosystem in Congo. 1. Chemical composition of rainfall and stem flow solutions. *For. Ecol. Manag.* **2003**, *176*, 105–119.
2. Laclau, J.P.; Ranger, J.; Boullet, J.P.; Nzila, J.D.; Deleporte, P. Nutrient cycling in a clonal stand of Eucalyptus and an adjacent savanna ecosystem in Congo. 2. Chemical composition of soil solutions. *For. Ecol. Manag.* **2003**, *180*, 527–544.
3. Pate, J.S.; Verboom, W.H. Contemporary biogenic formation of clay pavement by eucalypts: Further support for the phytotarium concept. *Ann. Bot.* **2009**, *103*, 673–685.
4. Phalan, B.; Bertzky, M.; Butchart, S.H.M.; Donald, P.F.; Scharlemann, J.P.W.; Stattersfield, A.J.; Balmfordet, A. Crop expansion and conservation priorities in tropical countries. *PLoS ONE* **2013**, *8*, e51759.
5. Eufrade Junior, H.J.; Melo, R.X.; Sartori, M.M.P.; Guerra, S.P.S.; Ballarin, A.W. Sustainable use of eucalypt biomass grown on short rotation coppice for bioenergy. *Biomass Bioenerg.* **2016**, *90*, 15–21.
6. Jagger, P.; Pender, J. The role of trees for sustainable management of less-favoured lands: The case of eucalyptus in Ethiopia. *For. Policy Econ.* **2003**, *5*, 83–95.
7. Nouvellon, Y.; Epron, D.; Marsden, C.; Kinana, A.; Maire, G.L.; Deleporte, P.; SaintAndre, L.; Bouillet, J.P.; Laclau, J.P. Age-related changes in litter inputs explain annual trends in soil CO_2 effluxes over a full Eucalyptus rotation after afforestation of a tropical savannah. *Biogeochemistry* **2012**, *111*, 515–533.
8. Zhang, Y.X.; Wang, X.J. Geographical spatial distribution and productivity dynamic change of Eucalyptus plantations in China. *Sci. Rep.* **2021**, *11*, 19764.
9. Laclau, J.P.; Deleporte, P.; Ranger, J.; Bouilleti, J.P.; Kazotti, G. Nutrient dynamics throughout the rotation of Eucalyptus clonal stands in Congo. *Ann. Bot.* **2003**, *91*, 879–892.
10. Li, H.F.; Fu, S.L.; Zhao, H.T.; Xia, H.P. Effects of understory removal and N-fixing species seeding on soil N_2O fluxes in four forest plantations in southern China. *Soil Sci. Plant Nut.* **2010**, *56*, 541–551.
11. Florentine, S.K.; Fox, J.E.D. Allelopathic effects of *Eucalyptus victrix* L. on *Eucalyptus* species and grasses. *Allelopath* **2003**, *11*, 77–83.
12. Zhang, C.L.; Fu, S.L. Allelopathic effects of eucalyptus and the establishment of mixed stands of eucalyptus and native species. *For. Ecol. Manag.* **2009**, *258*, 1391–1396.
13. Kara, O.; Bolat, I.; Çakıroğlu, K.; Öztürk, M. Plant canopy effects on litter accumulation and soil microbial biomass in two temperate forests. *Biol. Fertil. Soils* **2008**, *45*, 193–198.
14. Xue, Y.Y. *Research on the Characteristics and Influencing Factors of Soil Microbial Communities in Forest and Cultivated Land*; Northwest Institute of Eco-Environment and Resource, Chinese Academy of Science: Lanzhou, China, 2022.

15. Wang, Y.Z.; Zheng, J.Q.; Xu, Z.H.; Abdullah, K.M.; Zhou, Q.X. Effects of changed litter inputs on soil labile carbon and nitrogen pools in a eucalyptus-dominated forest of southeast Queensland, Australia. *J. Soil Sediment.* **2019**, *19*, 1661–1671.
16. Mercedes, M.O.; Michael, B.; Richard, J.P.D.; Mark, K.J.O.; Miriam, M.R. Fire and land use impact soil properties in a Mediterranean dry sclerophyll woodland. *J. Environ. Manag.* **2020**, *324*, 116245.
17. Jiang, Y.S.; Luo, C.L.; Zhang, D.Y.; Ostle, N.J.; Cheng, Z.N.; Ding, P.; Shen, C.D.; Zhang, G. Radiocarbon evidence of the impact of forest-to-plantation conversion on soil organic carbon dynamics on a tropical island. *Geoderma* **2020**, *375*, 114484.
18. Qu, Z.L.; Liu, B.; Ma, Y.; Sun, H. Differences in bacterial community structure and potential functions among Eucalyptus plantations with different ages and species of trees. *Appl. Soil Ecol.* **2020**, *149*, 103515.
19. Pereira, A.P.A.; Santana, M.C.; Zagatto, M.R.G.; Brandani, C.B.; Wang, J.T.; Verma, J.P.; Singh, B.K.; Cardoso, E.J.B.N. Nitrogen-fixing trees in mixed forest systems regulate the ecology of fungal community and phosphorus cycling. *Sci. Total Environ.* **2021**, *758*, 143711.
20. Huang, X.M.; Liu, S.R.; Wang, H.; Hu, Z.D.; Li, Z.G.; You, Y.M. Changes of soil microbial biomass carbon and community composition through mixing nitrogen-fixing species with Eucalyptus urophylla in subtropical China. *Soil Biol. Biochem.* **2014**, *73*, 42–48.
21. Xu, Y.X.; Du, A.; Wang, Z.C.; Zhu, W.K.; Li, C.; Wu, L.C. Effects of different rotation periods of Eucalyptus plantations on soil physiochemical properties, enzyme activities, microbial biomass and microbial community structure and diversity. *For. Ecol. Manag.* **2020**, *456*, 117683.
22. Liu, X.; Chen, C.R.; Wang, W.J.; Hughes, J.M.; Lewis, T.; Hou, E.Q.; Shen, J.P. Soil environmental factors rather than denitrification gene abundance control N2O fluxes in a wet sclerophyll forest with different burning frequency. *Soil Biol. Biochem.* **2013**, *57*, 292–300. [CrossRef]
23. Gunina, A.; Smith, A.R.; Godbold, D.L.; Jones, D.L.; Kuzyakov, Y. Response of soil microbial community to afforestation with pure and mixed species. *Plant Soil* **2017**, *412*, 357–368. [CrossRef]
24. Wang, D.J.; Abdullaha, K.M.; Xu, Z.H.; Wang, W.J. Water extractable organic C and total N: The most sensitive indicator of soil labile C and N pools in response to the prescribed burning in a suburban natural forest of subtropical Australia. *Geoderma* **2020**, *377*, 114586. [CrossRef]
25. Zhu, L.Y.; Wang, J.C.; Weng, Y.L.; Chen, X.L.; Wu, L.C. Soil characteristics of Eucalyptus urophylla×Eucalyptus grandis plantations under different management practices for harvest residues with soil depth gradient across time. *Ecol. Indic.* **2020**, *117*, 106530. [CrossRef]
26. Wang, C.; Liu, D.W.; Bai, E. Decreasing soil microbial diversity is associated with decreasing microbial biomass under nitrogen addition. *Soil Biol. Biochem.* **2018**, *120*, 126–133. [CrossRef]
27. Palmer, H.D.; Denham, A.J.; Ooi, M.K.J. Fire severity drives variation in post-fire recruitment and residual seed bank size of Acacia species. *Plant Ecol.* **2018**, *219*, 527–537. [CrossRef]
28. Wang, C.; Wang, G.; Wang, Y.; Rafique, R.; Ma, L.; Hu, L.; Luo, Y. Fire alters vegetation and soil microbial community in alpine meadow. *Land Degrad. Dev.* **2015**, *27*, 1379–1390. [CrossRef]
29. Bárcenas-Moreno, G.; García-Orenes, F.; Mataix-Solera, J.; Mataix-Beneyto, J. Plant community influence on soil microbial response after a wildfire in Sierra Nevada National Park (Spain). *Sci. Total Environ.* **2016**, *573*, 1265–1274. [CrossRef]
30. Mendham, D.S.; Sankaran, K.V.; OConnell, A.M.; Grove, T.S. Eucalyptus globulus harvest residue management effects on soil carbon and microbial biomass at 1 and 5 years after plantation establishment. *Soil Biol. Biochem.* **2002**, *34*, 1903–1912. [CrossRef]
31. Rachid, C.T.C.C.; Balieiro, F.C.; Peixoto, R.S.; Pinheiro, Y.A.S.; Piccolo, M.C.; Chaer, G.M.; Rosado, A.S. Mixed plantations can promote microbial integration and soil nitrate increases with changes in the N cycling genes. *Soil Biol. Biochem.* **2013**, *66*, 146–153. [CrossRef]
32. Pereira, A.P.A.; Durrer, A.; Gumierec, T.; Gonçalves, J.L.M.; Robina, A.; Bouillet, J.P.; Wang, J.T.; Vermab, J.P.; Singh, B.K.; Cardoso, E.J.B.N. Mixed Eucalyptus plantations induce changes in microbial communities and increase biological functions in the soil and litter layers. *For. Ecol. Manag.* **2019**, *433*, 332–342.
33. Xu, Y.X.; Ren, S.Q.; Liang, Y.F.; Du, A.; Li, C.; Wang, Z.C.; Zhu, W.K.; Wu, L.C. Soil nutrient supply and tree species drive changes in soil microbial communities during the transformation of a multi-generation Eucalyptus plantation. *Appl. Soil Ecol.* **2021**, *166*, 103991.
34. Bini, D.; dos Santos, C.A.; Bouillet, J.P.; Goncalves, J.L.D.M.; Cardoso, E.J.B.N. Eucalyptus grandis and Acacia mangium in monoculture and intercropped plantations: Evolution of soil and litter microbial and chemical attributes during early stages of plant development. *Appl. Soil Ecol.* **2013**, *63*, 57–66. [CrossRef]
35. Qian, Z.Z.; Zhu, K.X.; Zhuang, S.Y.; Tang, L.Z. Soil nutrient cycling and bacterial community structure in response to various green manures in a successive *Eucalyptus* (*Eucalyptus urophylla* × *Eucalyptus grandis*) plantation. *Land Degrad. Dev.* **2022**, *33*, 2809–2821.
36. Wang, Q.K.; Wang, S.L.; Liu, Y.X. Responses to N and P fertilization in a young *Eucalyptus dunnii* plantation: Microbial properties, enzyme activities and dissolved organic matter. *Appl. Soil Ecol.* **2008**, *40*, 484–490.
37. Kjoller, R.; Nilsson, L.O.; Hansen, K.; Schmidt, I.K.; Vesterdal, L.; Gundersen, P. Dramatic changes in ectomycorrhizal community composition, root tip abundance and mycelial production along a stand-scale nitrogen deposition gradient. *New Phytol.* **2012**, *194*, 278–286. [CrossRef] [PubMed]

38. Hasselquist, N.J.; Hgberg, P. Dosage and duration effects of nitrogen additions on ectomycorrhizal sporocarp production and functioning: An example from two N-limited boreal forests. *Ecol. Evol.* **2014**, *4*, 3015–3026. [CrossRef]
39. Asante, P.; Armstrong, G.W.; Adamowicz, W.L. Carbon sequestration and the optimal forest harvest decision: A dynamic programming approach considering biomass and dead organic matter. *J. For. Econ.* **2011**, *17*, 3–17.
40. Cavagnaroa, T.R.; Cunningham, S.C.; Fitzpatrick, S. Pastures to woodlands: Changes in soil microbial communities and carbon following reforestation. *Appl. Soil Ecol.* **2016**, *107*, 24–32. [CrossRef]
41. Heyde, M.V.D.; Bunce, M.; Dixon, K.; Wardell-Johnson, G.; White, N.E.; Nevill, P. Changes in soil microbial communities in post mine ecological restoration: Implications for monitoring using high throughput DNA sequencing. *Sci. Total Environ.* **2020**, *749*, 142262.
42. Trogisch, S.; He, J.S.; Hector, A.; Scherer-Lorenzen, M. Impact of species diversity, stand age and environmental factors on leaf litter decomposition in subtropical forests in China. *Plant Soil* **2015**, *400*, 337–350.
43. Sun, H.; Wang, Q.; Liu, N.; Li, L.; Zhang, C.; Liu, Z.; Zhang, Y. Effects of different leaf litters on the physicochemical properties and bacterial communities in Panax ginseng-growing soil. *Appl. Soil Ecol.* **2017**, *111*, 17–24.
44. Wang, J.; Yang, S.; Ma, T.F.; Raza, W.; Jing, L.; Howland, J.G.; Huang, Q.W.; Shen, Q.R. Impacts of inorganic and organic fertilization treatments on bacterial and fungal communities in a paddy soil. *Appl. Soil Ecol.* **2017**, *112*, 42–50.
45. Wang, S.; Li, T.; Zheng, Z. Distribution of microbial biomass and activity within soil aggregates as affected by tea plantation age. *CATENA* **2017**, *153*, 1–8.
46. Liu, D.; Huang, Y.; Sun, H.; An, S. The restoration age of Robinia pseudoacacia plantation impacts soil microbial biomass and microbial community structure in the Loess Plateau. *CATENA* **2018**, *165*, 192–200.
47. Zhu, L.; Wang, X.; Chen, F.; Li, C.; Wu, L. Effects of the successive planting of Eucalyptus urophylla on the soil bacterial and fungal community structure, diversity, microbial biomass, and enzyme activity. *Land Degrad. Dev.* **2019**, *30*, 636–646.
48. Guo, J.H.; Feng, H.L.; McNie, P.; Liu, Q.Y.; Xu, X.; Pan, C.; Yan, K.; Feng, L.; Goitom, E.A.; Yu, Y.C. Species mixing improves soil properties and enzymatic activities in Chinese fir plantations: A meta-analysis. *CATENA* **2023**, *220*, 106723.
49. Curtright, A.J.; Tiemann, L.K. Intercropping increases soil extracellular enzyme activity: A meta-analysis. *Agric. Ecosyst. Environ.* **2021**, *319*, 107489.
50. Zhou, Z.H.; Wang, C.K.; Zheng, M.H.; Jiang, L.F.; Luo, Y.Q. Patterns and mechanisms of responses by soil microbial communities to nitrogen addition. *Soil Biol. Biochem.* **2017**, *115*, 433–441.
51. Waring, B.; Gee, A.; Liang, G.P.; Adkins, S. A quantitative analysis of microbial community structure-function relationships in plant litter decay. *iScience* **2022**, *25*, 104523.
52. Achat, D.L.; Deleuze, C.; Landmann, G.; Pousse, N.; Ranger, J.; Augusto, L. Quantifying consequences of removing harvesting residues on forest soils and tree growth—A meta-analysis. *For. Ecol. Manag.* **2015**, *348*, 124–141.
53. Meng, C.; Tian, D.S.; Zeng, H.; Li, Z.L.; Yi, C.X.; Niu, S.L. Global soil acidification impacts on belowground processes. *Environ. Res. Lett.* **2019**, *14*, 074003.
54. Delgado-Baquerizo, M.; Maestre, F.T.; Reich, P.B.; Jeffries, T.C.; Gaitan, J.J.; Encinar, D.; Berdugo, M.; Campbell, C.D.; Singh, B.K. Microbial diversity drives multifunctionality in terrestrial ecosystems. *Nat. Commun.* **2016**, *7*, 10541. [PubMed]
55. Paliy, O.; Shankar, V. Application of multivariate statistical techniques in microbial ecology. *Mol. Ecol.* **2016**, *25*, 1032–1057.
56. Zhou, Z.H.; Wang, C.K.; Luo, Y.Q. Meta-analysis of the impacts of global change factors on soil microbial diversity and functionality. *Nat. Commun.* **2020**, *11*, 3072. [PubMed]
57. Oksanen, J.; Blanchet, F.G.; Friendly, M.; Kindt, R.; Legendre, P.; McGlinn, D.; Minchin, P.R.; OHara, R.B.; Simpson, G.L.; Solymos, P.; et al. Vegan: Community ecology package. *R. Package Version* **2019**, *2*, 5–6.
58. Yuan, C.P.; Gao, B.L.; Peng, Y.T.; Gao, X.; Fan, B.B.; Chen, Q. A meta-analysis of heavy metal bioavailability response to biochar aging: Importance of soil and biochar properties. *Sci. Total Environ.* **2021**, *756*, 144058.
59. Wang, J.; Ding, C.; Heino, J.; Jiang, X.; Tao, J.; Ding, L.; Su, W.; Huang, M.; He, D. What explains the variation in dam impacts on riverine macroinvertebrates? A global quantitative synthesis. *Environ. Res. Lett.* **2020**, *15*, 124028.
60. Viechtbauer, W. Conducting meta-analyses in R with the metafor package. *J. Stati. Softw.* **2010**, *36*, 1–48.
61. Calcagno, V.; de Mazancourt, C. glmulti: An R package for easy automated model selection with (generalized) linear models. *J. Stat. Softw.* **2010**, *34*, 1–29.
62. Luo, Y.J. Study on Soil Microorganism and Soil Enzyme Activity Along a Chronosequence of Eucalyptus plantation. Master's Thesis, Guangxi Normal University, Guilin, China, 2014.
63. Certini, G. Effects of fire on properties of forest soils: A review. *Oecologia* **2005**, *143*, 1–10. [PubMed]
64. Rousk, J.; Bååth, E.; Brookes, P.C.; Lauber, C.L.; Lozupone, C.; Caporaso, J.G.; Knight, R.; Fierer, N. Soil bacterial and fungal communities across a pH gradient in an arable soil. *ISME J.* **2010**, *4*, 1340–1351.
65. Chen, L.; Jiang, Y.; Liang, C.; Luo, Y.; Xu, Q.; Han, C.; Zhao, Q.; Sun, B. Competitive interaction with keystone taxa induced negative priming under biochar amendments. *Microbiome* **2019**, *7*, 77.
66. Xue, Y.Y.; Liu, W.; Feng, Q.; Zhu, M.; Wang, L.G.; Chen, Z.X.; Zhang, J.T. Variations and controls of soil microbial necromass carbon in grasslands along aridity gradients. *Ecol. Indic.* **2023**, *156*, 111188.

67. González-pérez, J.A.; González-Vila, F.J.; Almendros, G.; Knicker, H. The effect of fire on soil organic matter: A review. *Environ. Int.* **2004**, *30*, 855–870. [PubMed]
68. Hart, S.C.; DeLuca, T.H.; Newman, G.S.; MacKenzie, M.D.; Boyle, S.I. Post-fire vegetative dynamics as drives of microbial community structure and function in forest soil. *For. Ecol. Manag.* **2005**, *220*, 166–184.
69. Raison, R.J. Modification of the soil environment by vegetation fires, with particular reference to nitrogen transformations: A review. *Plant Soil* **1979**, *51*, 73–108.
70. Neary, D.G.; Klopatek, C.C.; DeBano, L.F.; Ffolliott, P.F. Fire effects on belowground sustainability: A review and synthesis. *For. Ecol. Manag.* **1999**, *122*, 51–71.
71. Chen, X.; Hu, Y.; Xia, Y.; Zheng, S.; Ma, C.; Rui, Y.; He, H.; Huang, D.; Zhang, Z.; Ge, T.; et al. Contrasting pathways of carbon sequestration in paddy and upland soils. *Glob. Change Biol.* **2021**, *27*, 2478–2490.
72. Paula, R.R.; Bouillet, J.P.; Gonçalves, J.L.M.; Trivelin, P.C.O.; Balieiro, F.C.; Nouvellon, Y.; Oliveira, J.C.; Júnior, J.C.D.; Bordron, B.; Laclau, J.P. Nitrogen fixation rate of Acacia mangium wild at mid rotation in Brazil is higher in mixed plantations with Eucalyptus grandis hill ex maiden than in monocultures. *Ann. Forest Sci.* **2018**, *75*, 14.
73. So, H.B.; Grabski, A.; Desborough, P. The impact of 14 years of conventional and no-till cultivation on the physical properties and crop yields of a loam soil at Grafton NSW, Australia. *Soil Tillage Res.* **2009**, *104*, 180–184.
74. Blagodatskaya, E.V.; Blagodatsky, S.A.; Anderson, T.H.; Kuzyakov, Y. Priming effects in Chernozem induced by glucose and N in relation to microbial growth strategies. *Appl. Soil Ecol.* **2007**, *37*, 95–105.
75. Demoling, F.; Figueroa, D.; Bååth, E. Comparison of factors limiting bacterial growth in different soils. *Soil Biol. Biochem.* **2007**, *39*, 2485–2495.
76. Cizungu, L.; Staelens, J.; Huygens, D.; Walangululu, J.; Muhindo, D.; Van Cleemput, O.; Boeckx, P. Litterfall and leaf litter decomposition in a central African tropical mountain forest and Eucalyptus plantation. *For. Ecol. Manag.* **2014**, *326*, 109–116.
77. Rottstock, T.; Joshi, J.; Kummer, V.; Fischer, M. Higher plant diversity promotes higher diversity of fungal pathogens, while it decreases pathogen infection per plant. *Ecology* **2014**, *95*, 1907–1917.
78. Wardle, D.A.; Jonsson, M.; Bansal, S.; Bardgett, R.D.; Gundale, M.J.; Metcalfe, D.B. Linking vegetation change, carbon sequestration and biodiversity: Insights from island ecosystems in a long term natural experiment. *J. Ecol.* **2012**, *100*, 16–30.
79. Zhong, J.D.; Pan, P.; Xiao, S.H.; Ouyang, X.Z. Influence of Eucalyptus plantation on soil organic carbon and its fractions in severely degraded soil in Leizhou Peninsula, China. *Forests* **2022**, *13*, 1606. [CrossRef]
80. Lombao, A.; Barreiro, A.; Carballas, T.; Fontúrbel, M.T.; Martín, A.; Vega, J.A.; Fernández, C.; Díaz-Raviña, M. Changes in soil properties after a wildfire in Fragas do Eume Natural Park (Galicia, NW Spain). *CATENA* **2015**, *135*, 409–418.
81. Oliveira, F.C.C.; Ferreira, G.W.D.; Dungait, J.A.J.; Araújo, E.F.; Soares, E.M.B.; Silva, I.R. Eucalypt harvest residue management influences microbial community structure and soil organic matter fractions in an afforested grassland. *Soil Till. Res.* **2021**, *205*, 104787.
82. Bhattacharyya, R.; Rabbi, S.M.; Zhang, Y.; Young, I.M.; Jones, A.R.; Dennis, P.G.; Menzies, N.W.; Kopittke, P.M.; Dalal, R.C. Soil organic carbon is significantly associated with the pore geometry, microbial diversity and enzyme activity of the macro-aggregates under different land uses. *Sci. Total Environ.* **2021**, *778*, 146286.
83. Flemming, H.C.; Wuertz, S. Bacteria and archaea on earth and their abundance in biofilms. *Nat. Rev. Microbiol.* **2019**, *17*, 247–260. [PubMed]
84. Sui, X.; Zhang, R.T.; Yang, L.B.; Xu, N.; Zhong, H.X.; Wang, J.F.; Ni, H.W. Bacterial community functional diversity of Calamagrostis angustifolia wetland in Sanjiang Plain. *Res. Environ. Sci.* **2016**, *29*, 1479–1486. (In Chinese)
85. Courty, P.E.; Buee, M.; Diedhiou, A.G.; Frey Klett, P.; Le Tacon, F.; Rineau, F.; Turpault, M.P.; Uroz, S.; Garbaye, J. The role of ectomycorrhizal communities in forest ecosystem processes: New perspectives and emerging concepts. *Soil Biol. Biochem.* **2010**, *42*, 679–698.
86. Cai, Y. *Distribution and Accumulation of Microbial-Derived Organic Carbon in Soils: The Regulatory Mechanism of Minerals and Substrates*; University of Chinese Academy of Sciences in Chinese: Beijing, China, 2020.

land

MDPI

Article

Spatial Distribution and Management of Trace Elements in Arid Agricultural Systems: A Geostatistical Assessment of the Jordan Valley

Mamoun A. Gharaibeh [1,*], Bernd Marschner [2], Nicolai Moos [3] and Nikolaos Monokrousos [4,*]

1 Department of Natural Resources and the Environment, Faculty of Agriculture, Jordan University of Science and Technology, P.O. Box 3030, Irbid 22110, Jordan
2 Department of Soil Science/Soil Ecology, Institute of Geography, Ruhr-University Bochum, 44801 Bochum, Germany; bernd.marschner@ruhr-uni-bochum.de
3 Department of Geodesy, Bochum University of Applied Sciences, 44801 Bochum, Germany; nicolai.moos@hs-bochum.de
4 University Center of International Programmes of Studies, International Hellenic University, 57001 Thessaloniki, Greece
* Correspondence: mamoun@just.edu.jo (M.A.G.); nmonokrousos@ihu.gr (N.M.)

Abstract: Sustainable land management in arid regions such as the Jordan Valley (JV) is essential as climate pressures and water shortages intensify. The extended use of treated wastewater (TWW) for irrigation, while necessary, brings potential risks related to the accumulation of trace elements and their impact on soil health and food safety. This study examined the spatial distribution, variability, and potential sources of five trace elements (Co, Hg, Mo, Mn, and Ni) in agricultural soils across a 305 km^2 area. A total of 127 surface soil samples were collected from fields irrigated with either TWW or freshwater (FW). Trace element concentrations were consistently higher in TWW-irrigated soils, although all values remained below WHO/FAO recommended thresholds for agricultural use. Spatial modeling was conducted using both ordinary kriging (OK) and empirical Bayesian kriging (EBK), with EBK showing greater prediction accuracy based on cross-validation statistics. To explore potential sources, semivariogram modeling, principal component analysis (PCA), and hierarchical clustering were employed. PCA, spatial distribution patterns, correlation analysis, and comparisons between TWW and FW sources suggest that Co, Mn, Mo, and Ni are primarily influenced by anthropogenic inputs, including TWW irrigation, chemical fertilizers, and organic amendments. Co exhibited a stronger association with TWW, whereas Mn, Mo, and Ni were more closely linked to fertilizer application. In contrast, Hg appears to originate predominantly from geogenic sources. These findings provide a foundation for improved irrigation management and fertilizer application strategies, contributing to long-term soil sustainability in water-limited environments like the JV.

Keywords: trace elements; treated wastewater; empirical Bayesian kriging; arid agriculture; land management

1. Introduction

Trace elements occur naturally in soils, primarily due to the mineral composition and weathering of parent geological materials. However, in modern agricultural systems, their levels are increasingly affected by human activities. The use of treated wastewater (TWW), various fertilizers, organic amendments, biosolids, and pesticides has introduced additional sources of these elements into the soil [1,2].

While many trace elements are stable and persist in the environment, their accumulation can become a concern, especially when concentrations rise to levels that may threaten ecological health or food safety. Elements like cobalt (Co), copper (Cu), manganese (Mn), molybdenum (Mo), and nickel (Ni) are vital for plant metabolism but can become harmful if they exceed threshold levels. Moreover, elements not directly required by plants, including chromium (Cr), selenium (Se), and vanadium (V), can still be absorbed into the food chain, posing risks to human and animal health [3,4].

Jordan, one of the most water-scarce countries globally, faces growing pressures on its agricultural sector. Climate variability, land degradation, and scarce renewable water resources are all contributing factors. The Jordan Valley (JV), a fertile strip extending over 35,000 hectares, plays a critical role in the nation's food production, particularly for fruits and vegetables. Due to persistent water shortages, especially in the central and southern areas, farmers have been relying on TWW combined with surface runoff for irrigation for over forty years. While TWW provides nutrients and organic matter that can enhance soil fertility, its prolonged use may also lead to unwanted trace element accumulation, potentially compromising soil quality and food safety [5,6]. The irrigation water, primarily sourced from the As-Samra wastewater treatment plant and King Talal Dam (KTD), lacks tertiary treatment or metal-specific removal steps, which, along with prolonged reuse practices, likely contribute to trace element accumulation in JV soils.

Recent studies conducted in the JV have reported elevated concentrations of Cr, Cu, Zn, and Pb in soils irrigated with TWW. Although most concentrations remained within international safety limits (e.g., WHO/FAO guidelines), these investigations revealed varying sources of contamination: Cd, Cr, and Pb were largely traced back to TWW; Zn and Cu were linked to chemical fertilizers; and As appeared to originate from natural geological formations. However, these studies often lacked spatial detail and did not adequately account for uncertainty in trace element distribution, limiting their utility for practical land management [5].

To address such gaps, spatial modeling techniques and multivariate statistical tools offer valuable insights. Interpolation methods like ordinary kriging (OK) and empirical Bayesian kriging (EBK) are commonly used to predict the spatial variability of soil contaminants [7–9]. Using EBK improves the robustness of spatial estimates by incorporating repeated simulations of semivariograms, thereby reducing predictive uncertainty. Meanwhile, multivariate analyses—such as principal component analysis (PCA) and hierarchical cluster analysis—can help distinguish between natural and anthropogenic sources of trace elements [10,11].

While similar patterns of trace element enrichment have been documented in other Mediterranean and semi-arid environments, differences in soil type, land use, and climate mean that local assessments remain essential. For example, elevated copper levels in vineyards of southern Italy were associated with fungicide applications [12], whereas high concentrations of Co, Ni, and Cr in parts of Spain [13] were linked to geological sources. In Egypt, cultivated soils showed higher levels of Cd, Co, Mo, and Pb compared to nearby desert soils, highlighting the influence of land use on element accumulation [14]. These international examples underscore the importance of region-specific studies, especially in vulnerable ecosystems like the JV.

Considering growing climatic pressures and persistent water scarcity, understanding how trace elements behave across space is essential, not only for assessing contamination risks but also for shaping adaptive land management strategies. Accurate spatial mapping of trace element concentrations enables site-specific interventions, such as identifying zones where TWW use should be minimized, targeting low-risk areas for intensive cultivation, and refining fertilizer practices based on accumulated soil burdens. This type of data-driven decision-making can support sustainable agricultural planning in the JV, helping balance productivity with environmental stewardship [15].

To the best of our knowledge, this is the first study in the JV to examine a specific group of trace elements (Co, Hg, Mo, Mn, and Ni) chosen for their environmental importance and agronomic relevance within arid, alkaline agricultural systems. Unlike the well-studied heavy metals such as Pb, Cd, and Zn, these elements have received relatively little attention in the JV. Mn, although commonly found in soils, often becomes deficient under the alkaline conditions prevalent in the JV, which can negatively impact crop yields. Hg, while not essential to plants, is a highly toxic contaminant primarily introduced through human activities such as pesticide application and biosolid amendments. Its behavior in arid, irrigated soils remains largely unexplored.

This research fills critical knowledge gaps by assessing the spatial distribution and associated uncertainties of these trace elements using geostatistical methods such as OK and, for the first time in the region, EBK. As an advancement over previous studies that relied solely on OK for classical heavy metals, EBK enhances spatial prediction accuracy and quantifies uncertainty by incorporating semivariogram variability [5]. Furthermore, multivariate statistical techniques, including PCA and cluster analysis, were used to identify possible sources of these elements. The study also evaluates the long-term effects of TWW irrigation on trace element accumulation in soils and offers recommendations to guide adaptive land and water management amid growing challenges from climate change and water scarcity in the JV.

2. Materials and Methods

2.1. Description and Climate of Study Area

The study area covers about 305 km^2 of intensive open-field and greenhouse agricultural farms extending from the northern to southern parts of the JV (32°19′ 32.45″ N, 35° 33′ 21.47″ E; 31°46′ 49.45″ N, 35° 32′ 47.76″ E) (Figure 1). Farms in the JV are organized in 0.3–0.4 ha units, with a total of 10,000 farm units extending from the northern borders to the Dead Sea. The total agricultural land in the JV produces more than half of the total food production and covers about 13 percent of the total agricultural land in Jordan.

The climate of the study area is arid, characterized by hot, dry summers and mild winters. Average annual evapotranspiration is about 2250 mm, with average annual precipitation ranging from 500 mm to less than 100 mm in the northern and middle to southern parts, respectively. Moreover, annual temperatures range from 21 to 27 °C during summer and from 10 to 12 °C during winter. Soils of the study area are Aridisols (Typic Camborthids, Typic Calciorthids) and Entisols (Typic Torriothents) [16].

Agricultural fields in the middle and south parts are irrigated with a blend of fresh and TWW, while the north parts are only irrigated with fresh water.

The total cultivated area in the JV is about 28,000 ha, about 42% is cultivated in the northern parts, 35% in the middle, and 23% in the south. Fruit (citrus trees) and vegetable crops are mainly grown in the northern and middle parts, whereas in the southern parts, banana and palm trees are mainly grown, in addition to vegetable crops. Farming areas between Kreymeh and Deir Alla are planted with greenhouse vegetables in small areas (0.4 ha). In Al-Muaddi, south of Deir Alla, large farms of palm and citrus trees are mainly found, in addition to small-scale greenhouse and open-field vegetable crops. In Dameih, south of Al-Muaddi stretching south to Karamah, open-field vegetables dominate [17].

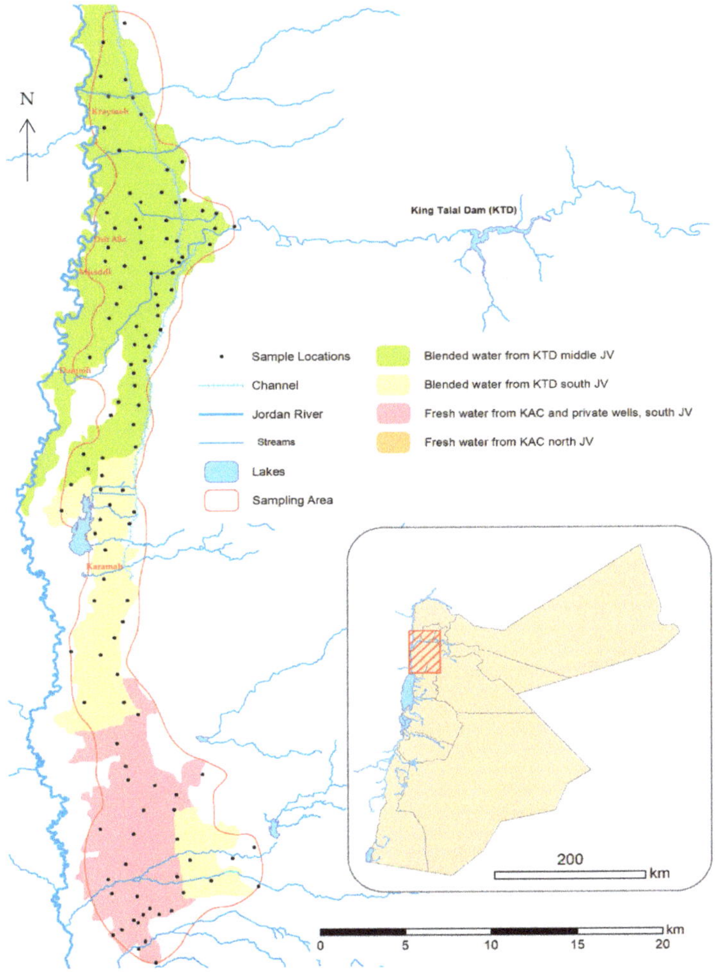

Figure 1. Location of collected soil samples.

In areas around the village of Karamah, special crops are planted (mint and parsley) in open fields and greenhouses. In the southern parts of the JV, banana farms and some vegetables are grown in open fields. These areas extend from the southern parts of Karamah to the north borders of the Dead Sea (South Shuna to Kafrein areas). Farms use water mainly from King Abdullah Canal (KAC) and Husban Wadi, in addition to private wells and some desalination plants. Water pumped from wells is salty; therefore, farmers mix this water with water from dams. Detailed information regarding sampling locations, soil classification, irrigation water sources, and grown crops are described in [5].

2.2. Collection of Soil Samples and Chemical Characterization

Soil samples were randomly collected from different farms. The sampling area covered an area of about 305 km^2 (70 km long × 4.7 km average width). Sample coordinates were recorded with a global positioning system receiver. From each location, four surface (0–25 cm) soil samples were collected and mixed, and one composite sample was made for each location, air dried, sieved to ≤2 mm, labeled, and stored for chemical analysis. A total of 127 composite surface soil samples (0–25 cm) were collected: 102 from fields irrigated with TWW and 25 from FW-irrigated fields.

Approximately 50 g of soil was ground to <250 microns using a ball mill, and after grinding, the mill was cleaned using clean quartz sand to reduce any possible cross-contamination. About 250 mg of finely ground soil from each sample was weighed into digestion vessels, followed by the addition of 10 mL of concentrated HNO_3. The samples were subjected to microwave-assisted digestion using the Mars Xpress system (CEM GmbH, Kamp-Lintfort, Germany) at a maximum temperature of 180 °C. The digestion protocol consisted of four sequential steps: (1) ramping the temperature from 25 °C to 90 °C over 4 min, (2) holding at 90 °C for 2 min, (3) increasing to 180 °C over 6 min, and (4) maintaining 180 °C for an additional 10 min. After digestion, samples were filtered using acid-washed Sartorius no. 640 filter paper and transferred into 50 mL polyethylene bottles for subsequent trace metal analysis [18].

The total concentrations of trace elements (Co, Mn, Mo, Ni, and Hg) in the digested soil samples were determined using Inductively Coupled Plasma Optical Emission Spectroscopy (ICP-OES) (Spectroblue, Ametek Materials Analysis Division, Kleve, Germany). The instrument was operated under the following conditions: RF power of 1300 W, plasma gas flow rate of 12 L/min, auxiliary gas flow rate of 1.0 L/min, nebulizer gas flow rate of 0.8 L/min, and sample uptake rate of 1.5 mL/min. The following analytical wavelengths were used: Co (228.616 nm), Mn (257.610 nm), Mo (202.030 nm), Ni (231.604 nm), and Hg (253.652 nm). Calibration curves were established using multi-element standard solutions at six concentrations: 0, 0.5, 1.0, 2.0, 5.0, and 10.0 mg/L, with correlation coefficients (R^2) above 0.999 for all analytes. The limits of detection (LOD) and limits of quantification (LOQ) were calculated as 3σ and 10σ, respectively, where σ is the standard deviation of ten blank measurements. LOD values ranged from 0.01 to 0.05 mg/L, and LOQ values ranged from 0.03 to 0.15 mg/L, depending on the element. For quality assurance, two replicate samples from each site were analyzed in duplicate, yielding four readings per location. The accuracy and precision of the analysis were evaluated using the certified reference material SPS-SW2 (Spectrapure Standards AS). Although originally certified for water matrices, the CRM was used due to the lack of a suitable soil-based reference material, and it showed good agreement with certified values, confirming acceptable analytical performance.

Furthermore, soil pH [19] and electrical conductivity (EC) [20], organic carbon (Org C) [21], total nitrogen-TN [22], and total phosphorus-TP (ICP-MS) were also determined for each soil sample.

2.3. Statistical Methods

Summary statistics were used to describe trace element concentrations and other soil chemical properties. Data were subjected to normality testing using Kolmogorov–Smirnov and Shapiro–Wilk tests, correlations between elements and other soil properties were assessed using the Pearson method, and source identification was performed using PCA and cluster analysis in SPSS 25. Varimax rotation was performed to extract components with eigenvalues greater than 1. Prior to PCA, data were z-transformed to normalize distribution, and the Kaiser–Meyer–Olkin (KMO) measure of adequacy and Bartlett's test of sphericity were conducted to test for homogeneity of variances and to check if data were well suited for factor analysis (KMO > 0.5). Furthermore, cluster analysis (dendrogram) was used to reduce studied variables into homogeneous groups of similar origin [23].

2.4. Semivariogram Analysis

Semivariogram analysis was used to study the spatial patterns of regionalized variables. Spatially related data were analyzed by plotting variance versus distance to create an experimental variogram to predict values of un-sampled locations (unmeasured data points) by kriging [24,25]. The semivariance is computed as half the average of the squared difference between points separated by distance h [26]:

$$\gamma(h) = \frac{1}{2N(h)} \sum_{i=1}^{N(h)} [(Z(x_i) - Z(x_i + h)]^2$$

where $N(h)$ is the number of sets of all data pairs separated by h, and $Z(x_i)$ and $Z(x_i + h)$ are the measured values at spatial locations x_i and $x_i + h$.

For the EBK, a semivariogram using a log-empirical transformation and K-Bessel de-trended type with 100 subsets and 1000 trials were performed in this study. Cross-validation was employed to evaluate the goodness of fit and to assess the reliability of the interpolation model (OK vs. EBK) for each element, using several performance indicators: the coefficient of determination (R^2), mean error (ME), root mean square error (RMSE), mean standardized error (MSE), root mean squared standardized error (RMSSE), and average standard error (avg. SE).

3. Results and Discussion

3.1. Trace Elements Contents in Soils

Summary statistics of trace elements concentrations, EC, Org C, TP, and TN in studied soils are shown in Table S1. Soils of the JV are slightly alkaline (pH 7.6) with a wide range of EC values (1.5–234 dS m^{-1}). Soils vary considerably in texture, with clay and clay loam textures dominating in the northern and middle parts, while loam, silt loam, and sandy loam dominate the southern parts of the study area. Levels of trace elements were compared with the maximum allowable concentration (MAC) reported by [27].

Average metal levels (mg kg^{-1}) in soils were as follows: Co (8.7), Hg (2.29), Mn (500.8), Mo (5.24), and Ni (21.9). The average contents of all analyzed elements were below the maximum permissible concentration (MPC) suggested by [28]. Soils irrigated with TWW contained elevated levels of trace elements compared to FW-irrigated soils (Figure 2). Summary statistics of trace elements in FW- and TWW-irrigated soils are provided in Tables S2 and S3.

Numerous studies have documented significant variability in trace element concentrations in agricultural soils across different regions. For instance, in Indonesia, average Ni levels were reported at 11.96 mg kg^{-1}, with a maximum of 19.5 mg kg^{-1}, while Mn concentrations averaged 1700 mg kg^{-1}, with a maximum of 2300 mg kg^{-1} [29]. In contrast, much higher Ni levels were found in Colombia's Sinú River Basin, where concentrations averaged 661 mg kg^{-1}, and Hg levels reached 0.159 mg kg^{-1}. These elevated values were attributed to both agricultural activities and ferronickel mining [30]. In another study, Hg concentrations ranged from 0.004 to 1.1 mg kg^{-1}, and Ni averaged 24.7 mg kg^{-1}, with a maximum of 55 mg kg^{-1}. The presence of Hg was linked to emissions from coal-fired power plants and the historical use of Hg-containing agrochemicals [31]. In Northeast China's Dehui region, the average Ni concentration in agricultural soils was reported at 20.8 mg kg^{-1} [32]. Meanwhile, in Shunde, Southeast China, average concentrations of Co, Hg, and Ni were measured at 16.76, 0.38, and 33.45 mg kg^{-1}, respectively [33]. Further data indicated Ni levels at 28.5 mg kg^{-1} and Mn at 550 mg kg^{-1}, emphasizing spatial disparities and potential risk patterns in heavy metal distribution [34]. For molybdenum (Mo), concentrations in Chinese topsoil ranged from 0.10 to 5.97 mg kg^{-1}, with an average of 0.66 mg kg^{-1} [35]. These figures are consistent with previously reported Mo levels of 0–44.1 mg kg^{-1} in the United States [36] and 0.026–14 mg kg^{-1} in European agricultural soils [37]. Overall, these data underscore the regional variability of trace elements and highlight the influence of both natural and anthropogenic sources critical for interpreting the metal concentrations observed in the current study.

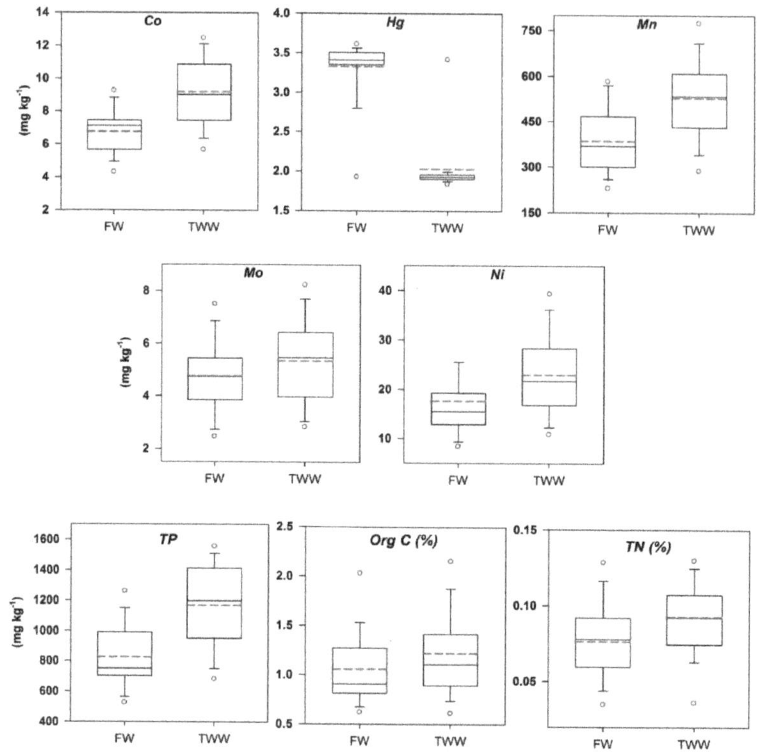

Figure 2. Box-plots for Co, Hg, Mo, Mn, Ni, TP, Org C, and TN contents in fresh (FW) and treated wastewater (TWW) irrigated soils.

The Mann–Whitney Rank Sum Test between medians was used if the normality test failed. Soils irrigated with TWW showed significantly ($p < 0.001$) higher Co, Mn, and Ni and lower Hg levels than FW-irrigated soils. Moreover, TWW-irrigated soils contained higher, but not significant, Mo levels as compared to FW-irrigated soils (Figure 2, Table S4).

For both FW and TWW, Co and Mo had the lowest skewness with a Guassian distribution using Shapiro–Wilk test, indicating non-point source (i.e., TWW and added fertilizers) loading for both elements. In addition, irrigation with TWW caused slight positive skewness and low to moderate relative standard deviation (CV) for Cr, indicating the presence of other sources of loading that might be coming from organic and inorganic fertilizer use. Crops in the JV are under intensive agricultural use, where more NPK fertilizers and organic manure have been used for improving yield quantity and quality. Furthermore, Hg had the lowest CV in both FW and TWW, while Ni had the highest CV value [38]. Low CV values indicate a natural source, while high values reflect anthropogenic sources [38].

3.2. Trace Elements in Main Soil Textural Classes

Concentrations of trace elements in the main textural classes under FW and TWW irrigation were investigated (Figure 3). Soils irrigated with FW had mainly loam (L), while textures of TWW-irrigated ones were mainly clay (C), clay loam (CL), and silt loam (SiL).

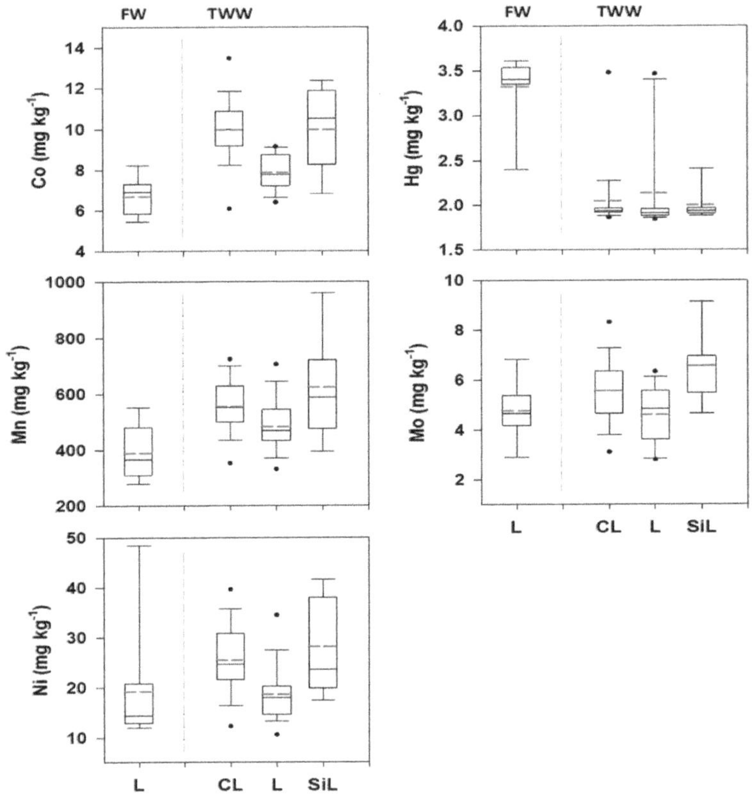

Figure 3. Box-and-whisker plots of trace element contents for main soil textures in FW- and TWW-irrigated soils.

ANOVA and Kruskal–Wallis tests were performed to determine whether the means of trace element concentrations significantly differ with soil texture in FW- and TWW-irrigated soils (Table S5). Generally, trace element concentrations in CL soils were significantly ($p < 0.05$) higher (except for Hg) than those in loam ones. Soil Co and Mn contents in loam TWW-irrigated sites were significantly higher ($p < 0.05$) than those in FW-irrigated sites. Except for Mo, no significant differences in element levels were observed between CL and SiL in TWW-irrigated soils, while SiL soils showed significantly higher ($p < 0.05$) levels of all elements than loam soils (except for Hg). In CL soils, element contents were significantly lower ($p < 0.05$) than those in loam soils (except for Hg).

The lower contents in TWW loam soils, mainly present in the southern parts of the JV, compared to other textures could be attributed to the use of blended waters from Shueib and Kafrein dams, private wells, and desalination units, which may contribute to greater dilution of used TWW. Concentrations of trace elements in soils under cropping land use (vegetable- vs. tree-planted soils) were also investigated (Figure S1). TWW vegetable irrigated soils showed significantly ($p < 0.05$) higher trace element contents than FW tree-cultivated soils (Table S6).

3.3. Spatial Mapping of Trace Elements

Spatial maps using OK (Figure 4a) and EBK (Figure 4b) show distinctively higher levels of Cr, Mo, Mn, and Ni in the west of Zarqa River (TWW-irrigated soils). These soils have fine texture (clay and clay loam) and are characterized by higher Org C, Tp, and

TN levels than FW-irrigated sites (southern parts). Furthermore, maps clearly show that the southern parts contained significantly ($p < 0.001$) higher Hg contents than the middle and northern parts of JV. This may indicate that Hg comes from a different source (parent material) than other elements.

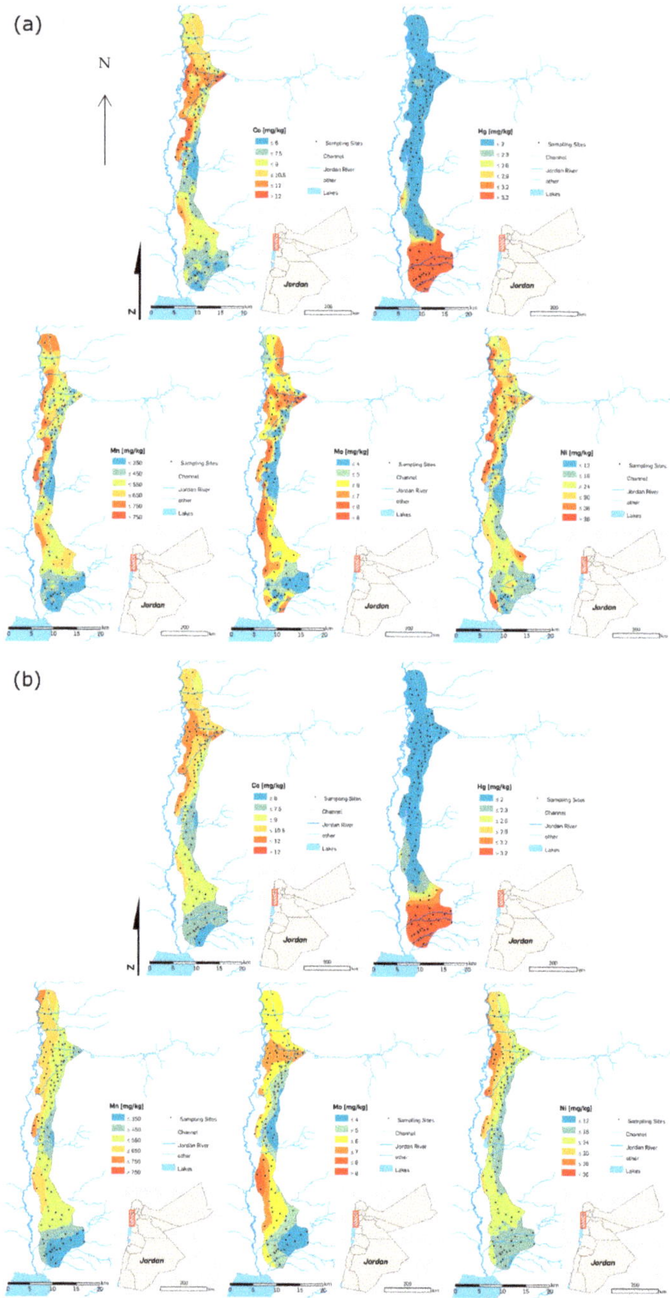

Figure 4. (**a**) Distribution of trace elements (mg kg^{-1}) using ordinary kriging (OK). (**b**) Distribution of trace elements (mg kg^{-1}) using empirical Bayesian kriging (EBK).

Spatial maps of Mo, Ni, and Mn using OK show patchy areas (variable elemental concentrations) in the northern and, to a lesser degree, in the southern parts of the JV (except for Hg), indicating that agricultural activities may primarily cause this variation. The relatively uniform spatial distribution of Hg across most sites, with distinct variability observed only in the southernmost areas, suggests a different origin compared to the other trace elements. This pattern is indicative of a lithogenic source, likely related to underlying parent material rather than anthropogenic inputs. Moreover, Co maps have a blocky distribution (Figure 4a) all over the studied area. The patchy distribution at shorter distances (especially Mo and Ni) indicates relatively higher inputs of these elements from anthropogenic activities (fertilizer and TWW use).

Using the EBK, the resulting maps (Figure 4b) display more continuous, block-like zones with fewer scattered hotspots compared to the OK maps. This smoother spatial pattern may be due to the reduced uncertainty inherent in the EBK approach relative to the OK method. Such uncertainties may result predominantly from spatial interpolation or spatial heterogeneity. In addition, the uncertainties of the spatial prediction of soil elements across the JV may be inherently linked to the complexity of interactions between natural processes (parent material) and associated agricultural activities (chemical and organic fertilizer use, in addition to irrigation with TWW). Therefore, the obtained results indicate that EBK is a better interpolator than the OK method for the estimation of trace element concentrations at unsampled locations. Several studies reported that metals had spatially weak dependence due to anthropogenic factors, while natural factors were attributed to spatially strong dependence in studied soils [13,39–43].

Examining the spatial maps reveals areas with high Co and Mo concentrations in the northern parts right across the Deir Allah area (western parts of Zarqa River). The high levels (red colored) may be attributed to the inputs from TWW. For Mn and Ni, maps show no distinct areas with high concentration (as observed in Co and Mo maps), indicating that both elements (Mn and Ni) may be influenced by a parent material origin.

3.4. Correlation Coefficient Analysis

Correlation is used to determine the association between two variables, and it forms the basis of other more sophisticated multivariate statistics such as factor analysis [44]. Pearson correlation coefficients were determined between Co, Hg, Mn, Mo, and Ni and other basic soil properties (org C, TP, TN, EC, pH, clay, sand, and CEC). In general, for all soils, coefficients revealed significant moderate to strong correlation between Co, Hg, Mn, Mo, and Ni, significant weak to moderate correlations between these elements and Org C, TN, and TP, significant strong correlations between pH and studied elements (except for Hg), and weak correlations between EC, CEC, and the rest of the soil tested properties (Table S7). Correlation coefficients between trace elements and tested soil properties in both FW and TWW and in main soil textural classes (CL, SiL, and L) are also provided in Tables S8–S13. Weak negative correlations coupled with low CV for Hg and other tested soil properties may indicate a different source (parent material). Strong positive correlations between tested elements may indicate similar possible sources. In addition, weak to moderate correlations between these elements and Org C, TP, and TN in all soils may also indicate possible multiple sources (chemical, organic fertilizers, and TWW) affecting trace element levels in soils.

3.5. Spatial Structure of Trace Elements

The spatial dependence of soil properties in the semivariogram is determined by the nugget-to-sill ratio. Three classifications are used for model explanation: A ratio of <25, 25–50, and <75% indicates strong, moderate, and weak spatial dependence, respectively [39]. For comparison purposes, both the isotropic and anisotropic spatial behavior of the attributes were explored. Results showed no differences in the spatial dependency of trace elements. The variability in measured soil properties can be caused by natural (inherent) and random factors. Anthropogenic activities such as the use of fertilizers and applying different farming and cropping systems result in a higher nugget-to-sill ratio, whereas a lower ratio may be related to the effect of climatic and soil-forming factors [45,46]. All trace elements were best described by the spherical model (Figure 5).

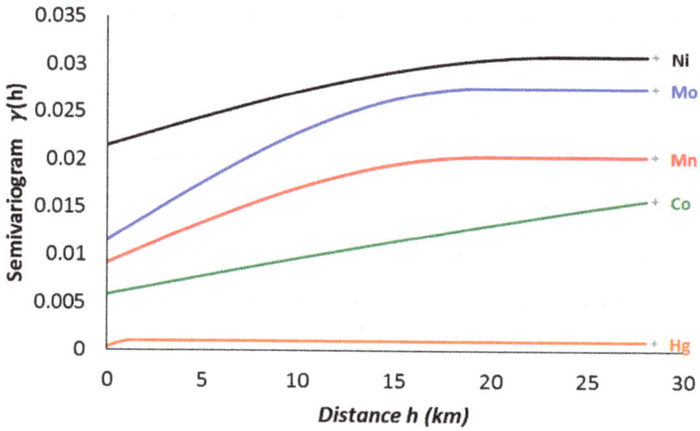

Figure 5. Fitted variogram models of Co, Hg, Mo, Mn, and Ni.

Semivariogram analysis showed that the nugget-to-sill ratios were larger than 0.25 and lower than 0.75 for all elements (excluding Hg), suggesting moderate spatial dependence (Table 1). A smaller nugget effect indicates enough collected samples to describe spatial variation of trace elements. Semivariogram analysis showed that the nugget-to-sill ratios were larger than 0.25 and lower than 0.75 for all elements (excluding Hg), suggesting moderate spatial dependence (Table 1). The result suggests that both natural and anthropogenic factors might be affecting Co, Mn, Mo, and Ni distribution. Variograms in Figure 5 also show similar shapes and slopes for Mn, Mo, and Ni, with higher nugget values for Ni, followed by Mo and Mn, indicating patchy distribution at shorter distances and relatively higher inputs of Ni from anthropogenic activities (fertilizer and TWW use) [47]. Furthermore, semivariogram results strongly indicate that Hg was of pedogenic origins.

Table 1. Nugget, sill, and range values of fitted variogram models for individual trace elements.

Element	Nugget	Par Sill	Sill	Nugget-to-Sill Ratio	R2	Range	Model
Co	0.005841	0.015188	0.021029	0.278	0.52	59,727	Spherical
Mn	0.009133	0.011193	0.020326	0.449	0.67	19,526	Spherical
Mo	0.011496	0.016039	0.027535	0.418	0.66	19,486	Spherical
Ni	0.021468	0.009403	0.030871	0.695	0.75	23,633	Spherical
Hg	0.000322	0.000672	0.000994	0.324	0.06	1305.4	Spherical

All trace elements showed relatively small nugget and sill values; however, the range values were quite different. For example, Mn and Mo had the lowest range, while Co had the highest range. Higher range values for Co and Ni indicate the presence of spatial dependency at longer distances than Mn and Mo [48,49]. For Co, a lower nugget value and a gradual increase in semivariogram slope at longer distances indicate lower steady inputs from fertilizers and TWW [50]. Best-fitted variogram models for isotropic directions are shown in (Table 1).

Cross-validation results (Table 2) demonstrate that EBK consistently outperformed OK in predicting trace element concentrations. EBK yielded lower RMSE and MSE values for most elements (Co, Mn, Mo, and Hg) and showed RMSSE values closer to 1, indicating improved model calibration. Lower average standard errors suggest reduced prediction uncertainty. These findings highlight EBK as the more reliable method for spatial interpolation of trace element distributions in the study area. For both RMSE and MSE, lower values or values close to zero indicate better fit, while for RMSSE, values close to one indicate a good estimation for the variability of prediction, and an MSE value nearest to the RMSE [24,51,52]. RMSE is a useful measure of accuracy to compare prediction errors of different models. A lower RMSE value is an indicator of higher precision of the interpolation method [53], while others considered a lower RMSSE value as indicating a stable model [54,55]. For Hg, both models were comparable; however, EBK showed less data dispersion when using EBK compared with the OK method [45,56,57]. Therefore, both natural and anthropogenic factors could be affecting Co, Mn, Mo, and Ni distribution.

Table 2. Cross-validation values of trace elements for evaluating the performance of ordinary kriging (OK) and empirical Bayesian kriging (EBK) models.

	Co		**Mn**		**Mo**		**Ni**		**Hg**	
	OK	**EBK**	**OK**	**EBK**	**OK**	**EBK**	**OK**	**EBK**	**OK**	**EBK**
ME	−0.00090	−0.00027	−0.00115	−0.00064	−0.00180	−0.00142	−0.00253	−0.00208	−0.00108	−0.00174
RMSE	0.07955	0.07463	0.10706	0.10157	0.12019	0.11321	0.15424	0.14205	0.03884	0.04178
MSE	0.00764	0.00192	−0.00553	0.00169	−0.01037	−0.00435	−0.01468	−0.00090	−0.02944	−0.01517
RMSSE	0.95200	0.97254	0.99866	0.97081	0.99073	0.98058	0.98642	1.0582	1.16246	0.98664
Avg. SE	0.08391	0.07772	0.10720	0.10443	0.12137	0.11482	0.15594	0.13399	0.03311	0.03931

3.6. Principal Component Analysis

PCA was performed to further delineate the sources of elements and relative contribution of irrigation (TWW vs. FW), agricultural practices (chemical fertilizers and organic manuring), and inherited soil characteristics in soils of the JV. The rotated component matrix of PCA for trace elements extracted components and loadings are shown for all, FW, and TWW-irrigated samples (Table 3). CEC was removed for not reaching the KMO criteria.

Table 3. Extracted components and loadings for trace elements (bold values indicate significant component loadings).

	All			**TWW**				**FW**		
Parameter	**PC1**	**PC2**	**PC3**	**PC1**	**PC2**	**PC3**	**PC4**	**PC1**	**PC2**	**PC3**
Co	0.883	0.156	−0.326	0.927	0.044	−0.148	−0.163	0.69	0.618	−0.121
Mn	0.768	0.335	−0.245	0.796	0.233	0.004	−0.23	0.832	0.368	−0.154
Mo	0.803	0.276	0.275	0.78	0.144	0.445	−0.083	0.394	0.778	0.208
Ni	0.85	0.02	−0.126	0.893	0.005	−0.037	−0.087	0.01	0.686	−0.072
Clay	0.677	−0.122	−0.423	0.732	0.009	−0.519	0.131	−0.119	0.078	−0.883
Org C	0.036	0.815	0.131	0.021	0.774	0.202	0.008	0.807	0.042	0.443
TN	0.194	0.767	−0.097	0.191	0.846	−0.096	0.164	0.752	0.101	−0.052
TP	0.221	0.681	−0.42	0.158	0.661	−0.073	−0.463	0.796	0.15	−0.344
EC_SPE	0.026	0.068	0.886	0.018	0.043	0.937	0.115	−0.305	0.468	0.55
Hg	−0.226	−0.287	0.638	−0.099	0.043	0.062	0.915	−0.369	−0.093	0.464

Table 3. *Cont.*

	All			TWW				FW		
Parameter	**PC1**	**PC2**	**PC3**	**PC1**	**PC2**	**PC3**	**PC4**	**PC1**	**PC2**	**PC3**
Sand	−0.86	−0.235	−0.083	−0.879	−0.209	−0.096	−0.126	−0.469	−0.687	0.162
Explained variance (%)	37.03	18.97	16.71	38.9	17.1	13.1	11.1	33.1	21.2	15.7
Cumulative	37.03	56	72.71	38.9	56	69.1	80.2	33.1	54.3	70
Eigen value	4.1	2.1	1.8	4.3	1.9	1.4	1.2	3.6	2.3	1.7

For all, TWW, and FW samples, PCA results explained 72.7, 80.2, and 70.0% of total variance, respectively. For all and TWW-irrigated soils, the first component (PC1) had strong positive loadings for Co, Mn, Mo, and Ni, moderate loadings for clay, and strong negative loadings for sand. PC2 showed strong loadings for Org C, TN, and TP; PC3 showed strong loadings for EC (all samples); and PC4 showed strong loadings for Hg (TWW).

For FW soils, PC1 showed strong loadings for Mn, Org C, TN, and TP; PC2 showed strong loadings for Mo and moderate to strong loadings for Ni and Co; and PC3 showed strong negative loadings for clay. Furthermore, Hg showed moderate loadings in PC3 and weak loadings in PC1 and PC2.

PC2 (with strong loadings in TWW soils for Org C, TN, and TP) indicates a clear signal of agricultural enrichment (TWW and fertilizer use, and composted organic material). In FW-irrigated soils, Org C, TN, and TP showed strong loadings in PC1, suggesting that elevated metal levels in these soils are linked to inputs from chemical fertilizers and composted organic materials. PC3 reflects saline inputs from TWW and the basic geochemical background, particularly from lacustrine sediments of the Pleistocene Lake Lisan [17]. Furthermore, in TWW-irrigated soils, Hg exhibited strong loadings in PC4, while only showing low to moderate loadings in PC3 for FW soils.

These patterns may strongly suggest an anthropogenic origin, primarily linked to the long-term use of TWW, which is known to introduce trace elements into soils. The association with clay may indicate that fine-textured soils act as sinks for these metals, enhancing their retention through sorption. Furthermore, TWW-irrigated soils exhibited consistently higher concentrations of trace metals compared to those irrigated with freshwater (FW), supporting the interpretation that these elements may be primarily derived from anthropogenic sources along with agricultural practices such as the use of organic manure and chemical fertilizers.

Furthermore, PCA results suggest that Hg accumulation is not significantly driven by fertilizers or organic amendments, but more likely reflects a lithogenic origin related to the geological characteristics of the region. The high association between Hg and EC suggests that Hg originates from basic rocks (residual parent material) found in that area [16]. Higher salinity in the southern parts of the JV was attributed to earlier huge floods occurring in the rift valley plain, which were considered a precursor for creating the Pleistocene Lake Lisan–Dead Sea [17]. As a result, layers of lacustrine sediments were left behind and the thickness of these sediments decreased, and therefore, soil salinity increased southward. Moreover, Hg shows a weak negative loading in PC1 and PC3; this may also confirm that chemical and organic fertilizers are weakly associated with Hg levels in the tested soils.

To further extrapolate the origins of trace elements; PCA was run excluding factors in PC2. Results showed that trace elements have moderate loading for clay (transported, alluvial parent material), moderate loading for TP, weak loadings for TN, and no loadings for Org C. Moderate loadings for trace elements and TP suggest the influence of chemical fertilizers and a greater possibility that p is complexed with these elements. Weak loadings for TN may indicate the influence of TWW and organic manure. Furthermore, PCA was run excluding factors in PC3. Results showed that EC and pH had very weak loading for all tested elements (except for Hg). Therefore, it could be postulated that Co, Mn, Mo, and Ni originate from multiple sources (pedogenic, chemical, and organic fertilizers).

PCA results were further clarified by the biplots using a different program (JMP software, JMP Version 11), where the angles formed between variable vectors and the horizontal (X) axis illustrate the strength and direction of their correlations (Figure 6). Acute angles (<90°) indicate a strong positive correlation, right angles (90°) suggest no correlation, and obtuse angles (>90°) denote a negative correlation [58]. These biplots indicate that these elements could have been derived primarily from weathering and genesis of soil parent rocks, whereas the upper graph indicates another secondary elemental source, which might be affected by agricultural practices (anthropogenic factors) such as chemical and organic fertilizers, as well as the use of TWW. Oblique angles (>0°–<90°) between vectors of soil trace elements strongly dominated by the same PC revealed strong correlations among these elements, suggesting that they were derived from similar sources [59].

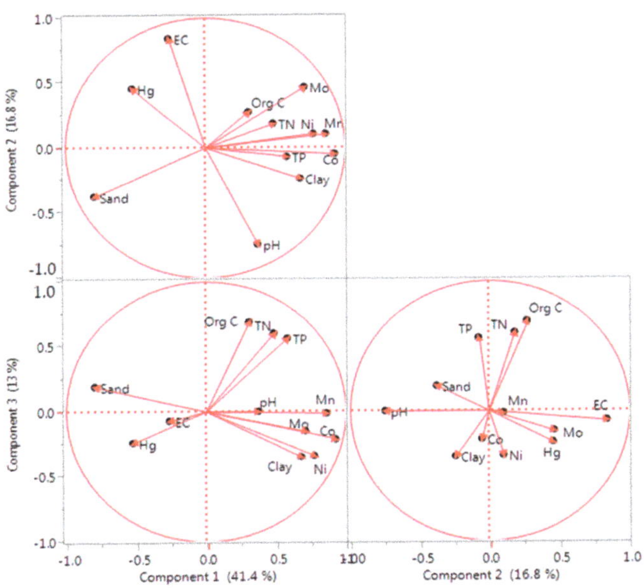

Figure 6. Principal component (PC) analysis biplot for trace element contents.

Results of this study agree with [60,61], which reported that Co, Ni, and Mn had positive loadings and belonged to one group (factor). In addition, Na and Si (sand) were negatively correlated with the other elements in this group.

In soils along Zarqa River, ref. [62] reported that Mn was mainly bound to iron-manganese oxides and calcite fractions, whereas Ni was found in the residual, iron-manganese oxides, and calcite fractions. Ref. [63] showed that Hg concentrations in Quanzhou Bay sediments were attributed to the variable geological distribution of their parent minerals.

Multivariate and geostatistical analyses in agricultural soils in Dehui, Northeast China, suggested that average Ni concentrations as high as 20.8 mg kg^{-1} had lithogenic origins [32]. Higher enrichment of Hg in the studied soils could be mainly attributed to the background value, which was determined only in the <2-micron fraction of 20 soil samples [64]. Moreover, higher explained variance in PC1 (34%) could be considered a strong indicator of geological or soil forming factors controlling Co, Mn, Mo, and Ni distribution in soil [65].

3.7. Cluster Analysis

Hierarchical cluster analysis using Ward's linkage (Figure 7) was used to group the tested parameters into main clusters. Results of the cluster analysis show three main groups (1) Co, Mn, Ni, Mo, and clay (red color); (2) Org C, TN, and TP (blue); and (3) Hg, EC, pH, and sand (black). Co and Mn were grouped tightly, Ni and Mo showed close association, and clay was linked with all four. The cluster analysis supports PCA outcomes by grouping Co, Mn, Mo, Ni, and clay together, Org C with TN and TP, and Hg with EC and pH. These findings may suggest that Co, Mn, Mo, and Ni could originate from mixed anthropogenic sources (TWW and fertilizers), while Hg could be largely attributed to geogenic factors [66].

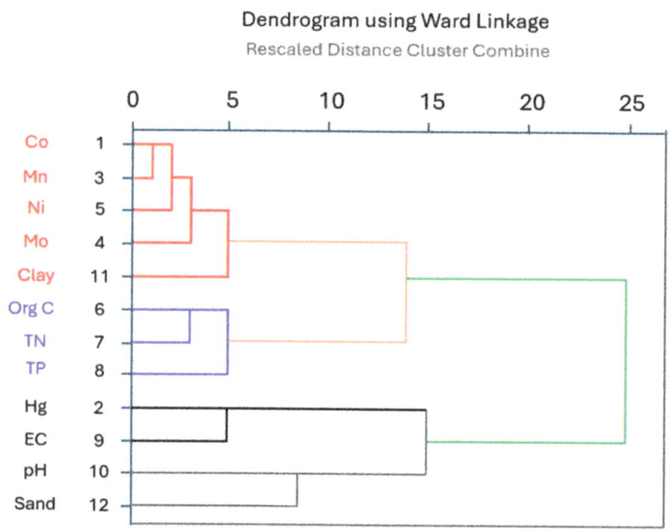

Figure 7. Hierarchical relationship (dendrogram) between trace elements and soil properties.

Within these clusters, Co and Mn were grouped tightly, Ni and Mo were clustered, and clay was associated with all four, indicating that soil texture modulates trace element retention but does not negate their anthropogenic source. Org C was more closely linked to TN, likely from wastewater and organic amendments, while TP diverged slightly, indicating a stronger association with chemical fertilizers. Hg's pairing with EC and pH, along with its weak correlation with other metals, supports its distinct geological origin.

Therefore, the combined evidence from PCA, hierarchical clustering, spatial mapping, correlation analysis, and comparisons of trace element levels in TWW versus FW suggests that the continued use of TWW for irrigation, along with fertilizer and manure application, is likely a contributor to the observed accumulation of Co, Mn, Ni, and Mo in soils. However, definitive source attribution would require further validation through isotopic fingerprinting or controlled experiments.

3.8. Implications for Land Management Under Water Scarcity

Studying the spatial variability and sources of trace elements in the JV offers critical insights for sustainable land management in arid and semi-arid agricultural systems. Although concentrations of Co, Mn, Mo, and Ni in TWW-irrigated areas remain below regulatory limits, their elevated presence underscores the need for proactive, site-specific strategies [67]. The kriging-based maps generated in this study serve as a decision support tool for identifying trace element hotspots, enabling more efficient freshwater allocation and guiding agroecological zoning through spatially informed crop planning to enhance crop safety and long-term sustainability. Farmers and extension services can use these spatial insights to implement selective irrigation—alternating TWW and freshwater based on localized trace element levels—and to develop long-term soil monitoring protocols that leverage model uncertainty ranges to detect early signs of accumulation in high-risk zones [68].

To further mitigate the accumulation of trace elements, land managers are advised to periodically dilute TWW with freshwater, particularly during dry periods, and to define region-specific threshold values that inform both irrigation scheduling and fertilizer management. Reducing the use of high-metal-content fertilizers and promoting slow-release or organic alternatives can also minimize trace element inputs [69]. Ultimately, integrating these targeted irrigation, monitoring, and input strategies within a broader agronomic planning framework will support resilient and adaptive land use in the face of growing climate and resource pressures.

4. Summary and Conclusions

This study investigated the spatial distribution and source attribution of five trace elements (Co, Hg, Mo, Mn, and Ni) in intensively cultivated soils of the JV, where irrigation with treated wastewater has been practiced for decades. Using both ordinary kriging (OK) and empirical Bayesian kriging (EBK), spatial variability and prediction uncertainty were quantified. EBK demonstrated superior predictive performance, providing more reliable spatial assessments essential for land monitoring programs. Multivariate analyses indicated that Hg likely originated from geogenic sources, while Co, Mn, Mo, and Ni are largely associated with anthropogenic activities—particularly the prolonged use of TWW, chemical fertilizers, and organic amendments.

Although none of the trace element concentrations exceeded international safety thresholds, the accumulation patterns emphasize the need for sustainable management practices to prevent future degradation. These include routine soil monitoring, the dilution of TWW with freshwater during sensitive growth periods, and more judicious fertilizer application. Site-specific maps produced from this study can support decision-makers and farmers in identifying at-risk zones and implementing precision agriculture techniques. Overall, this research provides a geostatistical framework for improving land resilience and ensuring soil sustainability under continued pressure from climate change and water scarcity in arid regions such as the JV.

Supplementary Materials: The following supporting information can be downloaded at: https://www.mdpi.com/article/10.3390/land14071325/s1. Figure S1. Box-and-whisker plots for trace element contents and related soil properties in FW- and TWW-irrigated soils under tree and vegetable cultivation in the Jordan valley; Table S1: Statistics of trace element concentrations and major soil properties of all samples; Table S2: Summary statistics of trace element contents and major soil properties for FW samples; Table S3: Summary statistics of trace element contents and major soil properties for TWW samples; Table S4: Pairwise multiple comparisons of trace element contents under FW and TWW irrigation; Table S5: Pairwise multiple comparisons of trace element contents in main soil textural classes under FW and TWW irrigation; Table S6: Pairwise multiple comparisons of trace element contents in FW and TWW soils under vegetable and tree cultivation; Table S7: Pearson correlation coefficients for all soils; Table S8: Pearson correlation coefficients for FW-irrigated soils; Table S9: Pearson correlation coefficients for TWW-irrigated soils; Table S10: Pearson correlation coefficients for FW loam soils; Table S11: Pearson correlation coefficients for TWW clay loam soils;

Table S12: Pearson correlation coefficients for TWW loam soils; Table S13: Pearson correlation coefficients for TWW silt loam soils.

Author Contributions: Conceptualization, M.A.G. and B.M.; methodology, M.A.G. and B.M.; software, M.A.G. and N.M. (Nicolai Moos); validation, M.A.G., B.M., and N.M. (Nicolai Moos); formal analysis, M.A.G. and B.M.; investigation, M.A.G. and B.M.; resources, M.A.G.; data curation, M.A.G., B.M., and N.M. (Nicolai Moos); writing—original draft preparation, M.A.G., B.M., N.M. (Nicolai Moos), and N.M. (Nikolaos Monokrousos); writing—review and editing, M.A.G., B.M., N.M. (Nicolai Moos), and N.M. (Nikolaos Monokrousos); visualization, M.A.G., B.M., and N.M. (Nicolai Moos); supervision, M.A.G. and B.M.; project administration, M.A.G.; funding acquisition, M.A.G. All authors have read and agreed to the published version of the manuscript.

Funding: This research was partially funded by the deanship of research at the Jordan University of Science and Technology (grant number 2018/180).

Institutional Review Board Statement: Not applicable.

Informed Consent Statement: Not applicable.

Data Availability Statement: Data are contained within the article or Supplementary Materials.

Acknowledgments: The authors gratefully acknowledge the external funding of the scientific research visit by the German Research Foundation (DFG), the deanship of research at the Jordan University of Science and Technology, and the access to laboratory facilities by the Department of Soil Science/Soil Ecology, Institute of Geography, Ruhr University Bochum, Germany.

Conflicts of Interest: The authors declare no conflicts of interest.

Abbreviations

The following abbreviations are used in this manuscript:

EC	electrical conductivity
EBK	empirical Bayesian kriging
FAO	Food and Agriculture Organization
ICP-AES	inductively coupled plasma atomic emission spectrometry
JV	Jordan Valley
KMO	Kaiser–Meyer–Olkin
OK	ordinary kriging
Org C	organic carbon
PCA	principal component analysis
SPSS	Statistical Package for Social Sciences
TN	total nitrogen
TP	total phosphorus
TWW	treated wastewater
WHO	World Health Organization

References

1. Antoniadis, V.; Levizou, E.; Shaheen, S.M.; Ok, Y.S.; Sebastian, A.; Baum, C.; Prasad, M.N.V.; Wenzel, W.W.; Rinklebe, J. Trace Elements in the Soil-Plant Interface: Phytoavailability, Translocation, and Phytoremediation—A Review. *Earth Sci. Rev.* **2017**, *171*, 621–645. [CrossRef]
2. Chittora, D. Harmful Impact of Synthetic Fertilizers on Growing Agriculture and Environment. *Glob. J. Pharm. Pharm. Sci.* **2023**, *11*, 555804. [CrossRef]
3. Alloway, B.J. Heavy Metals and Metalloids as Micronutrients for Plants and Animals. In *Heavy Metals in Soils*; Springer: Berlin/Heidelberg, Germany, 2013; pp. 195–209.
4. Mendieta-Mendoza, A.; Renteria-Villalobos, M.; Randall, H.; Ruíz-Gómez, S.; Ríos-López, M. Chemical Degradation of Agricultural Soil under Arid Conditions by the Accumulation of Potentially Toxic Elements and Salts. *Geoderma Reg.* **2023**, *35*, e00736. [CrossRef]

5. Gharaibeh, M.A.; Marschner, B.; Heinze, S.; Moos, N. Spatial Distribution of Metals in Soils under Agriculture in the Jordan Valley. *Geoderma Reg.* **2020**, *20*, e00245. [CrossRef]
6. FAO. *The State of the World's Land and Water Resources for Food and Agriculture—Systems at Breaking Point (SOLAW 2021)*; FAO: Rome, Italy, 2021.
7. Liu, H.; Zhang, Y.; Yang, J.; Wang, H.; Li, Y.; Shi, Y.; Li, D.; Holm, P.E.; Ou, Q.; Hu, W. Quantitative Source Apportionment, Risk Assessment and Distribution of Heavy Metals in Agricultural Soils from Southern Shandong Peninsula of China. *Sci. Total Environ.* **2021**, *767*, 144879. [CrossRef]
8. Ersoy, A.; Yünsel, T.Y. The Assessment of Soil Contamination by Heavy Metals Using Geostatistical Sequential Gaussian Simulation Method. *Hum. Ecol. Risk Assess.* **2018**, *24*, 2142–2161. [CrossRef]
9. Bednářová, Z.; Kalina, J.; Hájek, O.; Sáňka, M.; Komprdová, K. Spatial Distribution and Risk Assessment of Metals in Agricultural Soils. *Geoderma* **2016**, *284*, 113–121. [CrossRef]
10. Yuanan, H.; He, K.; Sun, Z.; Chen, G.; Cheng, H. Quantitative Source Apportionment of Heavy Metal(Loid)s in the Agricultural Soils of an Industrializing Region and Associated Model Uncertainty. *J. Hazard. Mater.* **2020**, *391*, 122244. [CrossRef]
11. Horta, A.; Azevedo, L.; Neves, J.; Soares, A.; Pozza, L. Integrating Portable X-Ray Fluorescence (PXRF) Measurement Uncertainty for Accurate Soil Contamination Mapping. *Geoderma* **2021**, *382*, 114712. [CrossRef]
12. Cicchella, D.; Giaccio, L.; Dinelli, E.; Albanese, S.; Lima, A.; Zuzolo, D.; Valera, P.; De Vivo, B. GEMAS: Spatial Distribution of Chemical Elements in Agricultural and Grazing Land Soil of Italy. *J. Geochem. Explor.* **2015**, *154*, 129–142. [CrossRef]
13. Rodríguez Martín, J.A.; Ramos-Miras, J.J.; Boluda, R.; Gil, C. Spatial Relations of Heavy Metals in Arable and Greenhouse Soils of a Mediterranean Environment Region (Spain). *Geoderma* **2013**, *200–201*, 180–188. [CrossRef]
14. Darwish, M.A.G.; Pöllmann, H. Trace Elements Assessment in Agricultural and Desert Soils of Aswan Area, South Egypt: Geochemical Characteristics and Environmental Impacts. *J. Afr. Earth Sci.* **2015**, *112*, 358–373. [CrossRef]
15. Savignan, L.; Lee, A.; Coynel, A.; Jalabert, S.; Faucher, S.; Lespes, G.; Chéry, P. Spatial Distribution of Trace Elements in the Soils of South-Western France and Identification of Natural and Anthropogenic Sources. *Catena* **2021**, *205*, 105446. [CrossRef]
16. Lucke, B.; Ziadat, F.; Taimeh, A. The Soils of Jordan. In *Atlas of Jordan*; Presses de l'Ifpo: Amman, Jordan, 2013; pp. 72–76.
17. Gharaibeh, M.A.; Albalasmeh, A.A.; El Hanandeh, A. Estimation of Saturated Paste Electrical Conductivity Using Three Modelling Approaches: Traditional Dilution Extracts; Saturation Percentage and Artificial Neural Networks. *Catena* **2021**, *200*, 105141. [CrossRef]
18. Chander, K.; Hartmann, G.; Joergensen, R.G.; Khan, K.S.; Lamersdorf, N. Comparison of Methods for Measuring Heavy Metals and Total Phosphorus in Soils Contaminated by Different Sources. *Arch. Agron. Soil. Sci.* **2008**, *54*, 413–422. [CrossRef]
19. Mclean, E.O. Soil PH and Lime Requirement. In *Methods of Soil Analysis*; Agronomy Monographs; Wiley: Hoboken, NJ, USA, 1982; pp. 199–224, ISBN 9780891189770.
20. Rhoades, J.D. Soluble Salts. In *Methods of Soil Analysis. Part 2. Chemical and Microbiological Properties*; Agronomy Monograph SV-9.2; American Society of Agronomy, Soil Science Society of America: Madison, WI, USA, 2015; pp. 167–179. ISBN 978-0-89118-204-7.
21. Nelson, D.W.; Sommers, L.E. Total Carbon, Organic Carbon, and Organic Matter. BT—Methods of Soil Analysis. Part 3. Chemical Methods. In *Methods of Soil Analysis. Part 3. Chemical Methods*; Wiley: Hoboken, NJ, USA, 1996; pp. 961–1010. ISBN 0-89118-825-8.
22. Bremner, J.; Mulvaney, C. Nitrogen-Total. In *Methods of Soil Analysis. Part 2*; Page, A.L., Ed.; Agronomy Monographs; American Society of Agronomy, Soil Science Society of America: Madison, WI, USA, 2015; pp. 595–624.
23. Morgan, G.A.; Barrett, K.C.; Leech, N.L.; Gloeckner, G.W. *IBM SPSS for Introductory Statistics: Use and Interpretation*; Routledge: New York, NY, USA, 2019.
24. Oliver, A.M.; Webster, R. *Basic Steps in Geostatistics: The Variogram and Kriging*; 2211-808X; Springer: Cham, Switzerland, 2015; Volume 1, ISBN 9783319158648.
25. Grekousis, G. *Spatial Analysis Methods and Practice*, 1st ed.; Cambridge University Press: New York, NY, USA, 2020; ISBN 9781108614528.
26. Goovaerts, P. Geostatistics in Soil Science: State-of-the-Art and Perspectives. *Geoderma* **1999**, *89*, 1–45. [CrossRef]
27. Kabata-Pendias, A. *Trace Elements in Soils and Plants*; CRC Press: Boca Raton, FL, USA, 2010; ISBN 9781420093681.
28. Kabata-Pendias, A.; Mukherjee, A.B. *Trace Elements from Soil to Human*; Springer: Berlin/Heidelberg, Germany, 2007; ISBN 3540327134.
29. Kristamtini; Widyayanti, S.; Widodo, S.; Pustika, A.B.; Purwaningsih, H.; Hanifa, A.P.; Muazam, A.; Sutardi; Badia Ginting, R.C.; Mulia, S.; et al. Potentially Toxic Elements' (PTEs) Spatial Distribution in Agricultural Soils and Their Impact on Ecological and Health Risks. *Case Stud. Chem. Environ. Eng.* **2024**, *10*, 100936. [CrossRef]
30. Marrugo-Negrete, J.; Pinedo-Hernández, J.; Díez, S. Assessment of Heavy Metal Pollution, Spatial Distribution and Origin in Agricultural Soils along the Sinú River Basin, Colombia. *Environ. Res.* **2017**, *154*, 380–388. [CrossRef]
31. Fan, M.; Margenot, A.J.; Zhang, H.; Lal, R.; Wu, J.; Wu, P.; Chen, F.; Gao, C. Distribution and Source Identification of Potentially Toxic Elements in Agricultural Soils through High-Resolution Sampling. *Environ. Pollut.* **2020**, *263*, 114527. [CrossRef]

32. Sun, C.; Liu, J.; Wang, Y.; Sun, L.; Yu, H. Multivariate and Geostatistical Analyses of the Spatial Distribution and Sources of Heavy Metals in Agricultural Soil in Dehui, Northeast China. *Chemosphere* **2013**, *92*, 517–523. [CrossRef]
33. Cai, L.; Xu, Z.; Bao, P.; He, M.; Dou, L.; Chen, L.; Zhou, Y.; Zhu, Y.G. Multivariate and Geostatistical Analyses of the Spatial Distribution and Source of Arsenic and Heavy Metals in the Agricultural Soils in Shunde, Southeast China. *J. Geochem. Explor.* **2015**, *148*, 189–195. [CrossRef]
34. Xu, X.; Xu, Z.; Liang, L.; Han, J.; Wu, G.; Lu, Q.; Liu, L.; Li, P.; Han, Q.; Wang, L.; et al. Risk Hotspots and Influencing Factors Identification of Heavy Metal(Loid)s in Agricultural Soils Using Spatial Bivariate Analysis and Random Forest. *Sci. Total Environ.* **2024**, *954*, 176359. [CrossRef]
35. Wu, Z.; Hou, Q.; Yang, Z.; Yu, T.; Li, D.; Lin, K.; Ma, X. Identification of Factors Driving the Spatial Distribution of Molybdenum (Mo) in Topsoil in the Longitudinal Range-Gorge Region of Southwestern China Using the Geodetector Model. *Ecotoxicol. Environ. Saf.* **2024**, *271*, 115846. [CrossRef]
36. Smith, D.B.; Solano, F.; Woodruff, L.G.; Cannon, W.F.; Ellefsen, K.J. Geochemical and Mineralogical Maps, with Interpretation, for Soils of the Conterminous United States. *USGS Sci. Investig. Rep.* **2019**, *2017–5118*. [CrossRef]
37. Reimann, C.; Fabian, K.; Birke, M.; Filzmoser, P.; Demetriades, A.; Négrel, P.; Oorts, K.; Matschullat, J.; de Caritat, P.; Albanese, S.; et al. GEMAS: Establishing Geochemical Background and Threshold for 53 Chemical Elements in European Agricultural Soil. *Appl. Geochem.* **2018**, *88*, 302–318. [CrossRef]
38. Baran, A.; Wieczorek, J.; Mazurek, R.; Urbański, K.; Klimkowicz-Pawlas, A. Potential Ecological Risk Assessment and Predicting Zinc Accumulation in Soils. *Environ. Geochem. Health* **2018**, *40*, 435–450. [CrossRef]
39. Cambardella, C.A.; Moorman, T.B.; Novak, J.M.; Parkin, T.B.; Karlen, D.L.; Turco, R.F.; Konopka, A.E. Field-Scale Variability of Soil Properties in Central Iowa Soils. *Soil Sci. Soc. Am. J.* **1994**, *58*, 1501–1511. [CrossRef]
40. Rodríguez Martín, J.A.; Arias, M.L.; Grau Corbí, J.M. Heavy Metals Contents in Agricultural Topsoils in the Ebro Basin (Spain). Application of the Multivariate Geoestatistical Methods to Study Spatial Variations. *Environ. Pollut.* **2006**, *144*, 1001–1012. [CrossRef]
41. Algül, F.; Beyhan, M. Concentrations and Sources of Heavy Metals in Shallow Sediments in Lake Bafa, Turkey. *Sci. Rep.* **2020**, *10*, 11782. [CrossRef]
42. Zhang, X.; Wei, S.; Sun, Q.; Wadood, S.A.; Guo, B. Source Identification and Spatial Distribution of Arsenic and Heavy Metals in Agricultural Soil around Hunan Industrial Estate by Positive Matrix Factorization Model, Principle Components Analysis and Geo Statistical Analysis. *Ecotoxicol. Environ. Saf.* **2018**, *159*, 354–362. [CrossRef]
43. Zhang, J.; Zhou, F.; Chen, C.; Sun, X.; Shi, Y.; Zhao, H.; Chen, F. Spatial Distribution and Correlation Characteristics of Heavy Metals in the Seawater, Suspended Particulate Matter and Sediments in Zhanjiang Bay, China. *PLoS ONE* **2018**, *13*, e0201414. [CrossRef] [PubMed]
44. Douaik, A.; van Meirvenne, M.; Toth, T. Statistical Methods for the Analysis of Soil Spatial and Temporal Variability. In *Principles, Application and Assessment in Soil Science*; IntechOpen: Rijeka, Croatia, 2011.
45. Verma, R.R.; Srivastava, T.K.; Singh, P.; Manjunath, B.L.; Kumar, A. Spatial Mapping of Soil Properties in Konkan Region of India Experiencing Anthropogenic Onslaught. *PLoS ONE* **2021**, *16*, e0247177. [CrossRef]
46. Bhunia, G.S.; Shit, P.K.; Chattopadhyay, R. Assessment of Spatial Variability of Soil Properties Using Geostatistical Approach of Lateritic Soil (West Bengal, India). *Ann. Agrar. Sci.* **2018**, *16*, 436–443. [CrossRef]
47. Xu, L.; Zhao, Y.; Wang, M.; Jiang, Y. Spatial Variability of Eight Hazardous Heavy Metals in Agricultural Topsoils: A Case Study from Xuzhou Area, Jiangsu, China. *Environ. Pollut. Bioavailab.* **2021**, *33*, 326–333. [CrossRef]
48. Huo, X.N.; Zhang, W.W.; Sun, D.F.; Li, H.; Zhou, L.D.; Li, B.G. Spatial Pattern Analysis of Heavy Metals in Beijing Agricultural Soils Based on Spatial Autocorrelation Statistics. *Int. J. Environ. Res. Public Health* **2011**, *8*, 2074–2089. [CrossRef]
49. Behrens, T.; Viscarra Rossel, R.A.; Kerry, R.; MacMillan, R.; Schmidt, K.; Lee, J.; Scholten, T.; Zhu, A.X. The Relevant Range of Scales for Multi-Scale Contextual Spatial Modelling. *Sci. Rep.* **2019**, *9*, 14800. [CrossRef]
50. Zhang, B.; Yang, Y. Spatiotemporal Modeling and Prediction of Soil Heavy Metals Based on Spatiotemporal Cokriging. *Sci. Rep.* **2017**, *7*, 16750. [CrossRef] [PubMed]
51. Zakeri, F.; Mariethoz, G. A Review of Geostatistical Simulation Models Applied to Satellite Remote Sensing: Methods and Applications. *Remote Sens. Environ.* **2021**, *259*, 112381. [CrossRef]
52. Özkan, B.; Dengiz, O.; Turan, İ.D. Site Suitability Analysis for Potential Agricultural Land with Spatial Fuzzy Multi-Criteria Decision Analysis in Regional Scale under Semi-Arid Terrestrial Ecosystem. *Sci. Rep.* **2020**, *10*, 22074. [CrossRef]
53. Gumiere, S.J.; Lafond, J.A.; Hallema, D.W.; Périard, Y.; Caron, J.; Gallichand, J. Mapping Soil Hydraulic Conductivity and Matric Potential for Water Management of Cranberry: Characterisation and Spatial Interpolation Methods. *Biosyst. Eng.* **2014**, *128*, 29–40. [CrossRef]
54. Shi, G. Kriging. In *Data Mining and Knowledge Discovery for Geoscientists*; Elsevier: Amsterdam, The Netherlands, 2014; pp. 238–274.
55. Tziachris, P.; Metaxa, E.; Papadopoulos, F.; Papadopoulou, M. Spatial Modelling and Prediction Assessment of Soil Iron Using Kriging Interpolation with PH as Auxiliary Information. *ISPRS Int. J. Geoinf.* **2017**, *6*, 283. [CrossRef]

56. Manjarrez-Domínguez, C.B.; Prieto-Amparán, J.A.; Valles-Aragón, M.C.; Delgado-Caballero, M.D.R.; Alarcón-Herrera, M.T.; Nevarez-Rodríguez, M.C.; Vázquez-Quintero, G.; Berzoza-Gaytan, C.A. Arsenic Distribution Assessment in a Residential Area Polluted with Mining Residues. *Int. J. Environ. Res. Public Health* **2019**, *16*, 375. [CrossRef]
57. Gupta, A.; Kamble, T.; Machiwal, D. Comparison of Ordinary and Bayesian Kriging Techniques in Depicting Rainfall Variability in Arid and Semi-Arid Regions of North-West India. *Environ. Earth Sci.* **2017**, *76*, 512. [CrossRef]
58. Hothorn, T.; Everitt, B.S. *A Handbook of Statistical Analyses Using R*; CRC Press: Boca Raton, FL, USA, 2014.
59. Shangguan, Y.; Cheng, B.; Zhao, L.; Hou, H.; Ma, J.; Sun, Z.; Xu, Y.; Zhao, R.; Zhang, Y.; Hua, X.; et al. Distribution Assessment and Source Identification Using Multivariate Statistical Analyses and Artificial Neutral Networks for Trace Elements in Agricultural Soils in Xinzhou of Shanxi Province, China. *Pedosphere* **2018**, *28*, 542–554. [CrossRef]
60. Han, Y.M.; Cao, J.J.; Wu, F.; Zhang, B.C.; Zhan, C.L.; Wei, C.; Zhao, Z.Z. Geochemistry and Environmental Assessment of Major and Trace Elements in the Surface Sediments of the Wei River, China. *J. Environ. Monit.* **2012**, *14*, 2762–2771. [CrossRef] [PubMed]
61. Shukla, K.; Kumar, B.; Agrawal, R.; Priyanka, K.; Venkatesh, M. Anshumali Assessment of Cr, Ni and Pb Pollution in Rural Agricultural Soils of Tonalite–Trondjhemite Series in Central India. *Bull. Environ. Contam. Toxicol.* **2017**, *98*, 856–866. [CrossRef]
62. Ghrefat, H.A.; Yusuf, N.; Jamarh, A.; Nazzal, J. Fractionation and Risk Assessment of Heavy Metals in Soil Samples Collected along Zerqa River, Jordan. *Environ. Earth Sci.* **2012**, *66*, 199–208. [CrossRef]
63. Yu, R.; Yuan, X.; Zhao, Y.; Hu, G.; Tu, X. Heavy Metal Pollution in Intertidal Sediments from Quanzhou Bay, China. *J. Environ. Sci.* **2008**, *20*, 664–669. [CrossRef]
64. Banat, K.M.; Howari, F.M.; To'mah, M.M. Chemical Fractionation and Heavy Metal Distribution in Agricultural Soils, North of Jordan Valley. *Soil Sediment Contam.* **2007**, *16*, 89–107. [CrossRef]
65. Franco-Uría, A.; López-Mateo, C.; Roca, E.; Fernández-Marcos, M.L. Source Identification of Heavy Metals in Pastureland by Multivariate Analysis in NW Spain. *J. Hazard. Mater.* **2009**, *165*, 1008–1015. [CrossRef]
66. Bux, R.K.; Batool, M.; Shah, S.M.; Solangi, A.R.; Shaikh, A.A.; Haider, S.I.; Shah, Z.-u.-H. Mapping the Spatial Distribution of Soil Heavy Metals Pollution by Principal Component Analysis and Cluster Analyses. *Water Air Soil Pollut.* **2023**, *234*, 330. [CrossRef]
67. Al-Kharabsheh, N.M.; Al-Zboon, K.K. Wastewater Treatment and Reuse in Jordan, 10 Years of Development. *Desalination Water Treat* **2021**, *238*, 15–27. [CrossRef]
68. Wu, Z.; Zhang, S.; Shan, B.; Zhang, F.; Chen, X. Optimizing Crop Spatial Structure to Improve Water Use Efficiency and Ecological Sustainability in Inland River Basin. *Agronomy* **2024**, *14*, 1645. [CrossRef]
69. Mishra, S.; Kumar, R.; Kumar, M. Use of Treated Sewage or Wastewater as an Irrigation Water for Agricultural Purposes-Environmental, Health, and Economic Impacts. *Total Environ. Res. Themes* **2023**, *6*, 100051. [CrossRef]

land

MDPI

Article

Edaphic Diversity, Polychemical Soil Status of the Prinevskaya Lowland and Prospects for Soils Use

Ekaterina Yu. Chebykina [1,*], Timur I. Nizamutdinov [1], Evgeny V. Abakumov [1] and Natalia V. Dinkelaker [2]

[1] Department of Applied Ecology, St. Petersburg State University, 199034 Saint-Petersburg, Russia; t.nizamutdinov@spbu.ru (T.I.N.); e.abakumov@spbu.ru (E.V.A.)
[2] Faculty of Ecotechnologies, ITMO University, 197101 Saint-Petersburg, Russia; nvdinkelaker@mail.ru
* Correspondence: e.chebykina@spbu.ru; Tel.: +7-921-987-70-51

Abstract: There will be a significant increase in anthropogenic load on the soils of the Prinevskaya lowland in the nearest decade due to the fact that a significant territory is occupied by St. Petersburg. The main objective is a study of the sanitary-hygienic state and soil diversity of the Prinevskaya lowland in case of a high degree of agricultural soil development there and the significant role of the lithological factor. Soils were studied at the following land use and land cover: agricultural and fallow soils of agrolandscapes; forest soils; and soils of industrial areas. Studies were carried out using morphological descriptions and analyses of chemical, physical, and biological properties. The most vulnerable land use are forest and agricultural and fallow zones, where active accumulation of priority toxicants of anthropogenic origin can occur. Geochemical peculiarities of studied soils are deficit of Mn, Cu, Mo, and Zn in soil-forming rock materials and accumulation of strontium and lead in arable horizons. The soils examined show minimal contamination with trace elements, as verified by a range of individual and combined ecotoxicological indicators. Urban development planning, particularly in St. Petersburg, should prioritize the preservation of biodiversity and soil resources.

Keywords: edaphic diversity; soil contamination; pollution status indexes; lowland; anthropogenic impact; soil functioning; land use

1. Introduction

Urban soils play a crucial role in the urban ecosystem, serving as an essential component of the habitat for humans, plants, and animals, while also underpinning various economic activities [1–4]. The condition of these soils is vital for evaluating the ecological health of a specific area, as they are significant in multiple ways: they act as the primary link in the food chain, serve as a source of secondary pollution for air and water, and provide a consolidated measure of the overall ecological status of the environment [5]. As cities often expand into surrounding agricultural lands, they encounter various agronatural soils and agrozems with distinct agrogenic horizons [6–8].

A significant portion of St. Petersburg is situated on the Prinevskaya lowland plain, a terraced lacustrine–glacial landscape formed from the glaciolimnium of the Baltic glacial lake (Figure 1) [9–11].

This flat region lies between the Gulf of Finland and Lake Ladoga, with its formation history linked to the abrasion and accumulation processes of late- and post-glacial basins, which have contributed to the variety of soil-forming materials present [12–14]. Notably, a large part of the Prinevskaya lowland is occupied by St. Petersburg and its associated industrial and agricultural enterprises, highlighting the significant impact of human activity on the development of lowland soils over the past three centuries.

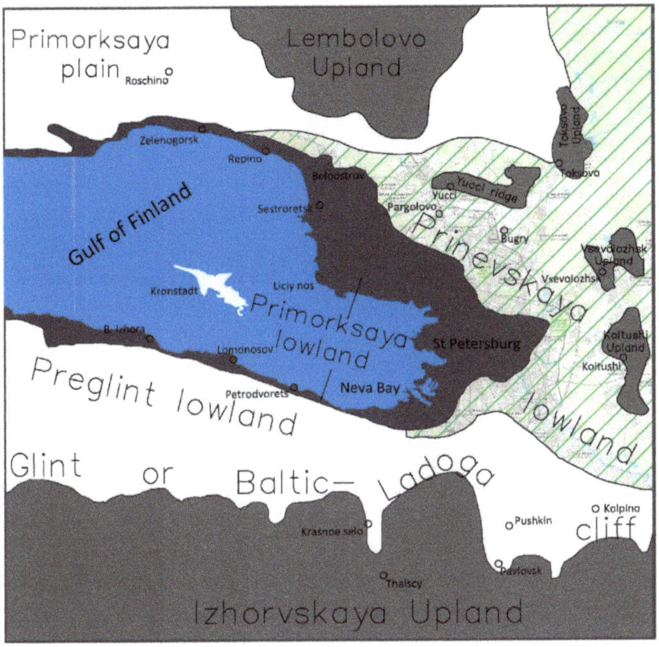

Figure 1. Prinevskaya lowland (marked with green shading).

The peculiarities of the soils of this territory were covered in the works [15–24] and a number of others. The works mentioned above were written in the last century and require updating. Additionally, the diversity of soils and their chemical properties are not well studied. There are challenges in assessing the sanitary and hygienic conditions of soils, as well as gaps in our understanding of soil development dynamics in agricultural landscapes. Furthermore, information regarding the sources of polychemical soil pollution is lacking. The current level of knowledge on the geochemistry of natural and anthropogenic landscapes in St. Petersburg and the Leningrad region is insufficient to address all issues related to the ecological and geochemical conditions of soils in the Prinevskaya lowland. Moreover, large cities such as St. Petersburg are important driving factors of environmental trends due to the increasing proportion of the global population living in urban areas and the high intensity of urban residents' activities. However, as the world urbanizes, people lose touch with the soil and the services it provides to sustain life. As developing cities and countries industrialize, soil pollution continues to increase and reaches a level that requires immediate action. Therefore, conducting a global assessment of the state of urban soils, starting from the local levels, such as the soils of the Prinevskaya lowland, to identify patterns, processes, and unique anthropogenic impacts is quite a relevant objective.

The lowland has a long history of agricultural development. The Prinevskaya lowland is characterized by the prevalence of humus–podzolic–gley soils with a thick peat horizon and drained sphagnum peat bogs [15]. In 1922, a soil survey conducted in the Shusharskaya farm area, situated in the heart of the Prinevskaya lowland, revealed that it can be challenging to reestablish the boundaries of natural soil types after years of cultivation and the application of urban waste and peat [16]. Today, the Prinevskaya lowland serves as the primary region for suburban agriculture, supplying Saint Petersburg with potatoes and vegetables, as well as functioning as a base for livestock feed. The fields were drained, treated with lime, and received large amounts of both organic and mineral fertilizers. Agricultural activities utilize 71–98% of all lowland areas, with over half of this land designated for arable farming (40–65%). A minor portion (ranging from 2 to 18%) is covered by secondary small-leaved forests [25,26].

Data from the third round of soil agrochemical surveys, conducted in the early 1980s during a period of intensive fertilizer application, indicated a significant increase in the average humus content of Leningrad region soils, reaching 3.5%. Additionally, there was a reduction in soil acidity and an enhancement in the levels of essential plant nutrients (the average content of mobile phosphorus and potassium in the region's soils attained average to above-average availability) [27,28]. Moreover, the Prinevskaya lowland stands out as one of the most intensively farmed regions in the Leningrad region. Consequently, areas with low pH, poor humification, and deficiencies in mineral nutrients occupy only a small fraction of their landscape. The practice of intensive agriculture has resulted in the emergence of a new soil component in the Prinevskaya lowland—agrosoils. These soils develop on various parent materials and are characterized by a thick (over 40 cm) organic layer that is well-structured and rich in plant nutrients [6].

Unfortunately, the economic decline in the Russian Federation during the 1990s and early 2000s was marked by inconsistencies in land use, resulting in a decrease in arable land. As a consequence of the changing economic landscape, a substantial portion of arable land in St. Petersburg has been left uncultivated [29–31].

The northwest region of the Russian Federation serves as a distinctive showcase of the diversity found in fallow lands. The quality of these lands, along with their biological characteristics and fertility levels, plays a crucial role in determining the quality of agricultural products, including seeds and grains [32,33].

In addition to the intentional effects of agriculture on the soils of the Prinevskaya lowland, there is also the unavoidable influence of the nearby city of Saint Petersburg, along with its industrial and transportation activities. Heavy metals and other toxic substances are released into the environment through the atmosphere, while sewage and urban waste serve as another major source of pollution. As a result of atmospheric transport, as well as surface and groundwater flow, a substantial suburban area is subjected to contamination. Therefore, the objectives of the study were: (a) to investigate the soil diversity of Prinevskaya lowland and describe the main types of urban, natural, and agrosoils at different land use and land cover (LULC), to determine their morphological features, and taxonomic position; (b) to evaluate the main chemical, physical and biological properties of soils of the different functional zones; (c) to assess heavy metal content and its geochemical distribution, and characterize the soil pollution status. The edaphic diversity and polychemical status of soils in the Prinevskaya lowland were studied on the example of the following objects (LULC):

1. Agricultural and fallow soils of agrolandscapes;
2. Forest soils;
3. Soils of industrial areas.

2. Materials and Methods

The territory of the Prinevskaya lowland is located in the northwest of the East European Plain in the southern part of the Karelia Isthmus. It is limited by the fluvio-glacial hills of the Koltushy upland in the north and by the Izhora upland in the south.

The climate of the Prinevskaya lowland is characterized as moderately cold and humid, influenced by the Atlantic Ocean, the Baltic Sea, and Lake Ladoga. In summer, the thermal regime is primarily affected by solar radiation, while in winter, it is largely determined by heat transfer from the Atlantic. Average temperatures in July range from

16.5 to 17.0 °C, while January sees averages of −8.0 to −8.5 °C, resulting in an average annual air temperature of 2.4 to 2.6 °C [34]. Precipitation patterns are mainly influenced by the topography; for instance, the lowland shores of the Gulf of Finland and Lake Ladoga receive the least rainfall. The average annual precipitation in the area is between 550 and 600 mm, with evapotranspiration rates of 400 to 500 mm, leading to a precipitation ratio of 1.8, indicating excessive moisture [35,36].

In terms of geomorphological zoning, the region falls within the Prinevsky–Estonian district of the Baltic–Valdai region and is part of the North Russian province of the Russian Plain. The Prinevskaya lowland is bordered to the north by the Central (Kotovskaya) Upland of the Karelia Isthmus and to the south by the Baltic–Ladoga escarpment. It features a terraced marshy plain with elevations ranging from 10–15 to 55–60 m [37]. The current landscape is primarily shaped by lake–glacial, glacial, lacustrine, and marine processes that occurred during the late Neopleistocene to Holocene periods. Common landforms include multi-aged lake and lake–glacial plains with coastal ramparts, sand spits, and abrasion scarps, all associated with Late Glacial–Holocene palaeobasins; rock outcrops and remnant uplands are more prevalent in the northern part of the study area. The border of the Prinevskaya lowland in the north is partly the abrasion ledges of the Rantolovsky plateau of the Toksovskaya kame upland.

The Prinevskaya lowland was formed in pre-glacial times. The contemporary landscape began to take shape as the last Valdai glaciation receded. The Prinevsky landscape emerged on sandy hills and loamy moraine deposits that constitute the Prinevskaya lowland, featuring granite boulders and banded clays. Most of the small boulders were removed from the soil during the process of territory development, while large boulders were blasted, crushed, and used in construction (for example, one of the boulders (Grom Stone) was used for the pedestal of the Bronze Horseman).

Moraine covers the bottom of the lowland and the surrounding areas, forming a flat surface (peneplain) and confirming that the lowland was formed before glaciation and is of tectonic origin. The geological column of Quaternary sediments is completed by marine, marsh, and eluvial sediments of the Holocene age.

The study area is characterized by middle-taiga forests, high and lowland swamps, and overgrown lakes. The primary element of the cropping system in the study area has been and continues to be perennial grasses, while annual grasses, vegetables, potatoes, and cereals cover significantly smaller areas.

Ten soil pits within Prinevskaya lowland were made in order to analyze soils of studied LULC (Figure 2).

Morphological soil diversity of the Prinevskaya lowland was studied in August–November 2023. These investigations were conducted utilizing standard soil characterization methods, including soil pits, morphological descriptions, and laboratory analyses focused on examining the chemical, physical, and biological properties of the soils.

Soil samples were collected from various depths of the soil horizons at each sampling location. Soil identification was performed in accordance with the "Classification and Diagnostics of Russian Soils" [38] and the World Reference Database of Soil Resources (FAO, 2015) [39].

All samples were air-dried at room temperature in the Department of Applied Ecology at St. Petersburg State University and subsequently passed through a 2 mm sieve. The analysis of soil properties was conducted on the fine earth fraction. The comprehensive analytical soil characteristics involved assessing chemical, physical, and physicochemical soil indicators through widely recognized methods [40,41].

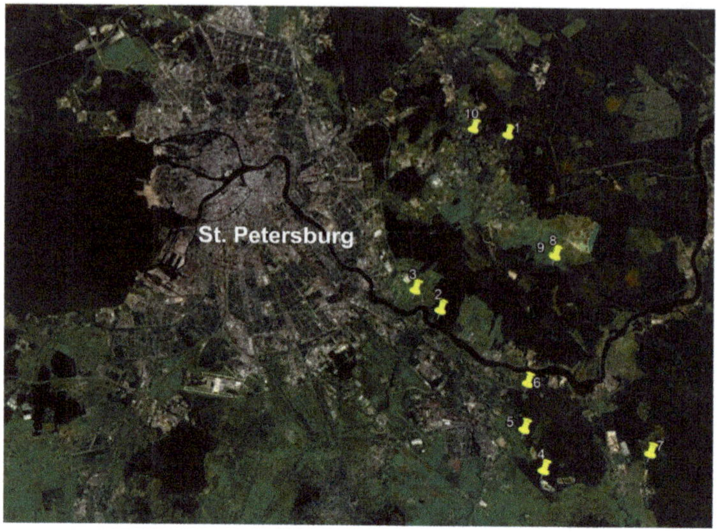

Figure 2. Map of studied plots and soil pits within the Prinevskaya lowland.

The particle size distribution was assessed using the Kachinsky pipette method, which involved the peptization of microaggregates with pyrophosphate. The analysis of mobile potassium compounds was performed following the Kirsanov method, as modified by the Central Scientific Research Institute of Agrochemical Service of Agriculture (CSRIASA) in accordance with Russian National Standard GOST R 54650-2011 [42]. The measurement of basal respiration (BR) was conducted according to the specified method [43]. Basal respiration is based on recording the CO_2 response in native soil. The pH values were determined in water and salt suspensions (soil-to-solution ratio 1:2.5). Substrate-induced soil respiration determined by the SID technique was also evaluated [44]. Carbon content was determined by the Tyurin method [40].

Soil samples in the solid phase were analyzed by X-ray fluorescence method using a portable X-ray spectrometer "Spectroscan" (M-049-P/16, 2016, Ekaterinburg, Russia) for the content of the following metals: Sr, Pb, As, Zn, Cu, Ni, Co, Fe_2O_3, MnO, Cr, V and TiO_2. Spectrometer "SPECTROSKAN" is designed for determination of elemental composition in the range from ^{11}Na to uranium (^{92}U), equipped with a vacuum-assisted scanning crystal-diffraction channel. The range of determinable contents from 0.0001% to 100% without concentrating depends on sample type, analyzed element, and matrix, and from 10^{-6} to 10^{-7}% to several proportions of percent with concentrating. The basic instrumental error does not exceed 0.5%. Calculation from oxide contents to element concentrations was carried out according to standard conversion coefficients [45].

Geochemical soil pollution by heavy metals was assessed by calculating the total soil pollution index Z_c, calculating exceeding the regional background values (single pollution index—*PI*) and maximum permissible concentrations specified in standard SanPiN 1.2.3685-21 [46]. The *PI* values are determined by taking the ratio of heavy metal concentrations (C_n) to their corresponding background regional values (B_n). The overall soil pollution index is computed using the following formula:

$$Z_c = \left(\sum_{i=1}^{n} PI\right) - (n-1) \tag{1}$$

Z_c is classified into four classes (Table 1).

The geoaccumulation index (I_{geo}) (proposed by Muller G. [47]), pollution load index (*PLI*) and potential environmental risk index (*RI*) were used to fully assess the pollution status of potentially toxic metals in soils of the Prinevskaya lowland.

The geoaccumulation index I_{geo} is used to determine the degree of contamination by trace metals relative to natural regional background values [47,48] and is calculated by the following formula:

$$I_{geo} = \log_2\left[\frac{C_n}{1.5\ B_n}\right], \tag{2}$$

where C_n represents the measured concentration of the element in the soil, B_n denotes the geochemical regional background value. Background values were determined according to Matinyan et al. 2007 [49]. A coefficient of 1.5 is used to minimize possible variations due to lithogenic variations [50]. I_{geo} is classified into seven classes (Table 1).

The Pollution Load Index (*PLI*) is determined as the geometric mean of the Pollution Index (*PI*) values [51–54]. This intricate index is calculated using the following formula:

$$PLI = \sqrt[n]{PI_1 \times PI_2 \times PI_3 \times \ldots PI_n} \tag{3}$$

PLI is divided into six categories (see Table 1).

The Potential Ecological Risk (*RI*) Index assesses the level of ecological risk associated with the harmful effects of trace metals [53,55,56]. This index is computed using the following formula:

$$RI = \sum_{i=1}^{n} E_r^i = \sum_{i=1}^{n} T_r^i \times P \tag{4}$$

where n represents the number of heavy metals, E_r^i indicates a single ecological risk factor index, and T_r^i refers to the toxicity response coefficient for each metal (As-10; Ni, Pb, Co-5; V, Cr-2; Zn-1) [55]. *PI* is calculated values for the Single Pollution Index. *RI* is categorized into five classes (Table 1).

All the indices utilized have their own evaluation scales, which are detailed in Table 1.

Table 1. Classification of pollution status indexes.

	The Total Soil Pollution Index (Z_c) [46]	
1	<16	Permissible pollution
2	16–32	Moderately dangerous pollution
3	32–128	Dangerous pollution
4	>128	Extremely dangerous pollution
	Geoaccumulation index (I_{geo}) [57]	
0	$I_{geo} \leq 0$	Absence of pollution
1	$0 < I_{geo} \leq 1$	From unpolluted to moderately polluted
2	$1 < I_{geo} \leq 2$	Moderately polluted
3	$2 < I_{geo} \leq 3$	From moderately to highly polluted
4	$3 < I_{geo} \leq 4$	Highly polluted
5	$4 < I_{geo} \leq 5$	From highly to extremely high polluted
6	$I_{geo} > 5$	Extremely high polluted
	Pollution load index (*PLI*) [53,58]	
0	PLI < 1	Absence of pollution
1	PLI = 1	Baseline levels of pollution
2	1 < PLI ≤ 2	Low pollution
3	2 < PLI ≤ 3	Moderate pollution
4	3 < PLI ≤ 5	High pollution
5	PLI > 5	Strong pollution

Table 1. *Cont.*

	Potential ecological risk (*RI*) [55,56]	
1	RI < 90	Low potential ecological risk
2	90 ≤ RI < 180	Moderate potential ecological risk
3	180 ≤ RI < 360	High potential ecological risk
4	360 ≤ RI < 720	Strong potential ecological risk
5	RI ≥ 720	Very strong potential ecological risk

The vertical electrical resistivity sounding (VERS) method, which enables the vertical division of soil layers into genetic layers with distinct properties and characteristics [59,60], was conducted using the portable LandMapper device (ERM-03, Landviser, LLC, League City, TX, USA). Measurements of apparent electrical resistance in the soil and strata were taken with electrode spacings of MN 10 and AB/2 at distances of 10, 20, 30, 40, 50, 60, 70, 80, 90, 100, 150, 200, 300, 400, and 500 cm. This approach allowed for the determination of apparent soil electrical resistance values at the corresponding depths.

The statistical data processing and analysis were performed using methods with the software packages MS Excel 2016, Past (version 3.20), and Statistica 64 (version 10).

3. Results and Discussion

3.1. Edaphic Soil Diversity of Prinevskaya Lowland

The soil cover of the Prinevskaya lowland was formed within the conditions of flat, poorly drained relief, with insignificant height variations. The cool, humid climate, along with this factor, leads to surface water stagnation and the occurrence of waterlogging processes in the region. The variety of soil-forming rocks in the lowland is a result of glacier and post-glacial water basins, which facilitated the erosion and redeposition of glacial sediments.

In the examined area of the Prinevskaya lowland, the following soil-forming materials were identified: (a) moraine loams and (b) fluvioglacial sands and sandy loams.

Additionally, the region contains banded clays and loams, limnoglacial sands and sandy loams, as well as sands and sandy loams of various origins, all underlain by loams and clays of lacustrine–glacial and moraine types.

The homogeneity of the relief in the Prinevskaya lowland means that soil diversity is largely influenced by lithological factors, which relate not only to the chemical composition and physical properties of the soil but also to their water availability. Additionally, human activities have significantly impacted the development of the soil cover in the study area over the past 200–300 years, leading to the formation of various agronatural soils and agrozems with distinct agrogenic horizons. Initially, this influence could be seen as beneficial—such as through land development and fertilizer application—but as anthropogenic impacts have intensified, their harmful effects have also become evident.

The diversity of soil-forming materials within the Prinevskaya lowland had a significant influence on the nature of soil-forming processes that determine the main soil types (Figure 3). Podzol formation associated with the impact of humus acids on the mineral soil part and further profile migration of decomposition products is most characteristic for soils formed on moraine loams. Gley soils develop on clay materials under conditions of impeded internal drainage and excessive moistening by surface waters. The formation of a lightened horizon in these soils is the result of reduction processes and the removal of elements of variable valence from upper horizons in a horizontal rather than vertical direction. Podzols and podburs (Entic Podzol) with illuvial–iron, and illuvial–humus horizons are formed on sandy rocks under the influence of podzolization and alfehumus processes. The alfehumus process consists of the removal by humus acids of aluminous-iron films from mineral grains and the formation of illuvial horizons with the content of semi-ferrous oxides and incrustive humus [23].

Agricultural and fallow soils of agrolandscapes:

Agrozem redoxymorphous light loamy formed on kama sandy loams (agricultural field) (Stagnic Anthrosol) N°. of study plot–3.

Agrozem texture-differentiated sub-eluvial loamy on cambrian clays (agricultural field) (Plaggic Anthrosol) N°. of study plot–5.

Sod-podzolic postagrogenic redoxymorphic loamy soil on cambrian clays (forest massif) (Umbric Plaggic Albeluvisol) N°. of study plot–7.

Postagrogenic grey humus fallow redoxymorphic sandy loam soil on fluvioglacial sediments (forest massif of the Prikoltushsky part of the Prinevskaya lowland) (Stagnic Anthrosol) N°. of study plot–8.

Agrozem ameliorative deep-turbid redoximorphic sandy loam on fluvioglacial sediments (agricultural field of the Prikoltushsky part of the Prinevskaya lowland) (Irragric Anthrosol) N°. of study plot–9.

Agrozem alfehumus postagrogenic sandy on kama sandy loams (Koltushy uplands nature protected area) (Plaggic Albic Anthrosol) N°. of study plot –10.

Forest soils:

Soddy podbur podzolized formed on kama sandy loams (border of Prinevskaya lowland and Koltushy upland *) (Umbric Entic Podzol) N°. of study plot–1.

Soddy podbur redoxymorphic sandy loam on kama sandy loams (forest park) (Umbric Entic Podzol) N°. of study plot–2.

Soils of industrial areas:

Lithostrata on cambrian clays (clay quarry) (Nudilithic Leptosol) N°. of study plot–4. Soil horizons were not identified there.

Lithostrata formed on overburden dumps (Spolic Technosol) N°. of study plot–6. Soil horizons were not identified there.

* Koltushy uplands is a unique natural object. This territory was not flooded by any of the seas and lakes that dominated the Prinevskaya lowland, not excluding the First Ioldian Sea—the deepest and largest waterbody: with depth of the First Ioldian Sea at 40–45 m, the Koltushy uplands was an archipelago of islands.

Figure 3. Sampling sites and soil profiles. Number of studying plots is given according to Figure 2.

The soil cover is mainly represented by patches of soddy podzolic soils (Podzol), podburs (Entic Podzol), and agrozems (Anthrosols). Relative microrelief rises, located in spots among the main massif, are occupied by automorphous soddy podzolic soils on moraine loams. Soddy podburs podzolized were found on sands with the thickness of the upper sediments exceeding 60 cm. The weak expression of the podzol process in soils at these plots can be explained by their anthropogenic transformation. Intensive farming leads to the formation of a peculiar component in the soil cover—agrozems. They develop on various types of rock materials and are distinguished by a thick organic layer (over 40 cm) that is well-structured and rich in humus and essential plant nutrients. It is important to mention that despite the long history of development in this region, soils with excessive moisture still cover a substantial area. This indicates the difficulty of soil drainage on clay rock materials in terms of low surface water runoff.

Currently, large areas of the Prinevskaya lowland are drained by an open and closed drainage network (e.g., agrozem ameliorative deep-turbid redoximorphic sandy loam on fluvioglacial sediments at the agricultural field of the Prikoltushsky part of the Prinevskaya lowland; number of study plot–9). In this case, a significant part of the soil cover belongs to soils without or with weak signs of gleization (72%). The water-air soil regime changes due to drainage: the moisture content of the arable horizon decreases by 1.5–2 times, aeration increases, and a zone of active aeration drops below the drainage boundary [61–64]. Amelioration causes the rise of redox potential and, therefore, promotes the oxidation processes in the soil profile: the content of ferrous oxide and mobile manganese decreases, and the content of ferric iron increases [65–67]. The processes of oxidogenesis (landscape geochemical processes of accumulation and transformation of iron oxides and hydroxides in soils and rock materials) are morphologically shown in the pronounced accumulation of ferruginous-manganese nodules in the upper horizons and heterogeneous color of aggregates [68–70].

Forest soils are characterized by the presence of litter horizon (AO), insignificant thickness, and fluffy structure of humus horizon, which consists mainly of coarse humus and unclear rudiments of podzol horizon—it is dense, whitish, and contains many ferruginous nodules.

Lands of industrial areas are characterized by the widespread distribution of technogenic surface formations—lithostratas.

During field studies measurements of electrophysical soil properties were performed. The measurements were extended to a depth of up to 3 m (Figure 4).

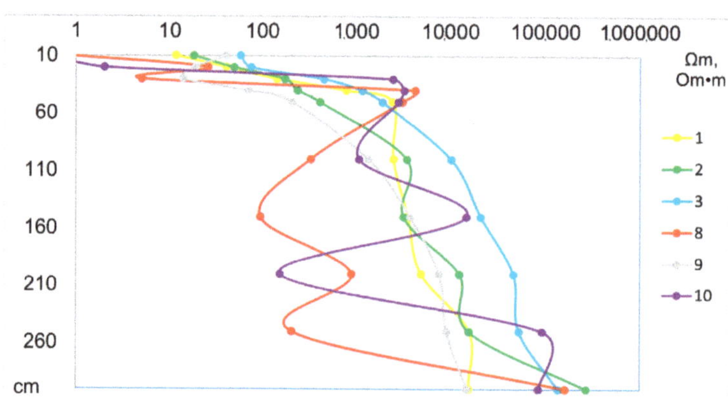

Figure 4. Electrical resistivity (ER, Ωm) of studied soils. Numbers indicate number of soil plots described in Figures 2 and 3.

The primary trend observed was an increase in the value of Ωm with depth, although there were some fluctuations noted at depths of 20–40 cm (Figure 4). This range was recognized as the transition zone between the arable horizon and the underlying layer. The first such fluctuations occur at a depth of 100 cm in non-agricultural soils. These disturbances are probably caused by changes in the structure and composition of the parent rock material. The method of VERS allowed the identification of five to seven heterogeneous layers in studied areas.

Studied soils were characterized by maximum values of temperature in the upper horizons, which smoothly decreased with depth.

3.2. The Physical and Chemical Soil Properties of Prinevskaya Lowland

The data on basic soil properties of studied soils are given in Table 2.

Table 2. Characteristics of studied soils.

LULC	Number of Study Plot *	Horizon, Depth, cm	Soil Moisture Content, %	pH_{H2O}	pH_{KCl}	C_{total}, %	Basal Respiration, μgC-CO_2/g Per hour	K_2O, mg/kg	Particle Size Distribution
Forest	1	AYe 0–10	1.60	6.3	3.9	1.74	1.62	36.9	Sandy loam
		AYe 10–15	1.00	5.6	3.9	1.13	2.55	39.1	Sandy loam
		BF 15–44	1.13	4.9	4.5	0.48	1.14	29.0	Sandy loam
		C 44–80	1.10	6.3	5.8	0.19	-	12.6	Medium-grained sand
Forest	2	AY 0–18	6.62	5.4	3.5	1.90	2.37	67.1	Coarse sandy loam
		BF 18–40	4.62	5.2	3.7	1.20	1.12	25.2	Coarse sandy loam
		Cff 40–80	3.33	5.5	3.5	0.24	0.73	22.6	Coarse sandy loam
Agricultural and fallow	3	P1 0–25	10.98	5.4	-	2.55	1.28	111.8	Light loam
		P2 25–42	5.79	5.7	5.4	1.12	0.45	206.8	Medium coarse loam
		Cff 42–75	3.03	6.2	5.3	0.05	0.04	177.1	Sandy loam
Industrial	4	0–65	2.50	5.2	-	0.83	1.12	450.1	Medium silty clay
		65–75	0.75	5.8	4.3	0.30	1.18	28.5	Light loam
		75–80	0.22	5.8	4.4	0.07	0.42	26.3	Fine-grained sand
		80–85	0.76	5.4	4.4	0.23	0.87	50.1	Sandy loam
		85–110	1.51	5.3	4.4	1.19	1.24	117.8	Fine clay
		Rock material (cambrian clay)	2.78	5.8	-	0.36	-	571.4	Fine clay
Agricultural and fallow	5	PY1 0–25	4.78	6.2	5.6	2.49	1.37	110.2	Medium coarse loam
		PY 25–40	4.79	5.9	5.1	2.18	0.69	119.4	Medium coarse loam
		BEL 40–65	4.05	6.1	5.3	0.67	1.54	15.4	Fine loam
		C 65–80	11.15	5.8	4.7	0.12	0.04	46.8	Medium coarse loam
		C 80–110	3.26	6.2	5.5	0.18	0.54	33.1	Light clay
Industrial	6	0–5	3.46	6.0	-	0.44	1.36	426.0	Medium coarse loam
		5–25	1.01	6.2	-	0.30	0.94	79.9	Sandy loam
		25–45	1.04	6.1	-	0.38	1.15	108.6	Sandy loam
Agricultural and fallow	7	AEL 0–20	13.54	6.0	4.9	4.94	1.46	70.4	Medium coarse loam
		EL 20–35	1.79	5.5	3.8	0.57	0.87	79.3	Light clay
		BEL 35–75	2.87	5.1	3.9	0.19	0.41	66.2	Medium silty clay
Agricultural and fallow	8	AYe,pa 0–25	21.68	4.9	3.1	10.61	2.27	112.0	Medium coarse loam
		C 25–60	1.12	6.4	4.2	0.12	0.86	9.1	Medium-grained sand
		D 60–100	10.49	5.8	4.3	0.15	-	12.5	Medium-grained sand
Agricultural and fallow	9	P 0–27	6.76	5.5	4.6	8.07	2.89	58.4	Light loam
		C 27–50	2.88	6.2	4.9	0.13	0.12	8.9	Sandy loam
		D 50–110	0.80	5.8	4.7	0.18	0.04	20.6	Sandy loam
Agricultural and fallow	10	AYpa 0–25	3.62	6.2	4.1	1.49	2.21	94.6	Medium-grained sand
		BF 25–35	0.98	5.9	4.8	0.22	0.08	21.9	Sandy loam
		BC 35–53	1.80	5.8	4.3	0.31	0.67	27.4	Light loam
		C 53–90	1.25	6.1	4.6	0.24	0.46	24.8	Fine loam
Post hoc test Forest–Agricultural–Industrial			0.20	$p << 0.05$	0.14	$p << 0.05$	$p << 0.05$	$p << 0.05$	
Significance of differences			Insign.	Sign.	Insign.	Sign.	Sign.	Sign.	

* Numbers indicate No. of soil plots described in Figures 2 and 3.

Soils of forest areas are characterized by acid pH throughout the soil profile, except for the upper humus horizon, which is close to neutral. The most acidic is the lower part of the soil profile. Postagrogenic and agricultural soils are characterized by higher pH values and have a slightly acidic pH in the lower part of the soil profile, while in the upper part, due to liming, it is close to neutral (pH_{H2O} 6.2).

Humus content in forest soils ranged from 1.9– to 10.61%. The high content of organic carbon in the accumulative part of the soil profile is explained by the time gap of litter decomposition from the intake of plant litter, leading to a significant accumulation of humified substances. At the same time, the qualitative composition of humus in these soils is characterized by a wide C/N ratio, a predominance of fulvic acids, and their mobile fractions [71–73]. The soil profile is divided into two or three parts by humus content. The upper part, the humus accumulative horizon, is characterized by the highest organic carbon content. The underlying horizon contains much less humus than the humus accumulative horizon. The organic matter content in it may either gradually decrease towards the rock material or have more or less close values in the whole profile. In the case of illuvial–humus process development, one more zone of soil profile (illuvial–humus horizon) is distinguished in sandy or sandy loam soils. There, the content of organic carbon is higher than in the neighboring mineral horizons. The humus profile of podburs (Entic Podzol) is characterized by a sharply decreasing distribution of humus. A separation of soil profile into humus-accumulative and mineral parts is very sharply expressed.

Soils of agricultural and fallow areas are characterized by the following features. The humus content in the upper soil profile is lower (4.4% on average), but the total humus stock is higher due to the higher density of the upper horizon and penetration of more significant humus amounts into the depth of the soil profile, i.e., its smoother distribution. Humus is more firmly connected with the mineral part, which is reflected in the crumby structure of the arable horizon (in the formation of which it participates along with calcium) and therefore in the higher content of the corresponding fraction of humic and fulvic acids. The vertical distribution of humus in agrozems is the smoothest. Humus content of 3–5% is maintained in the whole arable stratum, up to a depth of 40 cm.

The potassium distribution in the soil profile is influenced by the mineralogical composition, mainly of the silty fraction. Potassium is represented in clay soils mainly by forms that are difficult to access for plants and microorganisms. Mobile potassium compounds are less than 1% [74,75]. As a result of potassium fertilizer application, potassium accumulation in hard-to-access forms occurs in agrozems. Potassium fixation is also observed below the arable horizon, which is connected with potassium transport by soil colloids and the distribution of mica minerals. Forest soils are characterized by low potassium content (except for one soil at study plot number 8, which was located in a forest, and this forest, according to morphological description, was previously an agricultural field). The distribution of mobile potassium along the soil profile has the following peculiarities. The highest content is observed in the humus-accumulative horizon; the lowest is in the middle horizons. The content of mobile potassium in the rock material in forest soil is 1.1–1.5, and in agricultural soil, it is 3–4 times lower than in the upper soil horizon. It is explained by biogenic accumulation in forest soils, while in arable soils, among other things, it is the result of fertilizer application.

The rate of carbon dioxide release from soils and basal respiration is primarily influenced by two factors: the availability of nutrients and the quantity and physiological condition of the microbial community. The maximum values of basal respiration are characterized for upper humus horizons, where the greatest number of microorganisms and plant roots inhabit, for both forest soils (1.6–2.5 $\mu gC\text{-}CO_2/g$ per hour) and agricultural soils (0.45–2.89 $\mu gC\text{-}CO_2/g$ per hour).

Soils of industrial functional zones showed heterogeneity and the absence of any trend in the distribution of the main physical and chemical properties along the soil profile.

Particle size distribution is determined by rock material features. Studied soils represent all classes of particle size distribution, from sands to clay (Table 2). It could be noted a silt removal from the upper part of the soil profile in loamy soils and, on the contrary, an accumulation of fine particles in the arable layer in sandy soils. Silt is more distributed eluvial–illuvial with a maximum in the middle part of the middle horizon. An absolute predominance of fine sand fraction is observed in sandy soils throughout the soil profile. The arable layer in the agrozems is sandy loam or light loamy.

3.3. The Content of Trace Elements in Soils of Prinevskaya Lowland

Soils of St. Petersburg suburbs widespread at the Prinevskaya lowland have a significant anthropogenic impact. Trace metals are an essential part of the lithosphere [76]. As a result of anthropogenic and technogenic activities of industry and agriculture, the geochemical background of heavy metal content in soils can be changed. Due to the constant input of pollutants into the ecosystem, there is a destabilization of soil functioning, disturbance and change in its basic physical and chemical processes, and, as a consequence, further transformation with the formation of areas with high contents of heavy metals.

The heavy metals content in studied soils is given in Table 3, Figure 5. The average concentrations of trace elements varied, with Cu measuring less than 0.10 mg/kg and MnO reaching up to 591.76 mg/kg. Among all the metals analyzed, Cu exhibited the lowest concentration, often falling below the detection limit in most instances. The mean concentrations of heavy metals were distributed as follows: Sr > Cr > V > Zn > Pb > Ni > Co > As > Cu. Considering significant coefficients of variation in trace element concentrations, which ranged from 23.1% for As to 65.9–66.2% for Ni and Mn, further analysis of trace element concentrations was carried out in the soils of three different sampling sites. The industrial soils are characterized by the maximum heterogeneity, which is confirmed by the high coefficients of variation: from 17.08% for Sr to 67.34% for Co. The forest soils, on the contrary, are characterized by the lowest coefficients of variation.

Table 3. Trace elements content in soils of Prinevskaya lowland, mg/kg.

Trace Element	Mean	Max	Min	CV	SD
Sr	180.0	242.0	16.0	24.8	44.6
Pb	22.5	57.0	1.0	41.8	9.4
As	7.5	13.0	5.0	23.1	1.7
Zn	47.3	132.0	6.0	59.4	28.1
Cu	<0.10	<0.10	<0.10	0.0	0.0
Ni	21.1	72.0	4.0	65.4	13.8
Co	10.0	31.0	0.0	59.9	6.0
Fe_2O_3, %	3.0	6.9	0.9	48.8	1.4
MnO	591.8	1839.0	115.0	66.2	392.0
Cr	65.3	176.0	32.0	46.8	30.6
V	61.1	133.0	6.0	52.9	32.3
TiO_2, %	0.6	1.0	0.2	38.4	0.2

The highest indicators for Sr, As, Co, and Cr are observed among the studied soils. Manganese content is not included in Figure 5 due to high concentrations that did not fit on the graph with maximum coefficients of variation in the range. The MnO levels in forest soils range from 363 to 1839 mg/kg, with one sample exceeding the MPC. In industrial soils, the content varies from 151 to 1319 mg/kg, while agricultural soils show levels between 115 and 1236 mg/kg.

The distribution of heavy metal content is irregular across soil profiles. In industrial soils, there is generally a trend of decreasing content with depth or a fluctuation between high and low concentrations. Conversely, in forest soils, the concentration of trace metals tends to increase with soil depth, likely due to the sorption capacity of clay particles [77] (the increase in clay particles with depth was noted earlier).

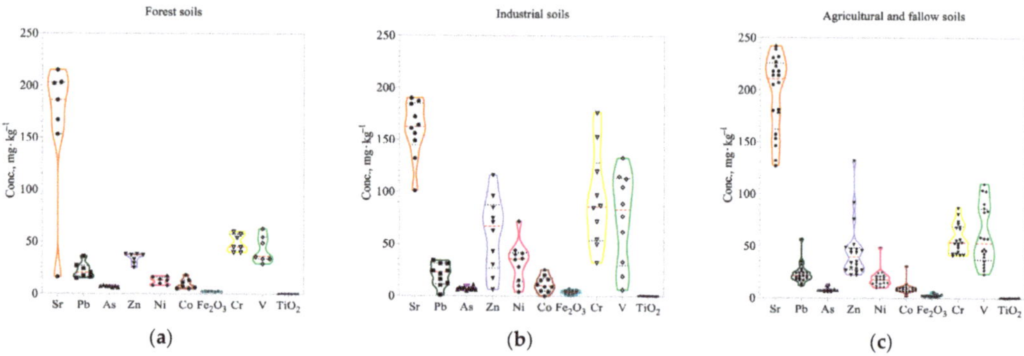

Figure 5. Variation of trace metals concentrations in studied soils s: (**a**)—forest soils; (**b**)—industrial soils; (**c**)—agricultural and fallow soils.

The primary factor influencing the concentration of trace elements in soils is their presence in the soil-forming rock materials. The literature indicates that banded clays have the highest concentration of trace elements among all soil-forming materials, with a total of 937 mg/kg for ten elements. In contrast, fluvioglacial and lake–glacial sands exhibit the lowest concentrations, measuring 626.7 mg/kg and 691.7 mg/kg, respectively. Sands have an increased concentration of strontium in contrast to the low content of other elements. The banded clays have higher concentrations of Cr, Pb, Mn, Zn, Ni, and Co compared to the other rock materials examined.

When comparing the rock materials from the study area to those of the northwest region overall [78], it is evident that clay rocks in this area contain more Ni, while sands have higher levels of Cr. Conversely, the rock materials in the Prinevskaya lowland exhibit lower concentrations of Mn, Zn, and Cu than those found across the northwest region as a whole.

The heavy metals content data were compared with the current standards for the content of trace elements in soil [46], as well as with geochemical background concentrations typical for the northwest of Russia, in particular, the Leningrad region (Figures 6 and 7).

The ecological soil state was assessed by calculating the concentration coefficients of trace elements, or PI (Figure 6). This indicator reflects the accumulation degree of heavy metals relative to their background content in the environment. Excess concentrations over the background content were recorded almost for studied elements, especially for Ni, Co, and Cr. The maximum accumulation was revealed for Ni; its content in all samples exceeds the background content on average from 2 to 7 times.

In general, the contents of trace elements in the analyzed samples were characterized as moderately dangerous (Figure 7) (up to 2 MPC) (for Sr, and separately for Pb, Zn, and Ni) and dangerous (from 2 to 5 MPC) (for As and Co (in case of industrial soils)) levels of soil contamination. An extremely dangerous level of soil contamination with Cr (over 5 MPC) was detected in all investigated soils with an excess of more than 1000 times.

The most polluted soils are found at the base of the quarry complex. This is primarily because the bottom of the quarry serves as a collection point for pollutants carried by rainwater. Elevated levels of these substances may also be linked to the operation of mining equipment, as transport activities were observed in the quarry complex during the material sampling.

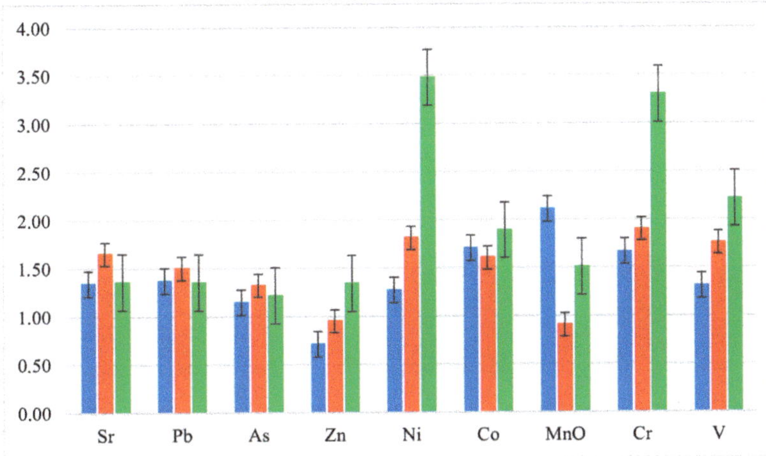

Figure 6. Single pollution index (PI) in studied soils according to LULC: blue color—forest soils; red color—agricultural and fallow soils; green color—industrial soils.

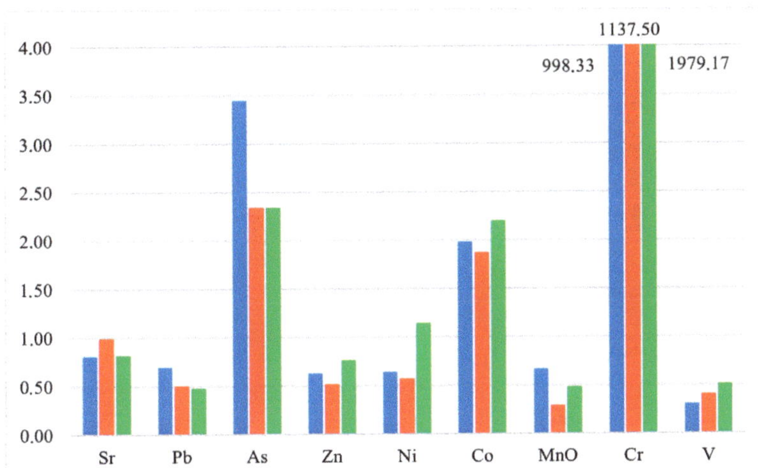

Figure 7. Heavy metal content relative to MPC in studied soils according to LULC: blue color—forest soils; red color—agricultural and fallow soils; green color—industrial soils.

The content of trace elements in arable soil horizon depends on a complex of factors: their content in the rock material, processes occurring in the soil, as well as on anthropogenic impacts. The accumulation of a trace element in the humus horizon relative to its content in the rock material can be explained by both biogenic accumulation and anthropogenic pollution from the surface. At the same time, there are data that biogenic accumulation cannot increase the trace element content more than 1.3–1.5 times in comparison with the rock material [79,80].

Sources of soil contamination by trace metals are as follows. Firstly, heavy metals can penetrate into soils from the atmosphere with emissions from industrial enterprises, transport, and thermal power plants. Secondly, it can be a by-effect of farming intensification. Heavy metals could be in the form of admixtures to chemical fertilizers, ameliorants, and as part of pesticides. Finally, heavy metals can enter the studied soils from solid and liquid urban wastes and through transfer from landfills [accidental or systematic (rubbish as fertilizer)].

Moreover, soil contamination with heavy metals (Pb, Co, Cd, Sr, and Hg) during fertilizer application has been noted by many researchers [81–83]. High doses of manure also contribute to the positive balance of Zn, Pb, Cd, Cu, and Fe [84].

To qualitatively assess the level of contamination of the studied soils with trace metals, the values of several individual and complex indexes were calculated. Its values are presented in Table 4.

Table 4. Qualitative assessment of contamination in studied soils. No. of study plots indicate number of soil plots described in Figures 2 and 3 and correspond to Table 2.

LULC	No. of Study Plot	Z_c		I_{geo}		PLI		RI	
		Value	Pollution Status	Value	Pollution Status	Value	Pollution Status	Value	Potential Ecological Risk
Forest	1	4.49	Permissible	0	Absence	0.99	Absence	36.44	Low
Forest	2	8.41	Permissible	0–1	Unpolluted to moderately polluted	1.63	Low	45.64	Low
Agricultural and fallow	3	8.47	Permissible	0–1	Unpolluted to moderately polluted	1.78	Low	56.80	Low
Industrial	4	16.33	Moderately dangerous	1–2	Moderately polluted	2.00	Moderate	75.59	Low
Agricultural and fallow	5	12.40	Permissible	0–1	Unpolluted to moderately polluted	1.56	Low	63.21	Low
Industrial	6	20.74	Moderately dangerous	1–2	Moderately polluted	2.85	Moderate	90.51	Moderate
Agricultural and fallow	7	7.47	Permissible	0–1	Unpolluted to moderately polluted	1.56	Low	46.53	Low
Agricultural and fallow	8	3.77	Permissible	0	Absence	0.84	Absence	51.09	Low
Agricultural and fallow	9	4.93	Permissible	0	Absence	1.40	Low	39.11	Low
Agricultural and fallow	10	5.09	Permissible	0	Absence	1.43	Low	42.66	Low

In order to assess the sanitary and ecological situation, the total soil pollution index Z_c was calculated. Its values are presented in Table 4. All investigated samples, except for soil samples from industrial areas (moderately hazardous), are characterized by the permissible category of total soil pollution by trace elements.

The contamination degree was also determined using a geoaccumulation index I_{geo}. The I_{geo} pollution degree for soils in industrial areas is characterized by rate No. 2 (moderately polluted); for some agricultural soils-rate No. 1 (from unpolluted to moderately polluted). At the same time, the I_{geo} pollution degree for forests and some agricultural soils was characterized by the rate No. 0 (absence of pollution).

Pollution load index PLI index was used to assess the degree of multiple contamination by toxic elements in studied soils (Table 4). PLI values in soils of three studied LULC ranged from 0.84 (absence of pollution) to 2.85 (moderate pollution).

The RI index provides an estimate of the potential ecological risk of pollution in the studied soils. The RI values ranged from 36.44 to 75.59, indicating a low potential ecological risk overall, except for one soil sample in the industrial zone, which had an RI index of 90.51, reflecting a moderate potential ecological risk.

Calculations of the PLI index reveal that nearly half of the investigated soils are experiencing a decline in soil quality. Despite this deterioration, the RI index remains low for all examined soils, with the exception of the one sample from the industrial area, which is classified as having a moderate environmental risk.

It can be anticipated that over the next decade, there will be a significant increase in anthropogenic pressures on the soils of the Prinevskaya lowland, largely due to the extensive urban area occupied by St. Petersburg. The forest and agricultural zones are the most susceptible LULC types, as they can actively accumulate priority toxicants of human origin. Currently, agricultural lands in St. Petersburg are utilized for crop cultivation and residential purposes, which may have potential implications for the health of the local population. These soils are likely to gather priority toxicants, polycyclic aromatic hydrocarbons, and petroleum products, which can persist in the environment for extended periods due to their low biological activity [85].

4. Conclusions

The Prinevskaya lowland features a complex and varied composition of soil cover components that have developed under various lithological, hydrological, and anthropogenic conditions. The prevailing soil combinations are variations of soddy podzols of different degrees of podzolization and a series of soddy podbur and agrozems. Automorphic soils have limited distribution, and their development is connected with human activity. The variety of soil-forming rock materials in several lowland regions with flat terrain results in complex combinations and mosaics of soddy podzols, exhibiting varying degrees of podzolization and the manifestation of alfehumus processes.

The primary factors driving the diversity of soil cover across different lowland areas differ. In regions dominated by clay, the variation in soil cover is influenced by the redistribution of water in microrelief features, while in areas with fluvio-glacial sandy deposits, it is determined by the level of soil moisture. Additionally, the lithological factor significantly impacts soil structure in regions characterized by a mosaic of soil-forming rock materials.

In comparison to other similar territories in the North-West, the geochemical characteristics of the Prinevskaya lowland are marked by a deficiency of Mn, Cu, and Zn in soil-forming rock materials, alongside an accumulation of Sr and Pb in arable horizons.

The geochemical soil characteristics of agricultural areas are primarily influenced by the underlying rock material from which they originate rather than by the extent of cultivation.

The analyzed soils show minimal contamination with trace elements, as confirmed by various individual and complex soil ecotoxicological indicators. Overall, the potential ecological risk across all areas is considered low, suggesting that the soils are in a good toxicological soil state and are currently suitable for agricultural use. An assessment of pollution status indexes (PI, PLI, RI) indicated that the quarry complex poses the highest level of threat. As anthropogenic impacts on the urban environment grow, industrial and agricultural zones are likely to become the most vulnerable areas of the city, primarily due to the low resilience of these soils to external influences.

According to our view, urban development planning in St. Petersburg should prioritize the principles of maximizing biodiversity conservation and protecting soil resources.

Author Contributions: Conceptualization, E.Y.C. and T.I.N.; methodology, E.Y.C., T.I.N. and N.V.D.; laboratory analysis, E.Y.C., T.I.N. and N.V.D.; field investigation, E.Y.C., T.I.N. and E.V.A.; data curation, E.Y.C., T.I.N. and E.V.A.; writing—original draft preparation, E.Y.C.; writing—review and editing, E.V.A.; visualization, E.Y.C.; supervision, E.V.A.; project administration, E.V.A.; funding acquisition, E.V.A. All authors have read and agreed to the published version of the manuscript.

Funding: This work was supported by the Russian Scientific Foundation, in accordance with the agreement from 20 April 2023 No. 23-16-20003, and the Saint Petersburg Scientific Foundation, in accordance with the agreement from 5 May 2023 No. 23-16-20003.

Data Availability Statement: The original contributions presented in the study are included in the article, further inquiries can be directed to the corresponding author.

Acknowledgments: This work is dedicated to the 300th anniversary of Saint Petersburg State University. Our acknowledgments are extended to the anonymous reviewers for their constructive reviews of the manuscript.

Conflicts of Interest: The authors declare no conflicts of interest.

References

1. Yang, J.-L.; Zhang, G.-L. Formation, characteristics and eco-environmental implications of urban soils—A review. *Soil Sci. Plant Nutr.* **2015**, *61*, 30–46. [CrossRef]
2. Pouyat, R.V. Urban Soils. In *Forest and Rangeland Soils of the United States Under Changing Conditions*; Pouyat, R., Page-Dumroese, D., Patel-Weynand, T., Geiser, L., Eds.; Springer: Cham, Switzerland, 2020. [CrossRef]
3. Pavao-Zuckerman, M. The nature of urban soils and their role in ecological restoration in cities. *Restor. Ecol.* **2008**, *16*, 642–649. [CrossRef]
4. Schröder, A.; Schloter, M.; Roccotiello, E.; Weisser, W.W.; Schulz, S. Improving ecosystem services of urban soils—How to manage the microbiome of Technosols? *Front. Environ. Sci.* **2024**, *12*, 1460099. [CrossRef]
5. Davies, F.T.; Garrett, B. Technology for sustainable urban food ecosystems in the developing world: Strengthening the nexus of food–water–energy–nutrition. *Front. Sustain. Food Syst.* **2018**, *2*, 84. [CrossRef]
6. Gerasimova, M.I.; Ananko, T.V.; Konyushkov, D.E. Agrogenic soils on the updated version of the soil map of Russia, 1:2.5 M scale: Classification diversity and position in the soil cover. *Eurasian Soil Sci.* **2023**, *56*, 122–131. [CrossRef]
7. Polyakov, V.; Abakumov, E.; Shamilishvily, G.; Chebykina, E.; Lavrishchev, A. Agrosoils in the city of St. Petersburg: Anthropogenic evolution and current state. In *Advances in Understanding Soil Degradation. Innovations in Landscape Research*; Saljnikov, E., Mueller, L., Lavrishchev, A., Eulenstein, F., Eds.; Springer: Cham, Switzerland, 2022. [CrossRef]
8. Alekseev, I.; Abakumov, E.; Maksimova, E. Agrochemical state and vertical organization of alluviated soils of Saint Petersburg's 300th anniversary park, Russia. In *Green Technologies and Infrastructure to Enhance Urban Ecosystem Services*; Springer: Cham, Switzerland, 2020; pp. 76–87. [CrossRef]
9. Dashko, R.E.; Aleksandrova, O.Y.; Kotyukov, P.V.; Shidlovskaya, A.V. Peculiarities of engineering-geological conditions of St. Petersburg. *Urban Dev. Geotech. Constr. Issue* **2011**, *1*, 1–47.
10. Paranin, R.V. Problems of water protection zones of the rivers of the Prineva lowland. *Geopolit. Ecogeodynamics Reg.* **2023**, *9*, 299–304.
11. Sheetov, M.V.; Dudanova, V.I.; Ruchkin, M.V.; Shukhvostov, R.S. New data on the quarter of Neva river area: Preliminary results of field work in 2021. *Relief Quat. Form. Arct. Subarct. North-West Russ.* **2021**, *8*, 359–364.
12. Bakhmatova, K.A. Agrogenetic Characterisation of Soils of Prinevskaya Lowland. Ph.D. Thesis, St. Petersburg University, St. Petersburg, Russia, 1997; p. 26.
13. Dolukhanov, P.M.; Subetto, D.A.; Arslanov, K.A.; Davydova, N.N.; Zaitseva, G.I.; Djinoridze, E.N.; Kuznetsov, D.D.; Ludikova, A.V.; Sapelko, T.V.; Savelieva, L.A. The Baltic Sea and Ladoga Lake transgressions and early human migrations in North-western Russia. *Quat. Int.* **2009**, *203*, 33–51. [CrossRef]
14. Subetto, D.A. History of the formation of Lake Ladoga and its connection with the Baltic Sea. *Soc. Environ. Dev. (Terra Humana)* **2007**, *1*, 111–120.
15. Rizpolozhensky, R.R. *Description of Petrograd Province in Soil Respect*; Printing House of the Imperial University: Kazan, Russia, 1922; p. 126.
16. Prasolov, L.I. *Natural Conditions of Agriculture in Petrogradskiy Sub-Stolichniy District. Soils of Shusharskaya Farm*; State Publishing House: Petrograd, Russia, 1922; p. 23.
17. Blagovidov, N.L. *Soils of the Leningrad Region*; Lenizdat: Leningrad, Russia, 1946.

18. Vladimirova, M.N. Changes in the content and composition of humus in heavy loamy sod-podzolic soils at application of high doses of organic fertilisers. *Notes Leningr. Agric. Inst.* **1973**, *206*, 30–34.
19. Pestryakov, V.K. *Soils of the Leningrad Region*; Lenizdat: Leningrad, Russia, 1973.
20. Orelskaya, N.G. Soddy-Weekly Podzolic Gley Soils of Prinevskaya Lowland and Diagnostics of Their Waterlogging Degree. Ph.D. Thesis, Leningrad State University, Leningrad, Russia, 1974.
21. Kozlov, A.V. Moisture and Density Regime of Sod-Podzolic Gley Loamy Soils of Prinevskaya Lowland in the Annual Cycle. Ph.D. Thesis, Leningrad State University, Leningrad, Russia, 1975.
22. Litvinovich, A.B. Changes in the Composition and Properties of Sod-Podzolic Gley Soils and Their Fine Fractions Under Drainage and Long-Term Agricultural Use. Ph.D. Thesis, Leningrad Agricultural Institute, Leningrad, Russia, 1985; p. 16.
23. Gagarina, E.I.; Matinyan, H.H.; Schastanaya, L.S.; Kasatkina, G.A. *Soils and Soil Cover of the North-West of Russia*; Publishing house of St. Petersburg State University: St. Petersburg, Russia, 1995; p. 236.
24. Gagarina, E.I.; Shelemina, A.N.; Abakumov, E.V. Ontogenesis of soils on earthen belligerent structures of the Leningrad region. *Vestn. St. Petersburg Univ.* **2011**, *3*, 100–107.
25. Administration of the Leningrad Region Committee for Natural Resources of the Leningrad Region. *Report on the Environmental Situation in the Leningrad Region in 2023*; Publishing house of the Committee for Natural Resources of the Leningrad Region: St. Petersburg, Russia, 2024; p. 216.
26. German, A.V.; Serebritsky, I.A. (Eds.) *Report on the Ecological Situation in St. Petersburg in 2023*; Publishing House of the Committee for Nature Management of St. Petersburg: St. Petersburg, Russia, 2024; p. 221.
27. Vedenin, O.L.; Ksenofontova, V.A. Changes in soil properties of the Leningrad region, under intensive farming. Scientific bases of soil protection of the Leningrad region. *Bull. V.V. Dokuchaev Soil Inst.* **1986**, *38*, 3–6.
28. Kashchenko, A.S. Change of fertility indicators of arable soils of the Leningrad region. *Bull. V.V. Dokuchaev Soil Inst.* **1986**, *38*, 6–10.
29. Lyuri, D.I.; Nekrich, A.S.; Karelin, D.V. Cropland dynamics in Russia in 1990–2015 and soil emission of carbon dioxide. *Bull. Mosc. Univ. Ser. Geogr.* **2018**, *5*, 70–76.
30. Nefedova, T.G. Main tendencies of changes in the rural space of Russia. *Izv. Russ. Acad. Sci. Ser. Geogr.* **2012**, *3*, 7–13.
31. Antropov, D.V.; Kirillov, R.A.; Chibirkina, E.A. Trends in the development of agricultural land use in Russia. *Int. J. Agric.* **2023**, *66*, 436–440.
32. Franco, M.F.S.; Delgado, E.U.A. Relation of fertilization and the quality of agricultural products. *Res. Soc. Dev.* **2022**, *11*, e36311427562. [CrossRef]
33. Komarova, A.A.; Lobasenko, B.A.; Pavsky, V.A.; Ivanova, S. Difficulties in determining the qualitative and quantitative characteristics of agricultural land. *Afr. J. Bio. Sci.* **2024**, *6*, 9046–9055.
34. Isachenko, A.G.; Dashkevich, Z.V.; Karnaukhova, E.V. *Physico-Geographical Zoning of the North-West of the USSR*; St. Petersburg University Publishing House: St. Petersburg, Russia, 1965; p. 248.
35. Guryanov, D.A. Variability of Climatic Seasons of the Year and Extreme Characteristics of Air Temperature in St. Petersburg and on the Territory of the Leningrad Region Under the Conditions of Modern Climate Change. Ph.D. Thesis, Russian State Hydrometeorological University (RSHU), St. Petersburg, Russia, 2016; p. 22.
36. Sidorenko, A.V. (Ed.) *Geology of the USSR. Leningrad, Pskov and Novgorod Regions. Part 1. Geological description*; Nedra: Moscow, Russia, 1971; Volume 1, p. 504.
37. Ananyev, G.S.; Andreeva, T.S.; Varushchenko, S.I.; Voskresensky, S.S.; Leontiev, O.K.; Lukyanova, S.A.; Spasskaya, I.I.; Spiridonov, A.I.; Ulyanova, N.S. *Geomorphological Zoning of the USSR and Adjacent Seas*; Vysshaya Shkola: Moscow, Russia, 1980; p. 343.
38. Shishov, L.L.; Tonkonogov, V.D.; Lebedeva, I.I.; Gerasimova, M.I. *Classification and Diagnostics of Soils in Russia*; Ojkumena: Smolensk, Russia, 2004; p. 342.
39. IUSS Working Group WRB. *World Reference Base for Soil Resources International Soil Classification System for Naming Soils and Creating Legends for Soil Maps*; World Soil Resources Reports No. 106; FAO: Rome, Italy, 2015.
40. Vorobyova, L.A. *Theory and Practice of Soil Chemical Analysis*; GEOS: Moscow, Russia, 2006; p. 400.
41. Rastvorova, O.G.; Andreev, D.P.; Gagarina, E.I. *Soil Chemical Analysis. Workbook*; Publishing House of Saint-Petersburg State University: Saint-Petersburg, Russia, 1995; p. 263.
42. *GOST R 54650-2011*; Soils. Determination of Mobile Compounds of Phosphorus and Potassium Using the Kirsanov Method as Modified by CINAO. The Federal Agency on Technical Regulating and Metrology (GOST R): Moscow, Russia, 2013.
43. Ananyeva, N.D. *Microbiological Aspects of Self-Purification and Soil Stability*; Nauka: Moscow, Russia, 2003; p. 222.
44. Anderson, J.P.E.; Domsch, K.H. A physiological method for the quantitative measurement of microbial biomass in soil. *Soil Biol. Biochem.* **1978**, *10*, 215–221. [CrossRef]
45. Panova, E.G.; Akhmedov, A.M. *Geochemical Indicators of Terrigenous Rocks Genesis: Textbook*; St. Petersburg State University: St. Petersburg, Russia, 2011; p. 64.
46. *SanPiN 1.2.3685-21*; Sanitary Regulations and Standards of Russian Federation. Hygienic Standards and Requirements for Ensuring the Safety and (or) Harmlessness of Habitat Factors to Humans. GOST R: Moscow, Russia, 2021.

47. Muller, G. Schwermetalle in den Sedimenten des Rheins: Veranderungen seit 1971. *Umschau* **1979**, *79*, 778–783.
48. Jiang, F.; Ren, B.; Hursthouse, A.; Deng, R.; Wang, Z. Distribution, source identification, and ecological-health risks of potentially toxic elements (PTEs) in soil of thallium mine area (southwestern Guizhou, China). *Environ. Sci. Pollut. Res.* **2019**, *26*, 16556–16567. [CrossRef] [PubMed]
49. Matinyan, N.N.; Reimann, K.; Bakhmatova, K.A.; Rusakov, A.V. Background content of heavy metals and arsenic in arable soils of North-West Russia (based on the materials of the International Geochemical Atlas). *Vestn. St.-Petersburg Univ.* **2007**, *3*, 123–134.
50. Taylor, S.R.; McLennan, S.M. The geochemical evolution of the continental crust. *Rev. Geophys.* **1995**, *33*, 241–265. [CrossRef]
51. Neyestani, M.R.; Bastami, K.D.; Esmaeilzadeh, M.; Shemirani, F.; Khazaali, A.; Molamohyeddin, N.; Afkhami, M.; Nourbakhsh, S.; Dehghani, M.; Aghaei, S.; et al. Geochemical speciation, ecological risk assessment of selected metals in the surface sediments of the northern Persian Gulf. *Mar. Pollut. Bull.* **2016**, *109*, 603–611. [CrossRef]
52. Tomlinson, D.L.; Wilson, J.G.; Harris, C.R.; Jeffrey, D.W. Problems in the assessment of heavy-metal levels in estuaries, the formation of a pollution index. *Helgol. Meeresunters.* **1980**, *33*, 566–575. [CrossRef]
53. Kowalska, J.B.; Mazurek, R.; Gasiorek, M.; Zaleski, T. Pollution indices as useful tools for the comprehensive evaluation of the degree of soil contamination—A review. *Environ. Geochem. Health* **2018**, *40*, 2395–2420. [CrossRef]
54. Varol, M. Assessment of heavy metal contamination in sediments of the Tigris River (Turkey) using pollution indices and multivariate statistical techniques. *J. Hazard. Mater.* **2011**, *195*, 355–364. [CrossRef]
55. Hakanson, L. An ecological risk index for aquatic pollution control—A sedimentological approach. *Water Res.* **1980**, *14*, 975–1001. [CrossRef]
56. Zhu, H.; Yuan, X.; Zeng, G.; Jiang, M.; Liang, J.; Zhang, C.; Yin, J.; Huang, H.; Liu, Z.; Jiang, H. Ecological risk assessment of heavy metals in sediments of Xiawan Port based on modified potential ecological risk index. *Trans. Nonferrous Met. Soc. China* **2012**, *22*, 1470–1477. [CrossRef]
57. Muller, G. Index of Geoaccumulation in Sediments of the Rhine River. *GeoJournal* **1969**, *2*, 108–118.
58. Jorfi, S.; Maleki, R.; Jaafarzadeh, N.; Ahmadi, M. Pollution load index for heavy metals in Mian-Ab plain soil, Khuzestan, Iran. *Data Brief* **2017**, *15*, 584–590. [CrossRef] [PubMed]
59. Pozdnyakov, A.I.; Pozdnyakova, L.A.; Pozdnyakova, D.A. *Constant Electric Fields in Soils*; KMK Scientific Press: Moscow, Russia, 1996; p. 360.
60. Pozdnyakov, A.I. Electrical parameters of soils and pedogenesis. *Eurasian Soil Sci.* **2008**, *10*, 1050–1058. [CrossRef]
61. Matinyan, N.N.; Rastvorova, O.G. Anthropogenic dynamics of soil processes on ribbon clays. In Proceedings of the Second Congress of the All-Russian Society of Soil Scientists, Saint-Petersburg, Russia, 27–30 June 1996; Volume 2, p. 314.
62. Antsiferova, O.A.; Safonova, D.N. Ecological and hydrological state and productivity of drained soils in the agricultural landscape of the Sambian plain. *Agrophysics* **2022**, *1*, 1–10.
63. Prudnikova, E.Y.; Savin, I.Y.; Lebedeva, M.P. Transformation of the surface layer in the arable soil horizon under the impact of atmospheric precipitation. *Eurasian Soil Sci.* **2021**, *54*, 1770–1781. [CrossRef]
64. Amézquita, E.; Rao, I.; Hoyos, P.; Molina, D.; Chavez, L.; Bernal, J. Development of an arable layer: A key concept for better management of infertile tropical savanna soils. In *Advances in Integrated Soil Fertility Management in Sub-Saharan Africa: Challenges and Opportunities*; Bationo, A., Waswa, B., Kihara, J., Kimetu, J., Eds.; Springer: Dordrecht, The Netherlands, 2007. [CrossRef]
65. Husson, O. Redox potential (Eh) and pH as drivers of soil/plant/microorganism systems: A transdisciplinary overview pointing to integrative opportunities for agronomy. *Plant Soil* **2013**, *362*, 389–417. [CrossRef]
66. Sharma, P.; Chouhan, R.; Bakshi, P.; Gandhi, S.G.; Kaur, R.; Sharma, A.; Bhardwaj, R. Amelioration of chromium-induced oxidative stress by combined treatment of selected plant-growth-promoting Rhizobacteria and earthworms via modulating the expression of genes related to reactive oxygen species metabolism in *Brassica juncea*. *Front. Microbiol.* **2022**, *13*, 802512. [CrossRef]
67. Abirami, R.; Jothimani, S.; Leninraja, D. Mechanism of redox reaction in soil chemistry. *Biol. Forum—Int. J.* **2023**, *15*, 1317–1321.
68. Schwertmann, U. Occurrence and formation of iron oxides in various pedoenvironments. In *Iron in Soils and Clay Minerals*; Stucki, J.W., Goodman, B.A., Schwertmann, U., Eds.; NATO ASI Series; Springer: Dordrecht, The Netherlands, 1988; Volume 217. [CrossRef]
69. Han, J.; Kim, M.; Ro, H.M. Factors modifying the structural configuration of oxyanions and organic acids adsorbed on iron (hydr)oxides in soils. A review. *Environ. Chem. Lett.* **2020**, *18*, 631–662. [CrossRef]
70. Wen, Y.; Xiao, J.; Goodman, B.A.; He, X. Effects of organic amendments on the transformation of Fe (oxyhydr)oxides and soil organic carbon storage. *Front. Earth Sci.* **2019**, *7*, 257. [CrossRef]
71. Korotkov, A.A. Humus substances in sod-podzolic soils. *Notes Leningr. Agric. Inst.* **1970**, *142*, 198–213.
72. Abakumov, E.V. Elemental composition and structural features of humic substances in young podzols developed on sand quarry dumps. *Eurasian Soil Sci.* **2009**, *42*, 616–622. [CrossRef]
73. Brocka, O.; Kalbitza, K.; Absalaha, S.; Jansena, B. Effects of development stage on organic matter transformation in Podzols. *Geoderma* **2020**, *378*, 114625. [CrossRef]

74. Leonicheva, E.V.; Stolyarov, M.E.; Roeva, T.A. The effect of soil nutrition and foliar fertilizers on the soil potassium regime and potassium status of apple trees in a rainfed Orchard. *Moscow Univ. Soil Sci. Bull.* **2024**, *79*, 65–77. [CrossRef]
75. Soumare, A.; Sarr, D.; Diédhiou, A.G. Potassium sources, microorganisms and plant nutrition: Challenges and future research directions. *Pedosphere* **2023**, *33*, 105–115. [CrossRef]
76. Kabata-Pendias, A.; Mukherjee, A.B. *Trace Elements from Soil to Human*; Springer: Berlin/Heidelberg, Germany, 2007. [CrossRef]
77. Lavado, R.S.; Rodríguez, M.B.; Scheiner, J.D.; Taboada, M.A.; Rubio, G.; Alvarez, R.; Alconada, M.; Zubillaga, M.S. Heavy metals in soils of Argentina: Comparison between urban and agricultural soils. *Commun. Soil Sci. Plant Anal.* **1998**, *29*, 1913–1917. [CrossRef]
78. Matinyan, N.N.; Gagarina, E.I.; Schastnaya, L.S.; Saprykin, F.Y.; Kulachkova, A.F. Geochemical characterization of the soil cover of the North-West of the Non-Black Earth Zone of the RSFSR. *Bull. Pushkin Leningr. State Univ.* **1985**, *10*, 91–99.
79. Dudás, F.Ö.; Zhang, H.; Shen, S.-Z.; Bowring, S.A. Major and trace element geochemistry of the permian-triassic boundary section at Meishan, South China. *Front. Earth Sci.* **2021**, *9*, 637102. [CrossRef]
80. Dupla, X.; Möller, B.; Baveye, P.C.; Grand, S. Potential accumulation of toxic trace elements in soils during enhanced rock weathering. *Eur. J. Soil Sci.* **2023**, *74*, e13343. [CrossRef]
81. Atafar, Z.; Mesdaghinia, A.; Nouri, J.; Homaee, M.; Yunesian, M.; Ahmadimoghaddam, M.; Mahvi, A.H. Effect of fertilizer application on soil heavy metal concentration. *Environ. Monit. Assess.* **2010**, *160*, 83–89. [CrossRef]
82. Srivastava, V.; Sarkar, A.; Singh, S.; Singh, P.; de Araujo, A.S.F.; Singh, R.P. Agroecological responses of heavy metal pollution with special emphasis on soil health and plant performances. *Front. Environ. Sci.* **2017**, *5*, 64. [CrossRef]
83. Pogrzeba, M.; Rusinowski, S.; Krzyżak, J. Macroelements and heavy metals content in energy crops cultivated on contaminated soil under different fertilization—Case studies on autumn harvest. *Environ. Sci. Pollut. Res.* **2018**, *25*, 12096–12106. [CrossRef] [PubMed]
84. Shahid, M.; Shukla, A.K.; Bhattacharyya, P.; Tripathi, R.; Mohanty, S.; Kumar, A.; Lal, B.; Gautam, P.; Raja, R.; Panda, B.B.; et al. Micronutrients (Fe, Mn, Zn and Cu) balance under long-term application of fertilizer and manure in a tropical rice-rice system. *J. Soils Sediments* **2016**, *16*, 737–747. [CrossRef]
85. Agapkina, G.I.; Chikov, P.A.; Shelepchikov, A.A.; Brodskii, E.S.; Feshin, D.B.; Bukhanko, N.G.; Balashova, S.P. Polycyclic aromatic hydrocarbons in soils of Moscow. *Mosc. Univ. Soil Sci. Bull.* **2007**, *62*, 149–158. [CrossRef]

land

MDPI

Article

Quality of Constructed Technogenic Soils in Urban Gardens Located on a Reclaimed Clay Pit

Dariusz Gruszka, Katarzyna Szopka and Cezary Kabala *

Institute of Soil Science Plant Nutrition and Environmental Protection, Wroclaw University of Environmental and Life Sciences, Grunwaldzka 53, 50375 Wroclaw, Poland; dariusz.gruszka@upwr.edu.pl (D.G.); katarzyna.szopka@upwr.edu.pl (K.S.)

* Correspondence: cezary.kabala@upwr.edu.pl; Tel.: +48-713201943

Abstract: Urban gardening plays diverse social, cultural and economic roles; its further development appears to be worthwhile, provided that soil contamination does not compromise ecosystem services. This study was conducted at a complex of urban gardens in Wroclaw (Poland) where topsoil screening indicated significant spatial differentiation of trace elements content, presumably related to the history of the site. Urbic Technosols cover the reclaimed section of the gardens, where industrial and urban waste materials, such as ash, slag, construction and demolition, and household waste, were used to fill former clay and sand mines. Although the topsoil layers, comprised of transported external soil, exhibited beneficial physicochemical properties and high fertility, they were seriously contaminated with trace elements (up to 1700, 920, 740, 5.1, 7.4, and 5.1 mg kg^{-1} zinc, lead, copper, cadmium, mercury, and arsenic, respectively). The trace elements were likely transferred from technogenic materials used for mine infilling, which now underlie the thin humus layers of the garden soils. The results suggest that the quality of soils in urban gardens located at reclaimed post-mining sites, while seemingly beneficial for horticulture based on physicochemical soil properties and fertility indices, can be seriously and permanently compromised by soil contamination from inappropriate materials used for site reclamation, thereby affecting soil quality and posing potential health and ecological risks.

Keywords: soil contamination; trace elements; soil health; garden soils; Technosols; reclaimed lands

1. Introduction

Although gardening has long existed in various forms in urban areas, it has recently received special attention in many countries [1,2]. Urban gardens, along with urban parks and forests, contribute to the urban systems of green areas, which influence the microclimate and mitigate various effects [3]; affect the water cycle by disrupting surface sealing [4]; sequester carbon dioxide and improve air quality [5]; and improve the biodiversity of urban ecosystems [6]. Community or allotment gardens provide an alternative to public parks for family recreation, cultivation-oriented physical activity, and local community-building interactions [7–9]. The most commonly noted function of gardens is the cultivation of fruits and vegetables, considered as local, safe, and healthy products [10,11].

However, numerous studies have reported on the contamination of soils and vegetables in urban gardens, particularly those located in large cities and urban–industrial agglomerations [12–17]. Xenobiotics in garden soils pose a general environmental risk [18], including direct and indirect risks to human health [19,20]. Urban traffic and transport, along with local or regional industrial emissions, are among the most frequently identified sources of trace metal contamination of urban soils [21,22]. Kabala et al. [23] reported on the relationship between soil pH and heavy metal content in garden soils and the long-term application of contaminated lime (of industrial/smelting origin), identified as the source of soil pollution. Some types of household or industrial waste, including composts and sludges, although considered beneficial and cheap fertilizers, even if aligned with the principles of the circular economy, can deliver excessive amounts of trace elements and other xenobiotics [24,25]. Furthermore, urban garden soils can contain an admixture of construction and demolition debris, particularly if the city experienced war damage [26–28]. Although crushed bricks and concrete can increase the availability of water for plants and improve the physicochemical properties of sandy soils [29], other construction materials or additions, such as paints, painted wood (including joinery), glazed pottery, crystals, stained glass, and metal products, can release Cd, Cu, Zn, Hg, Ni, and other toxic substances into the soil [30,31].

Most urban gardens are located on 'natural' soils, which often contain admixtures of anthropogenic materials [30,32,33]. However, little is known about gardens located on artificially constructed soils at reclaimed post-mining sites such as former sand and clay pits [32,34]. Many cities in Europe and around the world consumed large volumes of construction materials excavated locally. Such rapid growth in the nineteenth century contributed to the creation of numerous suburban sand and clay pits, the latter often accompanied by brickyards. As cities continued to grow, they absorbed previous suburbs, increasing the need to reclaim depleted and abandoned mines [35]. Depending on the type of mine, topography, hydrological conditions, and local needs, the pits were converted into water bodies or, after being filled with construction and other urban debris, transformed into parks, sports fields, or residential areas [36,37]. Some reclaimed sites were converted into urban gardens, as the demand for gardens in urban spaces could not be met with 'natural soils', already consumed by residential or industrial projects [13,35].

Urban gardens in Wroclaw, Poland were first established in the mid-19th century and were rented to poor people and factory workers [38,39]. Since the beginning of the 20th century, many school gardens and the so-called Schreber gardens (education plus health promotion) have been established, with these gardens becoming an important source of food for urban populations during the First World War (1914–1918) [38]. During the economic crisis of the 1920s and 1930s, urban gardens served not only for food production, but even as provisional homes for unemployed and homeless people [39]. Many of the gardens established in Wroclaw (then called Breslau) in the 1920s–1930s now exist as 'allotment gardens', with many of them situated in reclaimed sand and clay pits (Figure 1).

Although many other cities in Poland and Germany share a similar scenario, little is known about the properties of constructed garden soils at reclaimed sites or about the impact of the materials used for reclamation on soil quality. The objective of this study was to analyze the physicochemical properties of constructed technogenic soils, including the content of selected trace elements, as a possible source of risk to human health, in a complex of allotment urban gardens situated in a reclaimed post-mining area in Wroclaw. Understanding the scale of contamination and its sources is essential for sustainable urban gardening on reclaimed lands.

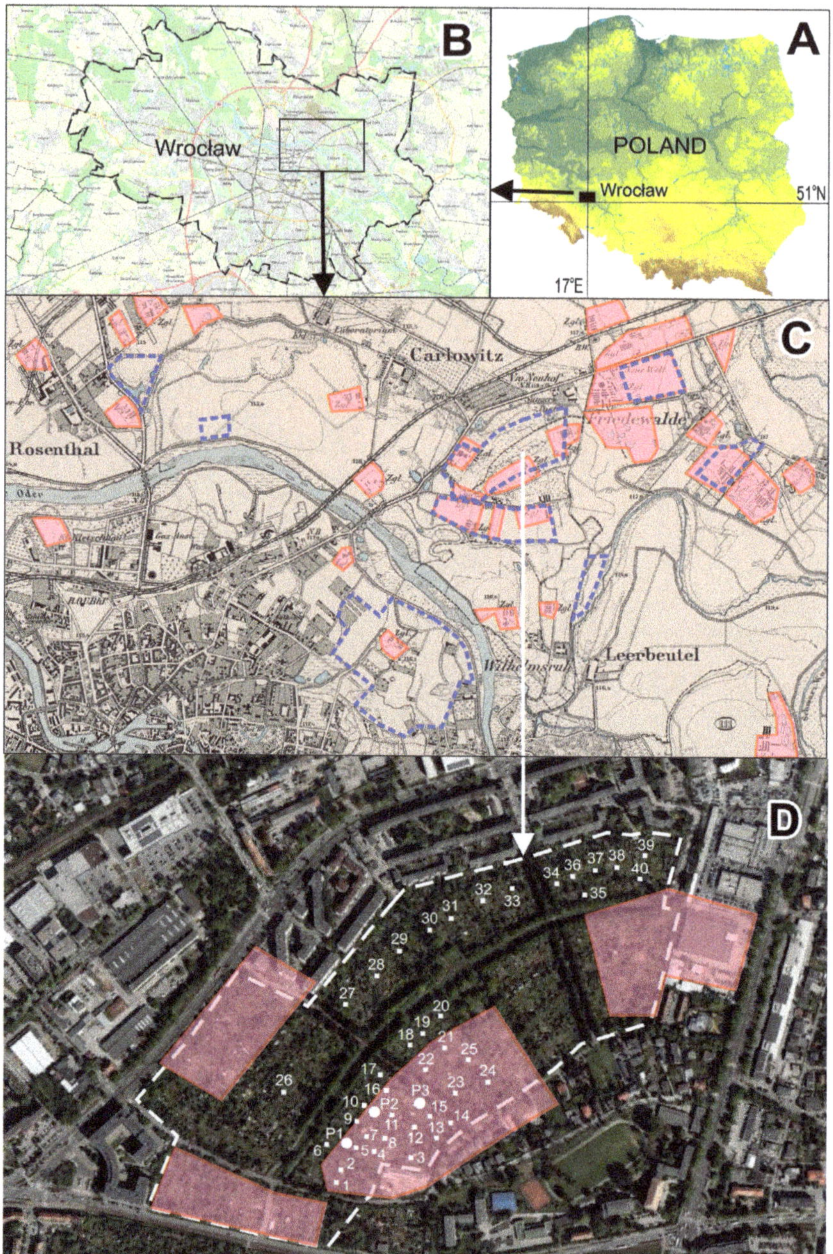

Figure 1. Location of the study site in Poland (**A**); study site in Wroclaw city (**B**); (**C**) historical map (1886 [40]) of the north-east forelands with marked sand and clay pits (pink areas). Present-day urban (allotment) gardens are bordered with blue dashed lines; and (**D**) the urban garden complex under study with marked reclaimed sand/clay pits (pink areas), topsoil (0–25 cm) sampling sites (small white dots, numbers 1–40), and soil profiles (large white dots, numbers P1–P3).

2. Materials and Methods

2.1. Study Area

The study was conducted at the 'Wytchnienie' ('Relaxation') complex of allotment gardens, located in the north-east of Wroclaw, south-west Poland (Figure 1), where urban gardens (allotments) occur throughout the city and are particularly popular among retired people and families with small children [39].

Typical natural soils in the floodplains of the Odra River and its tributaries are Brunic Fluvisols and Fluvic Cambisols developed from stratified loamy and sandy alluvial sediments [41]. Beyond the Odra Valley, Luvisols and Brunic Arenosols developed from glacial tills dominate in the northern part of the city, while Phaeozems and Chernozems developed from thin loess covers on glacial sediments dominate in the southern part of the city [23,41]. The northern outskirts of the city have been extensively exploited for the mining of sand and clay/loam, and for brick processing, as illustrated on the 1886 map [40] (Figure 1C). Many other small sand and loam excavation sites are believed to have existed in the 19th century; however, these were not mapped due to their early reclamation [39]. Some of these sites have been converted into urban gardens, particularly in the north-east of Wroclaw (Figure 1C). The area of the current 'Wytchnienie' complex was reclaimed in the 1920s [42]; in the 1930s, it was converted into urban gardens [39].

2.2. Field and Laboratory Methods

The original project involved only topsoil (0–25 cm) sampling at 40 plots of the garden complex (Figure 1D) before the planned sampling of vegetable and fruit. However, an initial analysis of the concentration of five trace elements expressed a large difference in soil contamination between the northern and southern sections of the gardens, which could not be explained by the impact of flooding (the entire area experienced a 'millennial flood' in 1997 [43]) or by different management practices [23]. It was assumed that the higher soil contamination in the southern section was related to the reclamation of clay mines that existed in this area at the end of the nineteenth century and the beginning of the twentieth century. Therefore, three soil pits (P1–P3) were excavated and sampled in the southern part of the complex (Figure 1D). The extent of the clay mines (a pink polygon in Figure 1D) was approximated based on the historical topographical map [40].

Soil profiles were characterized (Figure 2) and classified according to the international soil classification system, the World Reference Base for Soil Resources (WRB) [44]. All collected soil samples were air-dried, crushed, and sieved (mesh: 2 mm). The particle size distribution was determined using a hydrometer and sieving methods, after organic matter removal, where necessary, and sample dispersion with Na-hexametaphosphate. Soil texture classes were distinguished according to the WRB system [44]. The soil pH in distilled water (1:2.5 suspension, v/v) and the electrical conductivity (EC) in a saturated paste were analyzed potentiometrically [45]. Soil organic carbon (SOC) and total nitrogen (N_t) were determined by high-temperature catalytic combustion (Vario MACROcube, Elementar Analysensysteme GmbH, Langenselbold, Germany) after carbonate removal. Calcium carbonate, expressed as $CaCO_3$ equivalent, was analyzed by volumetric methods using 10% HCl. The exchangeable base cations (Ca_{ex}, Mg_{ex}, K_{ex}, and Na_{ex}) and cation exchange capacity (CEC) were extracted using 1 M ammonium acetate buffered at pH = 7 [45]. Element concentrations in the extracts, including Na as an indicator of CEC, were determined by microwave plasma atomic emission spectroscopy (MP-AES 4200, Agilent Technologies, Santa Clara, CA, USA).

Figure 2. Technogenic soil profiles P1, P2, P3 (**A–C**) situated in the reclaimed part of the allotment gardens; and (**D**) artefacts from the horizons 2BCu and 5ACu of the profile P1.

The plant-available forms of P_{av}, Kav, and Mg_{av} were extracted following the Mehlich-3 procedure [46] and determined by inductively coupled plasma spectrometry (ICP-OES, Thermo Scientific iCAP 7400, Waltham, MA, USA). Total concentrations of Zn, Cu, Pb, Cd, and Hg in topsoil samples from the reclaimed and unreclaimed sections of the gardens, as well as samples collected from soil profiles P1–P3, were determined by ICP-OES after sample digestion with aqua regia (ISO standard 54321:2020) [47]. Furthermore, total concentrations of Fe, Mn, Cr, Ni, and As were measured in dry, finely ground samples from soil profiles using portable X-ray fluorescence (pXRF; EDX Explorer 7000). The certified reference materials ISE 836, ISE 838, ISE 851, ISE 853, ISE 856, ISE 869, RTH 912, and RTH 953 (WEPAL, Wageningen, the Netherlands) were used for the validation of the ICP-OES analysis after aqua regia digestion and for the calibration and validation of the pXRF technique.

The geo-accumulation index (Igeo) was calculated for a comprehensive evaluation of soil pollution [48], following the formula: $Igeo = \log_2(Ci/1.5{*}GB)$, where: Ci is a concentration of an individual element, and GB is the geochemical background value.

Median concentrations of trace elements in the plough layers, based on a long-term monitoring of arable soils in Poland (Table 1 [49]), were adopted as an ambient geochemical background (GB) for the Igeo calculations.

Table 1. Total concentrations of Zn, Pb, Cu, Cd, and Hg in the topsoil layers of arable soils in Poland as an ambient geochemical background and the legal thresholds (permissible levels) of trace elements in arable soils in Poland.

Area	Total Concentration, mg kg^{-1}								
	Zn	**Pb**	**Cu**	**As**	**Cd**	**Hg**	**Cr**	**Ni**	**Mn**
Geochemical background for arable soils in Poland [49]	32.0	11.8	6.2	2.7	0.15	0.02	10.2	6.4	325
Legal permissible levels in medium-textured soils in Poland [50]	500	250	150	20	3	4	300	150	-

The legal concentration thresholds, the so-called 'permissible levels' of trace elements in the medium-textured arable soils of Poland [50], are given in Table 1 for further interpretation.

2.3. Statistical Analysis

Basic statistical calculations, including mean, median, and standard deviations were performed using the Statistica 13 software package (TIBCO Statistica, Santa Clara, CA, USA). The statistical significance of the differences between the mean values was verified using post hoc Tukey's test.

3. Results

3.1. Trace Element Concentrations in the Topsoil Layer

The concentrations of trace elements in the topsoil layer (0–25 cm) of the allotment gardens varied widely (Table S1—Supplementary Materials). However, even the lowest concentrations of Zn, Pb, Cu, Cd, and Hg—measured as 186, 47.6, 26.7, 0.78 and 1.11 mg kg^{-1}, respectively (Table 2)—were 4 to 5 times higher than their median values in arable soils of Poland (Table 1). Minimum, maximum, and mean concentrations of these elements were noticeably higher in the sites located in the reclaimed southern section (formerly clay pit) than in the unreclaimed northern section of the complex (Table 2). The 2.5–4-fold differences, statistically significant (at $p < 0.05$) for each element, were reported between the reclaimed and unreclaimed garden sections.

Table 2. Total concentrations (aqua-regia extractable) of Zn, Pb, Cu, Cd, and Hg and geo-accumulation index for the topsoil layers (0–25 cm) in the unreclaimed and reclaimed sections of the allotment gardens complex. Detailed results are presented in Table S1 (Supplementary Materials).

Site History	Parameter	Total Concentration					Igeo				
		Zn	Pb	Cu	Cd	Hg	Zn	Pb	Cu	Cd	Hg
		mg kg^{-1}									
unreclaimed area $n = 20$	minimum	186	47.4	26.7	0.78	1.11	2.0	1.4	1.5	1.8	5.2
	maximum	764	245	145	2.38	1.66	4.0	3.8	4.0	3.4	5.6
	mean	378 a	109 a	64.6 a	1.34 a	1.35 a	2.8 a	2.3 a	2.4 a	2.5 a	5.4 a
	SD	180	73.2	42.4	0.53	0.25	0.6	0.8	0.8	0.5	0.2
reclaimed mine $n = 20$	minimum	386	73.8	42.3	1.48	1.61	3.2	2.6	3.0	2.3	5.7
	maximum	2340	922	736	5.13	5.06	5.6	5.7	6.3	4.5	7.4
	mean	1120 b	474 b	276 b	3.28 b	3.30 b	4.5 b	4.6 b	4.7 b	3.8 b	6.5 b
	SD	420	209	145	0.89	1.53	0.5	0.7	0.7	0.5	0.7

Explanation: Igeo—geo-accumulation index; SD—standard deviation; a, b—indication of statistical difference between mean values for unreclaimed and reclaimed sites, checked by Tukey's post hoc test at $p < 0.05$.

Significantly (at $p < 0.05$) higher mean geoaccumulation indexes (Igeo) for all trace elements (Table 2) confirmed greater contamination in reclaimed sections than in the unreclaimed sections and allowed for a relative assessment of the pollution level. If Igeo values in an unreclaimed part indicated 'moderate to high soil pollution', the values for the reclaimed part indicated 'high to extremely high soil pollution' [48]. However, an assessment based solely on Hg suggested 'high pollution' and 'extremely high pollution', in the unreclaimed and reclaimed sections, respectively.

3.2. Morphological Characterization of Soil Profiles on Reclaimed Clay Pit

All three soil profiles, situated in the marginal part of the former clay pit, were artificially constructed above the stratified alluvial sediments (horizons 4Cl) starting at a depth of 80–150 cm (Figure 2). As evidenced by the recently introduced symbol 'τ' (tau) for transported materials [44], the topsoil layers constructed during the pit reclamation consist of local translocated materials, free of technogenic artefacts. The underlying layers consist of, or are affected by, the waste materials used to fill the pit and shape the land surface. These technogenic materials consist of ash, slag, charred wood, and coarse fragments of bricks, concrete, stones, pottery, ceramic tiles, glass, metals (rods, cables, tools), roofing felt, animal bones, fabrics (Figure 2D) and other unidentified substances, mixed with quarzitic sand. The addition of artefacts is reflected by the symbol 'u' [44] in the horizon designations (Figure 2A). Infilling materials of variable origins were deposited in subsequent layers, as reflected in their current stratification. The layers containing larger admixtures of ash are partly cemented, while the strata consisting mainly of construction debris mixed with household waste are loose.

The topsoil horizons are 21–40 cm thick, black or very dark brown (Munsell colours 10YR 2/1-3, moist), rich in humified organic matter (SOC content of 5–11%, Table 3) and biologically active, as evidenced by numerous plant roots, earthworm channels/casts, and a well-developed, fine-to-medium granular structure. The transitional ABu horizons are still rich in humus but also rich in artefacts. The colors of the ABu and BCu horizons are grey-brown or strong brown due to the weathering of the artefacts and the release of iron oxides [27,28]. The structure of these subsoil horizons is blocky subangular; however, the abundance, size, and durability of structural aggregates is affected by the content and kind of artefacts, and is, thus, it can be poorly developed in case of predominance of coarse fragments.

Table 3. Particle size distribution and selected physicochemical properties of soils in the reclaimed part the allotment gardens.

Soil Profile	Soil Horizon	Depth	Sand 2–0.05 mm	Silt 0.05–0.002	Clay <0.002	Texture Class	$CaCO_3$	pH	EC	SOC	N_t
		cm		%			%		dS m^{-1}	%	
P1	Apτ1	0–25	73	25	2	LS	3.6	7.4	1.1	5.9	0.368
	Apτ2	25–40	81	17	2	LS	3.2	7.6	1.4	5.0	0.281
	ABu	40–50	88	11	1	LS	2.4	7.9	1.8	4.6	0.279
	2BCu	50–65	80	19	1	LS	5.8	7.8	2.9	7.8	0.428
	3BCτ	65–90	84	10	6	LS	0	7.9	1.9	1.5	0.118
	4Cl	90–150	98	1	1	S	0	8.0	2.1	0.2	0.014
	5ACu	60–110	77	22	1	LS	7.5	7.8	2.6	11.6	0.393
P2	Apτ	0–21	73	26	1	LS	4.1	7.7	0.9	11.5	0.336
	BCu	21–47	81	18	1	LS	9.5	7.9	1.3	11.0	0.329
	2BCu	47–55	55	30	15	SL	0.7	8.0	1.1	5.5	0.256
	3BCτg	55–70	36	41	23	L	0	7.9	1.2	2.1	0.171
	4Cl	70–120	98	1	1	S	0	7.9	1.4	0.3	0.024
P3	Apτ1	0–20	70	29	1	SL	5.0	7.4	0.5	9.9	0.484
	Apτ2	20–27	67	32	1	SL	4.9	7.5	1.0	9.7	0.436
	ABu	27–35	76	23	1	LS	7.1	7.8	1.5	7.7	0.318
	2BCu	35–60	75	24	1	LS	9.2	7.8	2.3	9.3	0.302
	3ABu	60–100	74	24	2	LS	9.5	8.1	2.7	9.8	0.312

Explanation: EC—electrical conductivity, SOC—soil organic carbon, N_t—total nitrogen; soil texture classes according to WRB classification [44]: S—sand, LS—loamy sand, SL—sandy loam.

3.3. Physicochemical Soil Properties in the Profiles P1–P3

The texture of the technogenic soils varied between loamy sand and sandy loam, with a very low clay content (Table 3). A coarser texture (sand class) was found only in the natural alluvial bedrock (horizons 4Cl), while fine-textured loam was only identified as the horizon 3BC in the P2 profile and most probably represents the loam residues of loam exploited in the former clay mine.

All horizons of the constructed soils, excluding the natural bedrock layers, were rich in organic matter (4.6–11.5% SOC) and nitrogen (0.25–0.48% N). In soils relatively poor in artefacts, the SOC and N contents were highest in the topsoil Ap horizons and decreased with depth. In soils rich in artefacts, particularly charred wood particles (such as the 5ACu layer in the P1 profile), the SOC content was even higher than in the topsoil horizons (Table 3). The high N_t content in these layers suggests the impact of household debris.

Although noticeably higher than in the arable soils of Poland [49], EC did not reach thresholds for natural or human-affected saline soils [44]. EC increased from 0.5–1.4 dS m^{-1} in the topsoil layers to 2.7–2.9 dS m^{-1} in the subsoil layers rich in artefacts (Table 3). The bedrock layers in the P1 and P2 profiles were characterized by elevated CE values of 1.4 to 2.1 dS m^{-1}, probably affected by the overlying anthropogenic materials.

Calcium carbonate was present in the fine-earth fractions of all topsoil layers and in all layers containing artefacts. The highest $CaCO_3$ contents, up to 9.5%, occurred in layers containing ash and slag. The presence of carbonates and other alkaline waste affected the soil pH, which was slightly alkaline (7.4–8.1) throughout the profiles, including the carbonate-free bottom layers (Table 3). The sum of base cations (BC) reached very high values, of up to 63 $cmol_c$ kg^{-1} (Table 4), clearly affected by calcium carbonate and other substances dissolved in buffered ammonium acetate. Therefore, BC values should be analyzed with caution, as they reflect 'extractable' rather than purely 'exchangeable' forms of Ca, Mg, K, and Na. However, the high CEC values, of up to 45.4 $cmol_c\ kg^{-1}$ (Table 4), suggest the high sorption capacity of the constructed soils, likely due to their high organic matter content and porous mineral particles in the silt fraction [30,33]. The lowest BC and CEC values, while still indicating high base saturation, were found in the alluvial sandy bedrock layers (Table 4).

Table 4. Cation exchange capacity and plant-available macronutrients in soils of the reclaimed part of the allotment gardens. n.d.—not determined.

Soil Profile	Soil Horizon	Depth	Exchangeable Cations				BC	CEC	BS	Plant-Available Nutrients		
			Ca_{ex}	Mg_{ex}	K_{ex}	Na_{ex}				P_{av}	K_{av}	Mg_{av}
		cm	$cmol_c\ kg^{-1}$						%	$mg\ kg^{-1}$		
P1	Apτ1	0–25	20.7	1.35	0.36	0.07	22.4	21.0	100	318	242	416
	Apτ2	25–40	21.2	1.41	0.29	0.09	23.0	13.2	100	145	170	222
	ABu	40–50	17.8	1.29	0.21	0.12	19.4	11.7	100	50	111	159
	2BCu	50–65	27.6	2.29	0.35	0.23	30.5	11.5	100	n.d.	n.d.	n.d.
	3BCτ	65–90	14.1	0.84	0.13	0.10	15.10	7.80	100	n.d.	n.d.	n.d.
	4Cl	90–150	1.60	0.17	0.10	0.10	2.00	2.13	95	n.d.	n.d.	n.d.
	5ACu	60–110	42.3	4.61	1.03	0.38	48.3	29.1	100	n.d.	n.d.	n.d.
P2	Apτ	0–21	39.2	3.42	0.27	0.19	43.1	26.2	100	175	162	523
	BCu	21–47	44.7	4.58	0.20	0.31	49.8	25.0	100	19	178	1030
	2BCu	47–55	22.2	6.29	0.49	0.21	29.2	28.4	100	18	217	880
	3BCτg	55–70	10.0	5.89	0.47	0.18	16.5	16.6	99	n.d.	n.d.	n.d.
	4Cl	70–120	1.60	0.28	0.10	0.10	2.10	2.55	85	n.d.	n.d.	n.d.
P3	Apτ1	0–20	33.9	3.02	0.66	0.14	37.7	36.8	100	396	266	612
	Apτ2	20–27	34.1	2.89	0.53	0.15	37.6	36.2	100	373	249	542
	ABu	27–35	50.7	3.63	0.66	0.32	55.3	33.1	100	48	260	569
	2BCu	35–60	55.9	5.95	0.71	0.53	63.1	45.4	100	n.d.	n.d.	n.d.
	3ABu	60–100	52.9	6.31	0.98	0.53	60.8	30.8	100	n.d.	n.d.	n.d.

Explanation: Ca_{ex}, Mg_{ex}, K_{ex}, and Na_{ex}—exchangeable Ca, Mg, K and Na, BC—sum of exchangeable base cations, CEC—cation exchange capacity, BS—base saturation.

Taking into account the thresholds for arable soils in Poland [46], the topsoil layers in the constructed soils were very rich in plant-available phosphorus, magnesium and potassium, with concentrations reaching 396, 612 and 266 $mg\ kg^{-1}$, respectively (Table 4).

3.4. Trace Element Concentrations in the Profiles of Constructed Soils

Total concentrations (aqua regia-extractable) of Zn, Pb, Cu, and Cd in the natural alluvial 4Cl horizons did not exceed 50, 19, 16 and 0.4 $mg\ kg^{-1}$, respectively (Table 5). In contrast, the topsoil layers contained 1000–1700, 300–500, 200–360, 2.5–3.9 and 1.2–3.1 $mg\ kg^{-1}$ of Zn, Pb, Cu, Cd, and Hg, respectively. The concentrations of elements in the subsoil layers rich in ash and slag were at a similar level or higher than in the topsoil layers, reaching, as follows: 1500–2000 $mg\ kg^{-1}$ Zn; over 500 $mg\ kg^{-1}$ of Pb and Cu; and over 4 $mg\ kg^{-1}$ of Cd. The highest concentrations of elements were found in the layers containing mixtures of ash, construction debris and household waste with up to 3500 $mg\ kg^{-1}$ Zn; 1500 $mg\ kg^{-1}$ Pb; 690 $mg\ kg^{-1}$ Cu; 8 $mg\ kg^{-1}$ Cd; and 4.7 $mg\ kg^{-1}$ Hg (Table 5).

Furthermore, the total concentrations of Fe, Mn, Cr, Ni, and As, measured by pXRF (Table 5) were several times higher in layers containing artefacts than in the natural alluvial bedrock (horizons 4Cl). The Cr, Ni, and As concentrations in topsoil layers varied in the ranges of 60–90, 50–90, and 50–120 $mg\ kg^{-1}$, respectively. Although these values were lower than those in the technogenic subsoil layers, they were markedly higher than the median values for arable soils in Poland (Table 1).

The classification of soil pollution levels based on the geo-accumulation index (Igeo) (Table 6) indicated 'high to extremely high pollution' of the topsoil horizons, and 'extremely high pollution' for the subsoil horizons containing anthropogenic artefacts, due to Hg, Zn, Pb, Cu, and As (Igeo of 4–5 and ≥5, respectively). The assessment indicated 'moderate to high pollution' based on Igeo for Cr and Ni (Igeo of 2–4) and 'no pollution' based on Mn (Igeo of 0–2).

Table 5. Total concentrations of trace elements and iron in soil profiles of the reclaimed part of allotment gardens. n.d.—not determined.

Soil Profile	Soil Horizon	Depth	Total Concentration (Aqua Regia-Extractable)						Total Concentration (XRF Technique)			
			Zn	Cu	Pb	Cd	Hg	Mn	Cr_{xrf}	Ni_{xrf}	As_{xrf}	Fe_{xrf}
		cm	mg kg^{-1}									%
P1	Apτ1	0–25	1080	207	311	2.50	1.20	486	60	50	61	1.94
	Apτ2	25–40	1240	203	300	2.46	n.d.	459	73	52	54	2.11
	ABu	40–50	1530	341	535	3.55	n.d.	510	76	57	50	2.23
	2BCu	50–65	2460	541	1070	5.46	n.d.	1090	246	120	196	4.80
	3BCτ	65–90	295	62	115	1.46	1.61	394	46	15	23	1.66
	4Cl	90–150	18	6	5	0.25	n.d.	94	37	11	3	0.52
	5ACu	60–110	3580	689	1520	8.25	4.67	1050	125	83	189	4.87
P2	Apτ	0–21	1410	336	485	3.60	3.13	946	85	66	107	3.66
	BCu	21–47	1660	627	671	4.07	5.05	1270	146	143	153	5.2
	2BCu	47–55	380	116	129	2.10	n.d.	916	91	60	41	3.07
	3BCτg	55–70	103	33	92	1.32	n.d.	475	74	24	28	2.82
	4Cl	70–120	50	16	19	0.37	n.d.	137	39	9	3	0.57
P3	Apτ1	0–20	1640	335	506	3.81	1.66	913	93	96	108	3.30
	Apτ2	20–27	1700	363	499	3.89	2.06	917	96	90	120	3.38
	ABu	27–35	2000	545	1220	4.39	n.d.	1170	121	128	292	4.81
	2BCu	35–60	1700	386	673	4.87	n.d.	1810	216	447	199	6.92
	3ABu	60–100	1560	408	649	4.98	n.d.	1470	137	178	145	5.45

Table 6. Mean geo-accumulation index (Igeo) for trace elements in the topsoil and subsoils layers of soil profiles in the reclaimed part of the allotment gardens.

Soil Horizons	Zn	Cu	Pb	Cd	Hg	Mn	Cr	Ni	As
topsoil Ap horizons	4.9	4.9	4.5	3.8	6.0	0.5	2.4	2.8	4.4
subsoil with artefacts	5.3	5.7	5.5	4.5	7.3	1.2	3.1	3.7	5.0

4. Discussion

4.1. Classification of the Constructed Soils in Urban Gardens

The complete soil name in the WRB classification reflects the soil origin, its present functions and its health, as it is constructed based on soil morphology, physicochemical properties and contamination [44]. The WRB classification distinguishes two taxa at the highest classification level: Anthrosols for human-affected soils and Technosols for human-created soils. The name of the reference soil group is accompanied by principal and supplementary qualifiers that allow for a precise identification of their unique properties.

The soils studied have a well-developed structural, dark, rich in humus and nutrient Ap horizons, which meet the requirements for diagnostic hortic horizons. However, these hortic horizons are less than 50 cm thick, which is the minimum for Anthrosols [33,44]. The soils in profiles P1–P3 contained mineral waste materials, such as coal-ignition debris, construction and demolition debris, and household waste, which were used to fill the former clay pit. The content of these materials (artefacts) exceeded the required 20% (by volume) averaged over the upper 100 cm of soil profiles P1 and P3, or formed a layer ≥10 cm thick starting within 50 cm of the soil surface, with ≥80% artefacts (profile

P2). This allowed for the soils to be classified as Technosols [44]. The complete soil classification included the following qualifiers:

- Profile P1 (left side, Figure 2A): Urbic Technosol (Arenic, Calcaric, Hortic, Humic, Pyric, Endoskeletic, Transportic);
- Profile P1 (right side, Figure 2A): Urbic Technosol (Arenic, Calcaric, Hortic, Humic, Mahic, Endoraptic, Transportic);
- Profile P3: Urbic Technosol (Arenic, Calcaric, Hortic, Humic, Pyric, Skeletic, Transportic).

The principal qualifier Urbic reflects the prevalence of mineral artefacts of urban origin. Therefore, the garden soils represent the Urbic Technosols, the taxon commonly reported from urbanized areas in Europe [26,27,30,32,34]. Among the alphabetically listed supplementary qualifiers, Arenic indicates the prevailing sandy texture; Calcaric, the presence of calcium carbonate; Humic, the high SOC content ($\geq$1%) to a depth of 50 cm; Pyric, the high ($\geq$5%) black carbon (charcoal) content in a layer $\geq$ 10 cm thick; Skeletic, the high content ($\geq$40%) of coarse fragments. The important qualifier Hortic reflects the presence of a dark, base-saturated ($\geq$50%), biologically active topsoil layer rich in humus ($\geq$1%) and plant-available phosphorus ($\geq$120 mg kg^{-1}) [30,33]. The Transportic qualifier refers to the external origin of the subsoil and topsoil layers that are not enriched with artefacts. Additionally, Endoraptic refers to the lithological discontinuity observed in profiles P1 and P2, where technogenic and transported materials overlie natural alluvial sediments.

Despite the high content of trace elements, the Toxic qualifier [32,44] was omitted from the soil names due to the high biological activity observed in the topsoil layers and the lack of evidence of toxic impacts on vegetation.

4.2. *Quality of Reclaimed Garden Soils in Terms of Their Physicochemical Properties and Fertility*

Although the soils in the reclaimed part of the gardens are characterized by predominantly loamy sandy and sandy loamy textures (Table 3), the high organic matter content and the well-developed granular structure in the thick A horizons may compensate for the low clay content, support water retention, and improve water availability to plants [51]. The presence of calcium carbonate, accompanied by a slightly alkaline pH, ensures a high calcium ion concentration and effective cation exchange and also stabilizes organic matter and soil structure [52]. Both the relatively high pH and the humus content can reduce the solubility and availability of toxic elements for plants and soil fauna [53]. Furthermore, high or very high concentrations of plant-available P, K, and Mg indicate high fertility and the overall high quality of reclaimed garden soils in terms of their suitability for fruit and vegetable production [54]. This high horticultural quality is similar to that reported for other garden soils in Wroclaw and elsewhere [23,27], regardless of the initial status of soils and terrains converted into gardens [32,33].

4.3. *Quality of Reclaimed Garden Soils in Terms of Contamination with Trace Elements*

The total concentrations of trace elements in the unreclaimed part of the garden complex under study were similar to the mean concentrations recorded in the gardens of Wroclaw [23]. On the contrary, the concentrations in the reclaimed part were significantly higher—up to four times higher for copper and lead—than the mean values for Wroclaw [23]. Furthermore, the element concentrations in the reclaimed section were noticeably higher than those reported for most gardens in both smaller towns [12,16,24,55–58] and large urban agglomerations [13,59–61], and were comparable to the concentrations in highly contaminated garden and arable soils in industrial regions of Australia, northern Germany, and southern Poland [15,62,63]. The potential risk was also confirmed by the geo-accumulation index (Igeo), which indicated the 'high to extreme' pollution level, rarely reported in urban gardens in Poland [16,48].

The mean concentrations of Zn, Pb, Cu, Hg, and As in the reclaimed garden soils studied exceeded the legal threshold limits ('permissible levels') for medium-textured cultivated soils in Poland by 2–6 times [50]. According to formal interpretations, concentrations exceeding these limits are considered to pose a potential risk to human health and require respective action [50,64]. In this case, a detailed risk assessment, followed by a respective soil remediation or a change in land use, is required to minimize the impact on human health [64].

4.4. Maintaining Urban Gardening in Reclaimed Mining Sites

Waste materials rich in trace elements—such as ash, slag, construction and demolition debris, and household waste—used for pit infilling and reclamation, then covered with only a thin layer of non-contaminated soil, pose a potential but persistent risk of topsoil contamination. This risk arises from the continuous transfer of elements from the subsoil, driven by: (a) mechanical mixing during earthworks or construction in the gardens; (b) zooturbation—deep soil mixing by earthworms, moles and other burrowing animals; and (c) bioaccumulation—the extraction of elements by roots, followed by their transfer to above-ground plant parts and their eventual return to the topsoil via leaf litter, post-harvest remains or garden-made compost [65].

The continuous transfer of contaminants from the shallow subsoil can render ineffective the remediation techniques intended to remove metals from the topsoil layer [17,24,26]. Decreasing the mobility and plant-availability of metals by liming and organic fertilization is considered one of the simplest methods to reduce the risk of metal transfer from contaminated soil to cultivated plants [12,21]. However, these practices may also reduce the availability of macronutrients, particularly phosphorus [46,53].

Covering the present surface with an additional layer of uncontaminated soil to a thickness of at least 30–50 cm to create a new rooting layer can be considered an alternative method of reducing metal transfer from the subsoil to the rooting zone [32]. However, raising the surface level involves reconstruction of the entire garden infrastructure (roads, fencing, buildings, etc.) and can threaten existing trees. Applying this remediation method would effectively mean re-establishing the gardens. This approach appears to be generally unrealistic owing to its financial and social costs [32].

Thus, the present study highlights the urgent need for assessing the quality of fruits and vegetables and the risk to human health in reclaimed gardens created on urban post-mining sites. Confirmation of this risk may justify revising the functions (services) of urban gardens and limiting fruit and vegetable production in favor of recreation in many urban gardens on reclaimed sites in Poland and Central Europe.

5. Conclusions

Intentionally constructed anthropogenic soils, Urbic Technosols, predominate in a reclaimed section of urban gardens in north-eastern Wroclaw, where industrial and urban waste materials, such as ash, slag, construction and demolition debris, and household waste, were used to fill former clay and sand pits (open-cast mines). Although the topsoil layers, formed from transported soil, were characterized by beneficial physicochemical properties and a high abundance of humus and plant-available macronutrients, soils were found to be severely contaminated with zinc, lead, copper, arsenic, and mercury, along with elevated concentrations of other metals (including cadmium, nickel and chromium), transferred from the technogenic materials used for mine infilling. These materials form the subsoil layers directly beneath the humus-rich topsoil. The concentrations of trace elements were found to exceed legally permissible levels and pose a potential risk to human health, questioning the main function of the gardens. The results highlight the urgent need for a risk assessment of similarly reclaimed urban gardens in Poland and elsewhere, followed by soil remediation or changes in garden services in the case of a confirmed health or ecological risk.

Supplementary Materials: The following supporting information can be downloaded at: https://www.mdpi.com/article/10.3390/land14081613/s1, Table S1: Pseudo-total concentrations of Zn, Pb, Cu, Cd and Hg and the geoaccumulation index (Igeo) for a topsoil layer (0–25 cm) of the allotment gardens complex Wytchnienie in Wroclaw (in the unreclaimed and reclaimed sections).

Author Contributions: Conceptualization, D.G., K.S. and C.K.; methodology, K.S. and C.K.; investigation, D.G. and C.K.; writing—original draft preparation, D.G.; writing—review and editing, K.S. and C.K.; funding acquisition, D.G. All authors have read and agreed to the published version of the manuscript.

Funding: This research was partly supported by the Wroclaw University of Environmental and Life Sciences (Poland) as the Ph.D. research program "Bon doktoranta SD UPWr" project no. N020/0004/22 and partly by city of Wroclaw, within a Student Activity Fund, grant number BWU-1/2024/F6 (edition 2024).

Data Availability Statement: The original contributions presented in this study are included in the article. Further inquiries can be directed to the corresponding author.

Acknowledgments: The voluntary student research group (Studenckie Koło Naukowe Gleboznawstwa i Ochrony Środowiska) is acknowledged for extensive field and laboratory assistance.

Conflicts of Interest: The authors declare no conflicts of interest.

References

1. Delshad, A.B. Community gardens: An investment in social cohesion, public health, economic sustainability, and the urban environment. *Urban For. Urban Green.* **2022**, *70*, 127549. [CrossRef]
2. Kwartnik-Pruc, A.; Droj, G. The role of allotments and community gardens and the challenges facing their development in urban environments—A literature review. *Land* **2023**, *12*, 325. [CrossRef]
3. Silveira, C.; Dias, A.T.C.; Amaral, F.G.; de Gois, G.; Pistón, N. The importance of private gardens and their spatial composition and configuration to urban heat island mitigation. *Sustain. Cities Soc.* **2024**, *112*, 105589. [CrossRef]
4. Setiawan, W.; Gawryszewska, B. Urban garden as a water reservoir in an urban area–a literature review. *Sci. Rev. Eng. Environ. Sci.* **2023**, *32*, 221–237. [CrossRef]
5. Dobson, M.C.; Crispo, M.; Blevins, R.S.; Warren, P.H.; Edmondson, J.L. An assessment of urban horticultural soil quality in the United Kingdom and its contribution to carbon storage. *Sci. Total Environ.* **2021**, *777*, 146199. [CrossRef]
6. Royer, H.; Yengue, J.L.; Bech, N. Urban agriculture and its biodiversity: What is it and what lives in it? *Agric. Ecosyst. Environ.* **2023**, *346*, 108342. [CrossRef]
7. Milligan, C.; Gatrell, A.; Bingley, A. 'Cultivating health': Therapeutic landscapes and older people in northern England. *Soc. Sci. Med.* **2004**, *58*, 1781–1793. [CrossRef]
8. Armstrong, D. A survey of community gardens in upstate New York: Implications for health promotion and community development. *Health Place* **2000**, *6*, 319–327. [CrossRef]
9. Wakefield, S.; Yeudall, F.; Taron, C.; Reynolds, J.; Skinner, A. Growing urban health: Community gardening in South-East Toronto. *Health Promot. Int.* **2007**, *22*, 92–101. [CrossRef]
10. O'Riordan, R.; Davies, J.; Stevens, C.; Quinton, J.N.; Boyko, C. The ecosystem services of urban soils: A review. *Geoderma* **2021**, *395*, 115076. [CrossRef]
11. Corrigan, M.P. Growing what you eat: Developing community gardens in Baltimore, Maryland. *Appl. Geogr.* **2011**, *31*, 1232–1241. [CrossRef]
12. Sanchez-Camazano, M.; Sanchez-Martin, M.J.; Lorenzo, L.F. Lead and cadmium in soils and vegetables from urban gardens of Salamanca (Spain). *Sci. Total Environ.* **1994**, *146*, 163–168. [CrossRef] [PubMed]
13. Heymann, H. Schwermetall und Arsen in Hamburger Kleingaerten-Bodenbelastung und Pflanzenverfuegbarkeit. *Hambg. Bodenkd. Arb.* **1994**, *23*, 1–77.
14. Finster, M.E.; Gray, K.A.; Binns, H.J. Lead levels of edibles grown in contaminated residential soils: A field survey. *Sci. Total Environ.* **2004**, *320*, 245–257. [CrossRef]

15. Kachenko, A.G.; Singh, B. Heavy metals contamination in vegetables grown in urban and metal smelter contaminated sites in Australia. *Water Air Soil Pollut.* **2006**, *169*, 101–123. [CrossRef]
16. Walczak, B.; Kostecki, J.; Wasylewicz, R.; Lassota, T.; Greinert, A.; Drab, M. The content of lead in soils of allotment gardens in Zielona Góra, Poland. *Pol. J. Soil Sci.* **2016**, *48*, 41. [CrossRef]
17. Paltseva, A.A.; Cheng, Z.; Egendorf, S.P.; Groffman, P.M. Remediation of an urban garden with elevated levels of soil contamination. *Sci. Total Environ.* **2020**, *722*, 137965. [CrossRef]
18. Gruszka, D.; Gruss, I.; Szopka, K. Assessing environmental risks of local contamination of garden urban soils with heavy metals using ecotoxicological tests. *Toxics* **2024**, *12*, 873. [CrossRef]
19. Kubicz, J. Health risk assessment for allotment gardens in Wrocław. *Ekon. Sr.* **2014**, *1*, 154–163.
20. Taylor, M.P.; Isley, C.F.; Fry, K.L.; Liu, X.; Gillings, M.M.; Rouillon, M.; Filippelli, G.M. A citizen science approach to identifying trace metal contamination risks in urban gardens. *Environ. Int.* **2021**, *155*, 106582. [CrossRef]
21. Clark, H.F.; Brabander, D.J.; Erdil, R.M. Sources, sinks, and exposure pathways of lead in urban garden soil. *J. Environ. Qual.* **2006**, *35*, 2066–2074. [CrossRef]
22. Gmochowska, W.; Pietranik, A.; Tyszka, R.; Ettler, V.; Mihaljevič, M.; Długosz, M.; Walenczak, K. Sources of pollution and distribution of Pb, Cd and Hg in Wrocław soils: Insight from chemical and Pb isotope composition. *Geochemistry* **2019**, *79*, 434–445. [CrossRef]
23. Kabala, C.; Chodak, T.; Szerszen, L.; Karczewska, A.; Szopka, K.; Fratczak, U. Factors influencing the concentration of heavy metals in soils of allotment gardens in the city of Wroclaw, Poland. *Fresenius Environ. Bull.* **2009**, *18*, 1118–1124.
24. Szolnoki, Z.S.; Farsang, A.; Puskás, I. Cumulative impacts of human activities on urban garden soils: Origin and accumulation of metals. *Environ. Pollut.* **2013**, *177*, 106–115. [CrossRef] [PubMed]
25. Cao, X.; Williams, P.N.; Zhan, Y.; Coughlin, S.A.; McGrath, J.W.; Chin, J.P.; Xu, Y. Municipal solid waste compost: Global trends and biogeochemical cycling. *Soil Environ. Health* **2023**, *1*, 100038. [CrossRef]
26. Gambi, L.G.O.; Junior, R.A.; Vanzela, L.S.; Tagliaferro, E.R.; Vazquez, G.H.; da Silva Ueda, P.R.; Navarrete, A.A. Technosol made with urban and industrial waste: Potential for improving soil quality and growing tree seedlings. *Soil Adv.* **2024**, *2*, 100009. [CrossRef]
27. Schleuß, U.; Wu, Q.; Blume, H.P. Variability of soils in urban and periurban areas in Northern Germany. *CATENA* **1998**, *33*, 255–270. [CrossRef]
28. Kabala, C.; Greinert, A.; Charzyński, P.; Uzarowicz, L. Technogenic soils–soils of the year 2020 in Poland. Concept, properties and classification of technogenic soils in Poland. *Soil Sci. Annu.* **2021**, *71*, 267–280. [CrossRef]
29. Nehls, T.; Rokia, S.; Mekiffer, B.; Schwartz, C.; Wessolek, G. Contribution of bricks to urban soil properties. *J. Soils Sediments* **2013**, *13*, 575–584. [CrossRef]
30. Charzyński, P.; Bednarek, R.; Hudańska, P.; Świtoniak, M. Issues related to classification of garden soils from the urban area of Toruń, Poland. *Soil Sci. Plant Nutr.* **2018**, *64*, 132–137. [CrossRef]
31. Molla, A.S.; Tang, P.; Sher, W.; Bekele, D.N. Chemicals of concern in construction and demolition waste fine residues: A systematic literature review. *J. Environ. Manag.* **2021**, *299*, 113654. [CrossRef]
32. Levin, M.J.; Kim, K.H.J.; Morel, J.L.; Burghardt, W.; Charzynski, P.; Shaw, R.K. *Soils Within Cities Global Approaches to Their Sustainable Management*; Catena Soil Sciences: Stuttgart, Germany, 2017; pp. 177–196.
33. Krupski, M.; Kabala, C.; Sady, A.; Gliński, R.; Wojcieszak, J. Double-and triple-depth digging and Anthrosol formation in a medieval and modern-era city (Wrocław, SW Poland). Geoarchaeological research on past horticultural practices. *CATENA* **2017**, *153*, 9–20. [CrossRef]
34. Burghardt, W.; Morel, J.L.; Zhang, G.L. Development of the soil research about urban, industrial, traffic, mining and military areas (SUITMA). *Soil Sci. Plant Nutr.* **2015**, *61*, 3–21. [CrossRef]
35. Kobojek, E. How post-mining areas impact the spatial development and land use of cities. The case of the city of Łódź. *Space Soc. Econ.* **2021**, *32*, 111–130. [CrossRef]
36. Colten, C.E. Chicago's waste lands: Refuse disposal and urban growth, 1840–1990. *J. Hist. Geogr.* **1994**, *20*, 124–142. [CrossRef]
37. Orewere, E.; Owonubi, A.; Zainab, S. Landscape reclamation for abandoned mining site for outdoor recreation on the Jos Plateau. *J. Environ. Sci. Resour. Manag.* **2020**, *12*, 26–50.
38. Kononowicz, W.; Gryniewicz-Balińska, K. Historical allotment gardens in Wrocław-the need to protection. *Civ. Environ. Eng. Rep.* **2016**, *21*, 43–52. [CrossRef]
39. Gryniewicz-Balińska, K. Historical community gardens in Wroclaw as an element of the planned system of urban green areas. *Kwart. Archit. Urban.* **2015**, *60*, 45–79.
40. Königliche Preussische Landesaufnahme Berlin. *Topographishe Karte 1:25000 (Messtischblatt), Breslau Nord*; Königliche Preussische Landesaufnahme: Berlin, Germany, 1886.
41. Kabała, C.; Bekier, J.; Bińczycki, T.; Bogacz, A.; Bojko, O.; Cuske, M. *Soils of Lower Silesia: Origins, Diversity and Protection*; Polskie Towarzystwo Gleboznawcze, Polskie Towarzystwo Substancji Humusowych: Wrocław, Poland, 2015; pp. 25–46.

42. Reichsamt für Landesaufnahme. *Topographishe Karte 1:25000 (Messtischblatt), Breslau Nord A*; Reichsamt für Landesaufnahme: Berlin, Germany, 1831.
43. Chodak, T.; Szerszen, L.; Kabala, C. Effects of flood 1997 on soil environment of Wroclaw area. *Zesz. Probl. Postępów Nauk Rol.* **1999**, *467*, 51–57.
44. IUSS Working Group WRB. *World Reference Base for Soil Resources. International Soil Classification System for Naming Soils and Creating Legends for Soil Maps*, 4th ed.; International Union of Soil Sciences: Vienna, Austria, 2022; pp. 1–236.
45. Van Reeuwijk, L.P. *Procedures for Soil Analysis*; Technical paper 19; International Soil Reference and Information Centre: Wageningen, The Netherlands, 1992; pp. 1–114.
46. Kesik, K.; Jadczyszyn, T.; Lipinski, W.; Jurga, B. Adaptation of the Mehlich 3 procedure for routine determination of phosphorus, potassium and magnesium in soil. *Przem. Chem.* **2015**, *94*, 973–976.
47. *ISO standard 54321:2020*; Soil, Treated Biowaste, Sludge and Waste—Digestion of Aqua Regia Soluble Fractions of Elements. ISO: Geneva, Switzerland, 2020.
48. Kowalska, J.B.; Mazurek, R.; Gąsiorek, M.; Zaleski, T. Pollution indices as useful tools for the comprehensive evaluation of the degree of soil contamination–A review. *Environ. Geochem. Health* **2018**, *40*, 2395–2420. [CrossRef] [PubMed]
49. Monitoring Chemizmu Gleb Ornych Polski (Monitoring of the Chemistry of Arable Soils in Poland). Available online: https://www.gios.gov.pl/chemizm_gleb/ (accessed on 1 June 2025).
50. Regulation of the Ministry of Environment (Poland) on the Assessment of Soil Contamination, 2016. Available online: https://isap.sejm.gov.pl/isap.nsf/DocDetails.xsp?id=wdu20160001395 (accessed on 1 June 2025).
51. Lal, R. Soil organic matter and water retention. *Agron. J.* **2020**, *112*, 3265–3277. [CrossRef]
52. Šimanský, V.; Jonczak, J. Sorption capacity of sandy soil under long-term fertilisation. *Agriculture* **2019**, *65*, 164–171. [CrossRef]
53. Karczewska, A.; Gałka, B.; Dradrach, A.; Litak, K. Solubility of arsenic and its uptake by ryegrass from polluted soils amended with organic matter. *J. Geochem. Explor.* **2017**, *182*, 193–200. [CrossRef]
54. Salomon, M.J.; Cavagnaro, T.R. Healthy soils: The backbone of productive, safe and sustainable urban agriculture. *J. Clean. Prod.* **2022**, *341*, 130808. [CrossRef]
55. Mocek, A.; Owczarzak, W.; Tyksinski, W.; Kaczmarek, W. Heavy metals in soils of the allotment gardens in Polkowice. *Zesz. Probl. Postępów Nauk Rol.* **1995**, *418*, 299–304.
56. Ferri, R.; Hashim, D.; Smith, D.R.; Guazzetti, S.; Donna, F.; Ferretti, E.; Lucchini, R.G. Metal contamination of home garden soils and cultivated vegetables in the province of Brescia, Italy: Implications for human exposure. *Sci. Total Environ.* **2015**, *518*, 507–517. [CrossRef]
57. Joimel, S.; Cortet, J.; Consalès, J.N.; Branchu, P.; Haudin, C.S.; Morel, J.L.; Schwartz, C. Contribution of chemical inputs on the trace elements concentrations of surface soils in urban allotment gardens. *J. Soils Sediments* **2021**, *21*, 328–337. [CrossRef]
58. Kahle, P. Schwermetallstatus Rostocker Gartenböden. *J. Plant Nutr. Soil Sci.* **2000**, *163*, 191–196. [CrossRef]
59. Cheng, Z.; Paltseva, A.; Li, I.; Morin, T.; Huot, H.; Egendorf, S.; Shaw, R. Trace metal contamination in New York City garden soils. *Soil Sci.* **2015**, *180*, 167–174. [CrossRef]
60. Laidlaw, M.A.; Alankarage, D.H.; Reichman, S.M.; Taylor, M.P.; Ball, A.S. Assessment of soil metal concentrations in residential and community vegetable gardens in Melbourne, Australia. *Chemosphere* **2018**, *199*, 303–311. [CrossRef] [PubMed]
61. Mielke, H.W.; Anderson, J.C.; Berry, K.J.; Mielke, P.W.; Chaney, R.L.; Leech, M. Lead concentrations in inner-city soils as a factor in the child lead problem. *Am. J. Public Health* **1983**, *73*, 1366–1369. [CrossRef]
62. Antoniadis, V.; Shaheen, S.M.; Boersch, J.; Frohne, T.; Du Laing, G.; Rinklebe, J. Bioavailability and risk assessment of potentially toxic elements in garden edible vegetables and soils around a highly contaminated former mining area in Germany. *J. Environ. Manag.* **2017**, *186*, 192–200. [CrossRef]
63. Dziubanek, G.; Baranowska, R.; Oleksiuk, K. Heavy metals in the soils of Upper Silesia—A problem from the past or a present hazard? *J. Ecol. Health* **2012**, *16*, 169–176.
64. Karczewska, A.; Kabala, C. Environmental risk assessment as a new basis for evaluation of soil contamination in Polish law. *Soil Sci. Annu.* **2017**, *68*, 67. [CrossRef]
65. Tresch, S.; Moretti, M.; Le Bayon, R.C.; Mäder, P.; Zanetta, A.; Frey, D.; Fliessbach, A. A gardener's influence on urban soil quality. *Front. Environ. Sci.* **2018**, *6*, 25. [CrossRef]

MDPI AG
Grosspeteranlage 5
4052 Basel
Switzerland
Tel.: +41 61 683 77 34

Land Editorial Office
E-mail: land@mdpi.com
www.mdpi.com/journal/land

www.ingramcontent.com/pod-product-compliance
Lightning Source LLC
Chambersburg PA
CBHW061502210926
51850CB00008B/1790
9783725870103